Generative AI for Remote Sensing of the Environment

This book explores the cutting-edge integration of generative artificial intelligence (AI) techniques to enhance environmental remote sensing, providing a comprehensive guide from foundational algorithms to practical applications. It explains how advanced AI technology can be used to improve the way we monitor and understand the environment from a distance, such as through satellites or drones. It starts with an explanation of basic algorithms behind generative AI and gradually moves to complex algorithms, showing how they can be applied to real-world environmental issues, such as tracking climate change, monitoring deforestation, and predicting natural disasters.

Features

- Includes real-world examples and case studies showing how generative AI improves environmental monitoring.
- Provides step-by-step explanations of algorithms and their implementation.
- Explains complex concepts in simple and easy-to-understand language and introduces strategies to address environmental challenges using AI-driven solutions.
- Offers cutting-edge research and advancements in AI and remote sensing, including the application of generative AI models like GANs and VAEs.
- Applies to diverse fields such as urban planning, agriculture, and disaster management.
- Discusses ethical considerations and challenges when integrating AI with remote sensing.

This book is for researchers and practitioners in environmental monitoring, urban planning, agriculture, and disaster management using remote sensing technologies and AI to address environmental challenges and sustainability. It is also intended for university professors, graduate, and postgraduate students in environmental science, geospatial analysis, computer science, and data science working on projects related to AI and remote sensing.

Generative AI for Remote Sensing of the Environment

Algorithms and Applications

Edited by
Vishakha Sood
Arun Lal Srivastav
Ravneet Kaur
Neha Bhati

CRC Press
Taylor & Francis Group
Boca Raton London New York

CRC Press is an imprint of the
Taylor & Francis Group, an **informa** business

Designed cover image: © Getty Images

First edition published 2026
by CRC Press
2385 NW Executive Center Drive, Suite 320, Boca Raton FL 33431

and by CRC Press
4 Park Square, Milton Park, Abingdon, Oxon, OX14 4RN

CRC Press is an imprint of Taylor & Francis Group, LLC

ISBN: 978-1-041-01744-8 (hbk)
ISBN: 978-1-041-01745-5 (pbk)
ISBN: 978-1-003-61620-7 (ebk)

DOI: 10.1201/9781003616207

Typeset in Times
by SPi Technologies India Pvt Ltd (Straive)

Contents

Preface

Generative AI for Remote Sensing of the Environment: Algorithms and Applications appears at a time when environmental problems are on the rise, and novel technologies are transforming how we observe, anticipate, and react to environmental change. Generative AI has quickly evolved from a novel computational approach into a revolutionary method for generating synthetic datasets, enhancing image details, and modeling ecological phenomena and impacts with high accuracy. Remote sensing has a long history in environmental sciences, providing a vast range of data across various spatiotemporal scales. Therefore, the convergence of generative AI and remote sensing presents new opportunities for sustainable environmental monitoring and management. This book attempts to cover these converging domains, starting from the basics of generative AI, progressing through methodological advances, and concluding with several applied case studies and applications. This volume bridges the gap between theory and application by bringing together perspectives from academia, industry, and policymakers.

The foundational chapters, "Introduction to Generative AI and Its Role in Remote Sensing" and "Core Concepts of Understanding Generative AI Algorithms and Models," establish the necessary conceptual basis for a diverse audience. The former provides an overview of the fundamental principles and intuitive understanding behind generative models, including Generative Adversarial Networks (GANs), Variational Autoencoders (VAEs), and Diffusion Models. Based on this theoretical foundation, the volume then moves into applied contexts, where specific remote sensing technologies are examined in combination with AI-based approaches. Topics such as "Diffusion Model and Transformers for Enriching Remote Sensing Datasets" and "Satellite Imagery Processing with Multimodal or Vision-Language Models" showcase how these AI models are transforming the way we analyze remote sensing data. The conceptual discourse is furthered by examples of agricultural applications, monitoring climate change, and environmental prediction that illustrate how generative AI demonstrates its role in addressing pressing ecological and social problems.

What sets this book apart is its focus on practical implementation and end-to-end integration. "Tools and Software Essential Resources for AI Integration" provides readers not only with the theoretical background but also the technical skills needed to design, test, and deploy AI-driven models in remote sensing workflows. In addition to covering data analysis broadly, the book dedicates its third section to scenarios where advanced data handling skills are essential. The book acknowledges that extensive datasets are often disparate, incomplete, and noisy, which significantly impacts their practical use. It includes an entire section on how to work with such data, "Data Preparation, Collecting, Cleaning, and Managing Datasets in Generative AI." This addresses typical real-world constraints that are actual impediments to effective model deployment, ensuring that practitioners can use the resulting models. Such an applied orientation renders the volume valuable not only to researchers and academicians but also to practitioners and industry stakeholders interested in generative AI applications in the field. The book provides a theoretical rigor and technical clarity that allows the reader to confidently transition from a conceptual understanding to knowing how to implement it.

In addition to technical and methodological innovation, this book emphasizes the importance of responsibility and ethics in the application of generative AI for environmental purposes. As AI systems become increasingly autonomous, ensuring that environmental monitoring is both fair and trustworthy is not just a technical challenge but a moral obligation. Moreover, the chapters on "Challenges with Practical Solutions and Case Studies" and "Future Trends and Innovations in AI and Remote Sensing" extend beyond the current scope of AI in RS, prompting readers to consider the future direction of this rapidly expanding field. The book strikes a responsible balance between creative possibilities and new dangers, establishing a strong foundation for advancing generative AI for the betterment of people and the planet. These reflections are essential for policymakers, development organizations, and regulatory agencies that rely on their data-driven insights, which are ethical and trustworthy.

The contributors to this volume come from diverse disciplines and institutions worldwide, offering a stimulating blend of scholarly work, cutting-edge research, and practical knowledge. The book incorporates knowledge from diverse domains, including agriculture, climate science, ecology, and policy, ensuring a wide range of perspectives are represented. As editors, we hope that this will become a reliable resource for students seeking fundamental concepts, researchers interested in emerging trends, and practitioners applying generative AI in practice to address environmental challenges. We view it as a starting point for new collaborations across disciplines and borders. We offer this volume to the community as a timely addition to the emerging literature at the crossroads of generative AI and remote sensing, with our sincere thanks to all authors, reviewers, and project supporters.

Dr. Vishakha Sood
Project Scientist, Indian Institute of Technology (IIT) Ropar, Punjab, India
Dr. Arun Lal Srivastav
Associate Professor, School of Engineering and Technology,
Chitkara University, Himachal Pradesh, India
Ravneet Kaur
Assistant Professor, Chandigarh University, Gharuan, Punjab, India
Neha Bhati
Research Associate, Flexxited, Bengaluru, India

About the Editors

Dr. Vishakha Sood is working as a project scientist at Centre of Excellence in Socio-Environmental Sustainability for River Sand Mining at the Indian Institute of Technology, Ropar, Punjab. Earlier she worked as Women Scientist (WOS) by the Department of Science & Technology (DST), Govt. of India. She is also the founder of a company named Aiotronics Automation Pvt. Ltd, supported under the Himachal Pradesh CM Startup Scheme. She has more than 14 years of experience in the field of academics and research. She received her PhD in Electronics and Communication Engineering from Chitkara University, Punjab, in 2020. She has done B.Tech from Himachal Pradesh University (HPU), 2008, and M.Tech from Punjab Technical University (PTU) in Electronics and Communication Engineering, 2011. She also earned an MBA in Human Resources in 2010. She has authored more than 15 SCI-indexed articles (IEEE, T&F, Elsevier, and Springer) and Scopus-indexed book chapters, and holds many inventions. Her research interests include satellite sensors, remote sensing, and digital image analysis. She is an IEEE senior member and an active member of various societies, such as the Indian Society of Remote Sensing (ISRS), Indian Society of Technical Education (ISTE), Punjab Science Congress (Punjab Academy of Sciences), IEEE Sensors Council, Geoscience and Remote Sensing Society Digital Library, IEEE Women in Engineering, and European Geoscience Union (EGU).

Dr. Arun Lal Srivastav is working as an associate professor at the School of Engineering and Technology, Chitkara University, Himachal Pradesh, India. He received a PhD from the Indian Institute of Technology (BHU), Varanasi. He has also done post-doctoral research at National Chung Hsing University, Taiwan. He is currently involved in the teaching of Environmental Science, Environmental Engineering, Disaster Management, and Design Thinking to undergraduate engineering students. His research interests include water quality surveillance, climate change, water treatment, river ecosystem, soil health maintenance, engineering education, phytoremediation, and waste management. Under his supervision, three PhD degrees have been awarded. He has published >120 research papers in various prestigious journals (Elsevier, Springer, IWA, Taylor & Francis, etc.), including books, book chapters, and conference publications. He has edited many books with Elsevier, Springer, Nova, IGI Global, and Wiley. Additionally, he is also one of the series editors of Elsevier and Nova publishers on energy sustainability and e-waste management, respectively. He is also working on four government-sponsored projects (worth ~16 million INR) on phytoremediation, adsorption, capacity building, organic farming, leachate treatment, agro-waste management, etc. He received the prestigious "Teachers Associateship for Research Excellence (TARE) Fellowship," followed by a research grant from the Science and Engineering Research Board (SERB), Govt. of India, in 2022. He has also filed 25 patents on multidisciplinary topics, 12 of which have been granted by the Government of India. He has also been recognized among the 2% worldwide scientists in Earth and Environmental Sciences jointly by Stanford University and Elsevier in the years 2023 and 2024. Recently, his name has been included in the top 0.5% scholars of the year 2024 by Scholars in Environmental Sciences.

Ravneet Kaur is working as an assistant professor and serving the post of Program Leader of Cyber Security, Information Security, and Blockchain at Apex Institute of Technology, Chandigarh University, Gharuan. She holds a Master of Technology in Computer Science and Engineering with a specialization in Satellite Image Processing, a Master of Business Administration in Information Technology from SCDL, Pune, and a Bachelor of Technology in Information Technology from Punjab Technical University. She has published research papers in reputed journals (SCI/Scopus indexed) such as Elsevier and IEEE conferences. She has vast teaching experience of more than 18 years in engineering colleges. She is an active professional member of the Association for Computing Machinery, the Institution of Engineers, the Indian Society of Technical Education, and an honorary member of the Indian Engineering Teachers Association. She also served as an organizing secretary of the BBSBEC ACM Student Chapter for the past five years.

Neha Bhati is a researcher specializing in remote sensing, Geographic Information Systems (GIS), and AI-driven geospatial intelligence, currently serving as a research associate at Flexxited in Bengaluru, India. She focuses on developing innovative solutions for environmental monitoring, climate assessment, and resource management using tools such as Google Earth Engine, QGIS, ENVI, and ERDAS Imagine. She holds postgraduate degrees in Computer Science and Applications, blending technical and environmental domains. Neha has authored several articles published in SCI and Scopus-indexed journals and conferences, covering topics such as agricultural change detection, the Internet of Things (IoT), and machine learning applications in healthcare and sustainability. In addition to her research, she serves as an editor for books currently in production with renowned international publishers, including CRC Press, Wiley, and Apple Academic Press (AAP). Her work reflects a strong commitment to interdisciplinary collaboration, scientific rigor, and technological advancement in the geospatial domain.

Contributors

Mohamed A.E. AbdelRahman
Land Use Department
Division of Environmental Studies and Land Use National Authority for Remote Sensing and Space Sciences (NARSS)
Cairo 11769, Egypt

Fathi Ben Amar
Institute De L'olivier De Sfax
Sfax, Tunisia

R. Ayswarya
Sri Ramakrishna College of Arts and Science
Coimbatore, India

Sanket Babar
Department of MCA
Government College of Engineering
Karad, Maharashtra, India

Neha Bhati
Department of Research and Development
Flexxited
Bengaluru, India

Vaibhavi Chavan
School of Business Management and Commerce
Ravensbourne University
London, United Kingdom

Adrian David Cheok
Nanjing University of Information Science and Technology
Nanjing, China

Souvik Das
Department of Genetics and Plant Breeding
Bidhan Chandra Krishi Viswavidyalaya
Nadia, India

Sharbari Deb
Poornima University
Jaipur, India

Ruma Debnath
CMR Institute of Technology
Bengaluru, India

Akash Deep
Graphic Era Deemed to be University
Dehradun, India

Vishal Dutt
Department of Computer Science
Punjabi University
Patiala, Punjab, India

Department of Computer Science and
Engineering Chandigarh University
Mohali, Punjab, India

Karim Ennouri
Institute De L'olivier De Sfax
Sfax, Tunisia

Monia Ennouri
Institute De L'olivier De Sfax
Sfax, Tunisia

Pratik Gunjalkar
Department of MCA
Government College of Engineering
Karad, Maharashtra, India

Kawaljit Kaur
University Institute of Computing
Chandigarh University
Mohali, India

Atiya Khan
G H Raisoni College of Engineering
Nagpur, India

Shalini Kumari
SBAS
SGRR University
Dehradun, India

Sandeep Kunwar
Graphic Era Deemed to be University
Dehradun, India

Ezil Sam Leni A
Alliance University
Bengaluru, India

Revanth Madamala
Machine Learning Engineer, Recommendations
TikTok, Bytedance Pvt. Ltd.
Mountain View, California

Mahesh R
Graduate school
ICAR-Indian Agricultural Research Institute
New Delhi, India

Shankar Mali
School of Computer Science
Dr. Vishwanath Karad MIT World Peace University
Pune, India

Santanu Mallik
Poornima College of Engineering
Jaipur, India

Varun Narayan Mishra
Amity Institute of Geoinformatics and Remote Sensing (AIGIRS)
Amity University
Noida, India

C H Patil
School of Computer Science
Dr. Vishwanath Karad MIT World Peace University
Pune, India

Sakshi Rawat
Graduate school
ICAR-Indian Agricultural Statistics Research Institute
New Delhi, India

M Hemakumar Reddy
Department of Genetics and Plant Breeding
Bidhan Chandra Krishi Viswavidyalaya
Nadia, India

Revathi T
Alliance University
Bengaluru, India

Sridevi Sakhamuri
Department of IoT
Koneru Lakshmaiah Education Foundation
Guntur, India

Tanmay Sawant
Department of MCA
Government College of Engineering
Karad, Maharashtra, India

Ganesh Kumar Sethi
Department of Computer Science
Multani Mal Modi College
Patiala, India

Shalen S
Spaulding Ridge Advisory India Pvt Ltd
Bengaluru, India

Nisha Sharma
University Institute of Computing
Chandigarh University
Mohali, India

Priyanka P. Shinde
Department of MCA
Government College of Engineering
Karad, Maharashtra, India

Amanpreet Singh
Chitkara University Institute of Engineering and Technology
Chitkara University
Rajpura, India

Jarnail Singh
Swami Vivekanand Institute of Engineering and Technology
Banur, India

Jaswinder Singh
Chitkara University Institute of Engineering and Technology
Chitkara University
Rajpura, India

Rakesh Singh
Graphic Era Deemed to be University
Dehradun, India

Rupinder Singh
Chitkara University Institute of Engineering and Technology
Chitkara University
Rajpura, India

Sartajvir Singh
Indian Institute of Technology
Ropar, India

Thota Sivasankar
School of Computer Science
UPES
Dehradun, India

Mohamed Ali Triki
Institute De L'olivier De Sfax
Sfax, Tunisia

Amol D. Vibhute
School of Cyber Security and Digital Forensics
National Forensic Sciences University
Nagpur, Maharashtra, India

Narayan Vyas
Department of Computer Science and Engineering
Vivekananda Global University
Jaipur, India

Anurag Wazarkar
Department of MCA
Government College of Engineering
Karad, Maharashtra, India

Surendra Yadav
Department of Computer Science and Engineering
Vivekananda Global University
Jaipur, India

Part I

Introduction to Gen AI

1 Introduction to Generative AI and Its Role in Remote Sensing

Revanth Madamala and Sridevi Sakhamuri

1.1 INTRODUCTION TO GENERATIVE AI

The development of Artificial Intelligence (AI) during the last ten years has taken an extraordinary form, largely leaving the rule-based systems behind and focusing on the use of data-driven learning that is similar to how humans learn. Generative Artificial Intelligence (GenAI), a type of model that generates new information that is strongly correlated with a specific pattern discovered in reality, is at the forefront of this change. Those models create lifelike images, sensible text, and feasible environmental re-enactment, thus providing not only analytical assistance but also imaginative and forecasting capacity. Unlike standard AIs that are guided mainly by classification or interpretation of existing data, generative models actually generate new samples of data, making possible rather than ruling out more insightful results in various fields. The paradigm shift has mega potential in remote sensing, where the gathering of environmental data is highly troubled by occlusions, sparse data, and sensor restrictions. Generative AI provides data augmentation, super-resolution, and simulation, and expands the world of environmental modeling of the environment, precision agriculture, and real-time disaster management.

Developments in deep learning design have triggered the appearance of sound generative designs. The most notable one is Generative Adversarial Networks (GANs) proposed by Goodfellow et al. (2014), where a generator-discriminator architecture is used to create synthetically generated data that cannot be distinguished from real data. Variational Autoencoder (VAE) is an even greater extension and allows generative construction in latent spaces to provide structured and continuous variation. High-resolution diffusion models have proved effective in generating high-resolution images by denoising the inputs iteratively (Ma et al., 2019). Along the way, transformer-based architectures that were initially proposed to work on natural language tasks have demonstrated promising results in remote sensing solutions due to the ability to model both spectral and temporal properties with an attention mechanism. Although there are clear examples of successful applications of these models to language generation, healthcare imaging, and others, their combination with geospatial analytics has been an emerging and rather unexplored area.

DOI: 10.1201/9781003616207-2

Notwithstanding increased interest, a number of pertinent issues interfere with the popularization of GenAI in remote sensing. Access to quality labeled datasets with rare or episodic events to learn or train, for example, wildfires, floods, and pest outbreaks, is also one key limitation. Unpredictable cloud cover, sensor noise, and low rates of revisit further frustrate the consistency of data that is obtained by satellite. Various generative models are computationally demanding since they demand access to resource-intensive GPUs and other large-scale computing resources that are not readily available in resource-constrained areas. Also, the model outputs are still considered to be noninterpretable, particularly when such tools are used in important areas such as agriculture, climate response, and public policy decisions. Additionally, the matter of ownership of the synthetic data, regulatory considerations, and possible spread of biases generates ethical and legal concerns that need to be answered in the process of technological advancements.

In light of this background, the proposed study intends to conduct an inquiry into the application of GenAI to remote sensing in terms of environment, agriculture, and society. Its key goals are (a) to examine the contribution of advanced generative models like GANs, VAEs, diffusion models, and transformers to increase the usability of remote sensing data; (b) to identify what might be the contribution of GenAI to overcome the main limitations of data usability in remote sensing (cloud obstruction, sparsity of data, and data resolution); (c) to illustrate its real use in precision agriculture, environmental monitoring, and disaster response; and (d) to analyze whether and how GenAI-enabled remote sensing solutions support the UN Sustainable Development This work provides a starting point on how GenAI could foster more resistant, scalable, and equitable terrestrial systems of observation by determining its existing capabilities, challenges, and research gaps (Längkvist et al., 2014).

The rest of the chapter is organized as follows. An evolution of AI is given in Section 1.2, with particular attention to the generative models and their rise. Section 1.3 is a more in-depth, technical description of the most popular GenAI models, GANs, VAEs, diffusion models, and transformers, highlighting how they work and their flexibility for remote sensing tasks. Section 1.4 presents some practical applications to cloud removal, super-resolution, and seasonal image gap filling. Section 1.5 presents the pioneering case studies and numbers that show how to use it today in such fields as agriculture and weather tracking and, in case of emergency, disaster response. Limitations and future directions are assessed by Section 1.6 in terms of ethical, computational, and policy challenges. The chapter ends by establishing relative connections between GenAI capabilities and Sustainable Development Goals (SDGs) and then suggesting a roadmap for sustainable environmental intelligence.

Figure 1.1 shows the hierarchical relationship among key fields in intelligent technologies. At the core lies AI, encompassing the broader concept of intelligent systems. Machine Learning (ML) builds on this by enabling systems to learn from data. Deep learning, a subset of ML, uses neural networks for handling complex data tasks.

AI and its applications, such as in remote sensing, represent advanced, domain-specific uses built on these foundational layers. Remote sensing applications and output results together with the goals of discriminative and generative models are presented in Table 1.1. The generation models provide new analytical techniques that produce synthetic images for handling restricted datasets.

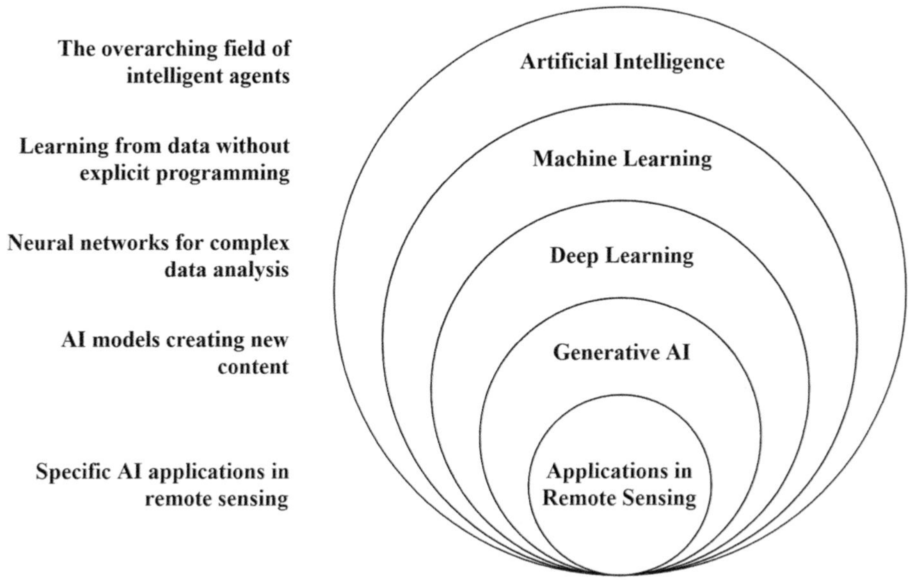

FIGURE 1.1 The evolution of AI to remote sensing applications.

TABLE 1.1
Comparison of Discriminative Versus Generative Models in Remote Sensing Applications

Feature	Discriminative Models	Generative Models
Goal	Classify or predict	Generate new data
Example Models	SVM, Logistic Regression, CNNs	GANs, VAEs, diffusion models
Output	Label or category	New data instances
Use Case in Remote Sensing	Land cover classification	Synthetic SAR/optical image creation

1.2 CORE GENERATIVE MODELS AND MECHANISMS

The GenAI system develops models that produce new data elements matching the statistical patterns observed in training data. These models have modernized the fields that need data extension, authentic simulations, and content invention. The application of generative models in remote sensing helps resolve three fundamental problems within the field involving scarce data resources and poor-quality imagery, as well as environmental simulation modeling. This section demonstrates an examination of the main generative models, GANs, VAEs, diffusion models and transformers, Large Language Models, evaluating their operational frameworks and functionality alongside remote sensing applications.

1.2.1 Generative Adversarial Networks

GANs were initially proposed with their dual architecture of the two components of a neural network: generator and discriminator that perform an ongoing adversarial procedure. The purpose of the generator is to produce information that seems to be as real as the actual data samples when pitted against the actual data samples by the discriminator. These dynamic forces are utilized by the generator to come up with increasingly real results.

The working of a GAN is shown in Figure 1.2. The model starts by sending data noise into a generator that generates a synthetic image. This fake image is then compared to a real one by a discriminator, which considers what is real and what is fake. The generator seeks to optimize its output by referring to the remarks of the discriminator. This circle goes on until the fake pictures can no longer be separated from the actual.

Applications in Remote Sensing

- **Image Dehazing**: Remote sensing images receive haze removal through GAN techniques, which enhance their clarity and make details more visible. The TRPC-GAN implements texture restoration with physical assessment methods, which results in improved haze-removed images that enhance downstream image recognition capabilities.
- **Cloud Removal**: The SpT-GAN operates through a sparse transformer-based system that removes thin clouds from satellite imagery while both maintaining ground feature detection and improving visual output (Jin et al., 2024).
- **Super-Resolution**: Models like ARISGAN generate high-resolution arctic surface imagery from low-resolution inputs, aiding in detailed environmental monitoring.

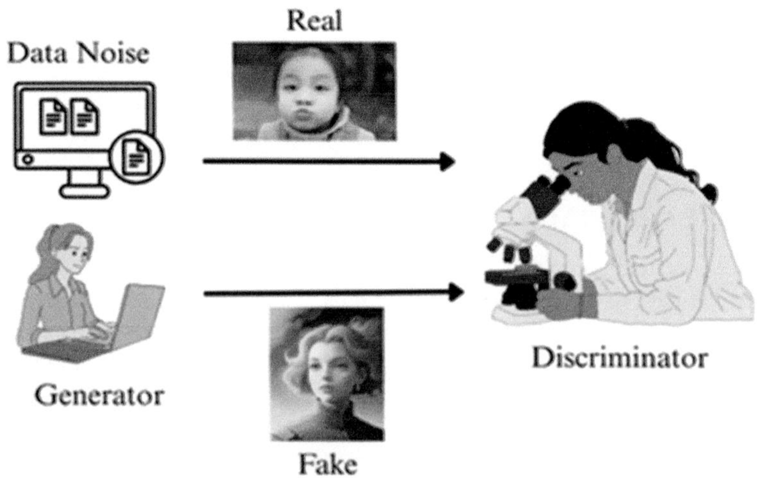

FIGURE 1.2 Architecture of GAN.

1.2.2 Variational Autoencoders

VAEs are probabilistic generative models that aim to pick up data representation by latent spaces so that they are able to generate new data points by sampling them. They are composed of an encoder, which projects raw input data to a latent distribution and a decoder, which transforms latent variable samples to data (Khanna et al., 2023).

The architecture of a VAE is shown in Figure 1.3. The encoder is used to map input data to a latent space distribution, and the decoder is used to reconstruct the data using Gaussian samples of this distribution. This architecture allows effective data generation and learning of representation.

1.2.3 Diffusion Models

Diffusion models are a type of generative model in which their task is to learn to reverse a diffusion process, converting a simple noise distribution to a complex data distribution. They have become renowned due to their stability and high quality of production generation.

Equation 1.1: Diffusion Process

$$q\left(x_t \middle| x_{t-1}\right) = N\left(x_t; \sqrt{}\left(1-\beta_t\right) x_{t-1}, \beta_t I\right) \tag{1.1}$$

Explanation

This defines the forward process in a diffusion model. At each step t, we add a small amount of Gaussian noise to the previous data sample x_{t-1}, resulting in a progressively noisier sample x_t.β_t: The noise schedule (how much noise is added at each time step) and N: Denotes a Gaussian distribution.

The reverse process, which the model learns, is used to generate clean data from pure noise, a kind of denoising-based generation.

Figure 1.4 depicts the diffusion model process, where data undergoes gradual noise addition in the forward diffusion and is denoised step-by-step during the reverse process. This enables the generation of realistic data samples from pure noise. The EDiffSR model leverages diffusion probabilistic methods for efficient

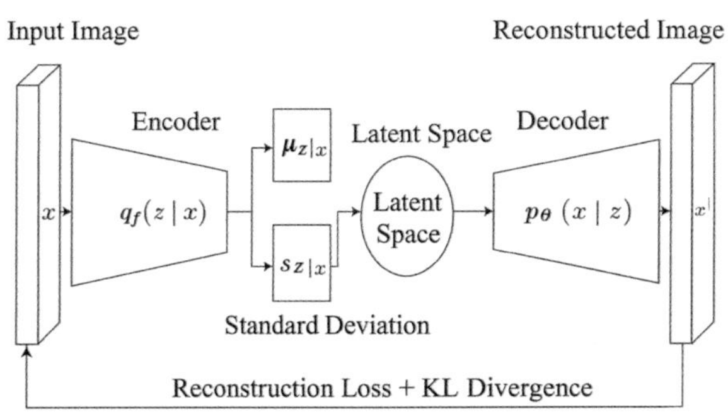

FIGURE 1.3 The structure of VAE, highlighting the latent space.

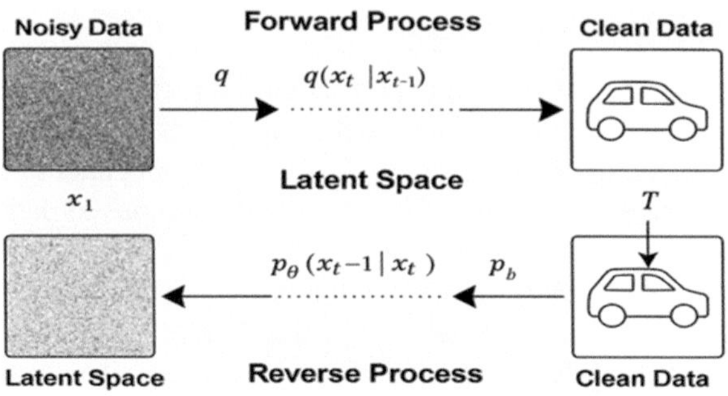

FIGURE 1.4 The forward and reverse diffusion processes in a diffusion mode.

super-resolution of remote sensing images, enhancing visual detail, and aiding in applications requiring high-resolution data.

1.2.4 Transformers and Large Language Models

Transformers, which are defined by self-attention mechanisms, have changed the landscape of sequence modeling works and opened doors to large language models (LLMs) that can understand and generate text that resembles humans. Due to their versatility, they are used outside natural language processing (Aleissaee et al., 2022).

Equation 1.2: Self-Attention Mechanism

$$\text{Attention}(Q,K,V) = \text{softmax}\left(\frac{QK^T}{\sqrt{d_k}}\right)V \tag{1.2}$$

Explanation

This formula powers the **self-attention** mechanism in transformers:

- ***Q*** (Query), ***K*** (Key), and ***V*** (Value) are all learned representations of the input.
- The dot product QK^T calculates how much attention one word (or pixel, in images) should pay to another.
- Dividing by $\sqrt{d_k}$ stabilizes gradients during training.

Figure 1.5, which represents the transformer architecture, is constituted by encoder and decoder that are linked through attention mechanisms. Self-attention and feed-forward layers contained in each block allow parallel access and contextual awareness of input sequences.

Applications in Remote Sensing

- **Spectral Data Modeling:** Spectral GPT, a foundation model based on transformers, has been developed to handle spectral remote sensing images, facilitating tasks like scene classification and change detection.
- **Cloud Removal:** The SpT-GAN integrates sparse transformer modules to enhance cloud removal processes in satellite imagery, demonstrating the versatility of transformer architectures in remote sensing applications.

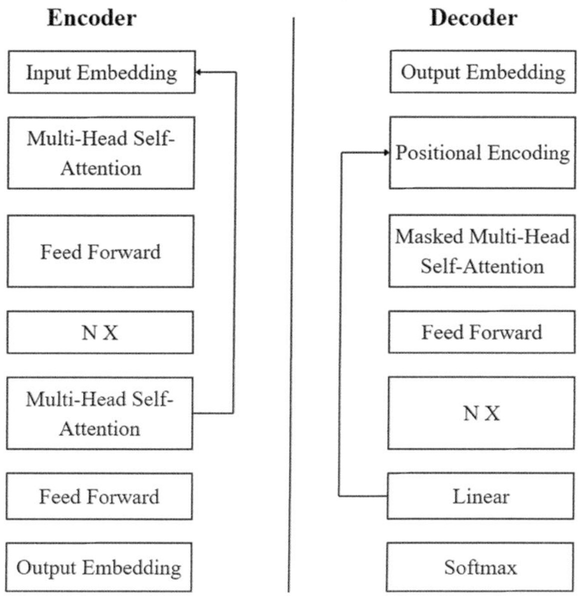

FIGURE 1.5 Transformer architecture overview.

1.3 METHODOLOGY

In order to overcome the disadvantages of remote sensing imagery (i.e., cloud cover, irregular revisit periods, and low spatial resolution), an elaborate methodological pipeline was developed. The workflow combines four strong GenAI models, viz., VAEs, GANs, diffusion models, and transformer-based architectures. The initial stage was initiated by gathering and preprocessing of remote sensing data, which was mainly based on the Sentinel-2 and MODIS satellite data stores. The preprocessing procedures involved in atmospheric correction, geometric rectification, and noise filtering were carried out with several ESA SNAP and Python-based custom pipelines. Such pre-processed images were stacked in multi-spectral and multi-temporal mode, in order to maintain seasonal consistency and depth of spectral color. The fourth step was to put each of the generative models in its particular role. VAEs were used to encode the missing data and also to detect anomalies because of their strong latent space. Conversely, GANs found application in areas such as cloud removal, SAR-to-optical translation, as well as image de-occlusion, due to their image fidelity behaviors (Zhu et al., 2017).

The diffusion models were used to boost image resolution, particularly on cloud-free pieces of information where super-resolution was required. Such models are recursive noise reduction and sharpened imagery with remarkable spatial resolution. Transformers, both Vision Transformers (ViT) and Spatio-Temporal Transformers (SpT), were added to the pipeline and used to process multi-modal sensor fusion and long-range spectral connection. Hybrid architectures came into the fore in the third stage of the pipeline, in which they were used to enhance the performance of the model. Of note, VAE-GAN was created to realize seasonal gap filling in images to reach a higher degree of temporal consistency and semantic conservation (Xiao et al., 2023). Moreover, SpT-GAN models added transformer blocks to learn time-series dependencies in the sequence of remote sensing imageries. The combination of outputs of several models was performed on the decision level with a softmax-based ensemble strategy, which normalizes confidence scores of various architectures. This integration strategy served to soften the weaknesses of individual models, and it enhanced individual strengths.

The fourth stage was concerned with training and assessment. NVIDIA A100 GPUs were trained, and Adam optimizers and adaptive learning rates were used for all models. The correctness of the work of each model was tested with well-known measures such as Peak Signal-to-Noise Ratio (PSNR), Structural Similarity Index (SSIM), and a personal temporal consistency measure that shows a good change of frames in time-series. The training data was taken in agriculturally significant and vulnerable to disaster areas of India and the Midwest, United States. Lastly, model results were displayed in QGIS and subjected to vegetation indices, including Normalized Difference Vegetation Index (NDVI), to monitor crops in real-time, detect changes, and for disaster mapping. Benchmarking of these results was done against UN SDGs to determine how they assist in climate action, food security, and disaster resilience. Such a unified approach has enabled scalable, explainable, and practical use of GenAI tools in remote sensing, such that raw satellite imagery can be converted into usable geospatial intelligence.

Figure 1.6 illustrates the workflow of applying GenAI to remote sensing image enhancement. It includes stages such as data input, model training, generation, and final application output. Various models like GANs and transformers are integrated into the pipeline. The figure emphasizes how these methods improve image quality, clarity, and utility in practical use cases.

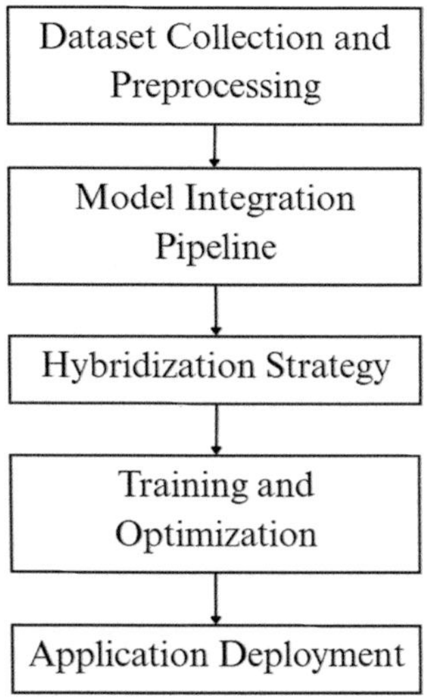

FIGURE 1.6 Methodology for GenAI-driven remote sensing enhancement.

1.4 REMOTE SENSING: CONCEPTS AND SIGNIFICANCE

Remote sensing refers to the acquisition of information about objects or phenomena without making physical contact with them. Earth surface and atmospheric studies through the utilization of sensors aboard space platforms and aircraft defines this method within environmental and agricultural fields. Space exploration technology serves as an essential base to track land usage, together with vegetation health status, and predict weather conditions, as well as find natural disasters and control natural resources.

Remote sensing that uses satellite technology with enhanced sensors and AI analytical tools has become essential for delivering multiple SDGs. This technology enables cost-effective observation and study of environmental systems across different spatial and temporal ranges using repeatable processes that are also non-invasive.

Figure 1.7 shows the remote sensing cycle, beginning with solar radiation reaching Earth's surface. Reflected signals from Earth's features are captured by a satellite sensor in orbit. The data is transmitted to a ground antenna station for further analysis. A computer system processes the received signals to extract meaningful insights. The final output supports various applications such as land monitoring and environmental management.

Table 1.2 provides an overview of remote sensing systems categorized by sensor types such as optical, SAR, Hyperspectral, and Thermal Infrared. Each sensor type is linked with specific platforms like satellites or drones and is distinguished by its unique capability, ranging from capturing light and heat to penetrating cloud cover. These systems support various applications, including vegetation monitoring, flood analysis, crop classification, and assessing drought stress.

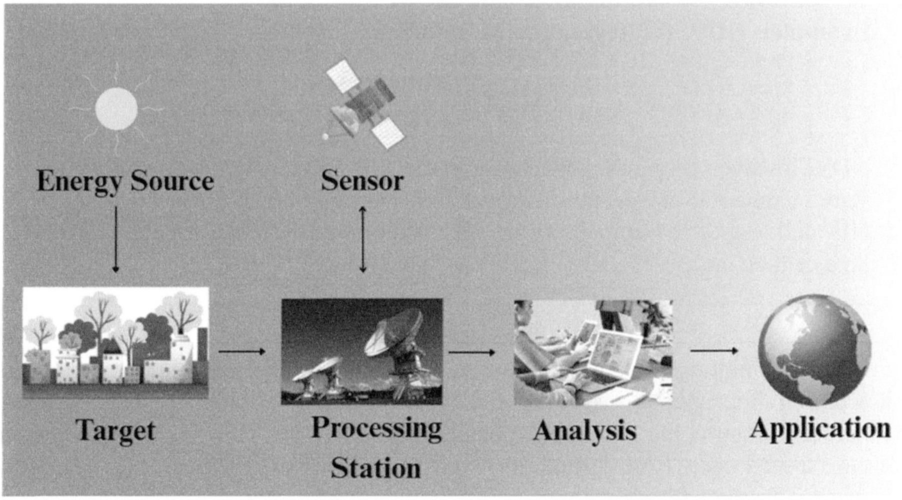

FIGURE 1.7 Remote sensing data flow diagram.

TABLE 1.2
Classification of Remote Sensing Systems by Sensor Type and Use Case

Sensor Type	Platform	Characteristics	Common Use Cases
Optical	Satellite/Drone	Captures visible and near-infrared light	Vegetation health, urban mapping
SAR	Satellite	Penetrates clouds, active radar sensing	Flood mapping, soil moisture analysis
Hyperspectral	Satellite/Drone	Captures hundreds of narrow spectral bands	Crop type classification, mineralogy
Thermal Infrared	Satellite/Drone	Detects heat emissions	Drought stress, water availability

1.4.1 Remote Sensing in Action

Remote sensing data undergoes several processing stages for standard operations.

1. **Data Acquisition:** This process obtains data through the utilization of satellites (Sentinel, Landsat, MODIS) and unmanned aerial vehicles (UAVs).
2. **Preprocessing:** It includes geometric adjustments together with radiometric calibration and atmospheric correction methods.
3. **Feature Extraction:** Satellite imagery analysis uses NDVI, LST (Land Surface Temperature), and texture measures as primary analytical indices.
4. **Classification and Analysis:** The identification of different land cover categories facilitates change monitoring and trend forecasting.
5. **Visualization and Decision Support:** Turning imagery into actionable maps, alerts, or insights.

Example: NDVI for Vegetation Monitoring

$$\mathrm{NDVI} = \frac{\mathrm{NIR} + \mathrm{RED}}{\mathrm{NIR} - \mathrm{RED}} \tag{1.3}$$

NDVI involves taking the reflectance in the NIR and red bands to measure the state of material or vegetation. Vegetation with good health reflects more of NIR and less of red lighting; it has high NDVI values (near 1), but a stressed land or land under desolation has a low value (near 0 or less).

Table 1.3 highlights key satellites widely used in environmental monitoring, showcasing their operators, resolution ranges, spectral bands, and application areas. It includes systems like Sentinel-2, Landsat 8/9, MODIS, and WorldView-3, each serving specific roles from agriculture and forestry to urban development and disaster tracking. These satellites vary in resolution and spectral capabilities, supporting both regional and global-scale observations.

TABLE 1.3
Characteristics of Popular Environmental Monitoring Satellites

Satellite	Operator	Resolution (m)	Spectral Bands	Primary Applications
Sentinel-2	ESA	10–60	13	Agriculture, water monitoring, forestry
Landsat 8/9	NASA/USGS	15–100	11	Land use, hydrology, urban mapping
MODIS	NASA	250–1000	36	Global-scale vegetation and fire tracking
WorldView-3	Maxar	0.31–7.5	29	Precision agriculture, urban analytics

1.5 GENERATIVE MODELS IN REMOTE SENSING

Generative models play a transformative role in remote sensing by reconstructing missing data, enhancing resolution, and simulating complex environmental patterns. These models compensate for satellite limitations such as cloud obstruction, low revisit frequencies, and spectral distortions. Unlike traditional interpolation techniques, GenAI approaches can infer fine-grained spatial or temporal details by learning underlying feature distributions from real datasets. GANs have become prominent for generating realistic satellite imagery and synthetic data augmentation. Their adversarial structure, pitting a generator against a discriminator, enables high-fidelity reconstructions. In particular, GANs are applied in cloud removal, urban reconstruction, and land cover synthesis. For example, models like SpT-GAN use spatial and temporal attention mechanisms to reconstruct cloud-covered images with improved feature preservation (Pang et al., 2024).

VAEs offer probabilistic modeling through latent space encoding, allowing generation of continuous variations in land cover maps and hyperspectral profiles. Their strength lies in generating semantically consistent samples, useful for anomaly detection and missing data recovery in sparsely monitored regions. Diffusion models iteratively refine noisy satellite input to generate high-resolution outputs, outperforming GANs in structural detail retention. Their growing use includes super-resolution of MODIS data and restoration of noisy SAR (Synthetic Aperture Radar) inputs. Although computationally intensive, they are favored for tasks requiring photorealistic output with minimal artifacts (Kong et al., 2022).

Transformers, with their self-attention mechanisms, are gaining traction in geospatial applications due to their ability to capture long-range spatial-temporal dependencies. Vision Transformers (ViTs) and Spatio-Temporal Transformers have been integrated with generative architectures for tasks such as synthetic sequence generation, NDVI prediction, and environmental forecasting. Collectively, these models support data generation in agricultural health monitoring, climate anomaly detection, and rapid disaster assessment (Prasad et al., 2022). Their effectiveness hinges on training data diversity, model architecture tuning, and computational capacity. GenAI is rapidly emerging as a game-changer in the field of remote sensing. Traditional remote sensing systems, while powerful, are often limited by cloud occlusion, sensor noise, low temporal resolution, or insufficient labeled data. GenAI models overcome these challenges by synthesizing missing data, enhancing image resolution, and generating entirely new datasets for training and simulation. These capabilities are critical for achieving high-precision, real-time environmental insights, especially in data-scarce regions.

Figure 1.8 illustrates GenAI in remote sensing pipeline, showing the sequential flow from raw input image through preprocessing and generative models to enhanced outputs. The final stage highlights practical applications in agriculture, disaster monitoring, and environmental analysis. Custom-designed flow based on the conceptual pipeline from.

1.5.1 Data Synthesis and Augmentation

In remote sensing, datasets often suffer from imbalance, particularly when events like floods or wildfires occur rarely but require accurate modeling (Wang et al., 2024). GenAI enables the creation of synthetic but realistic images to augment such limited data.

The details of addressing common remote sensing-related data handling issues are outlined in Table 1.4, and how these issues can be resolved using generative techniques. It solves such problems as unbalanced datasets, time lapsing, and expensive labeling. With the help of the GANs concept, diffusion models, and transformer-based systems, it is viable to generate synthetic data and labels to power model training.

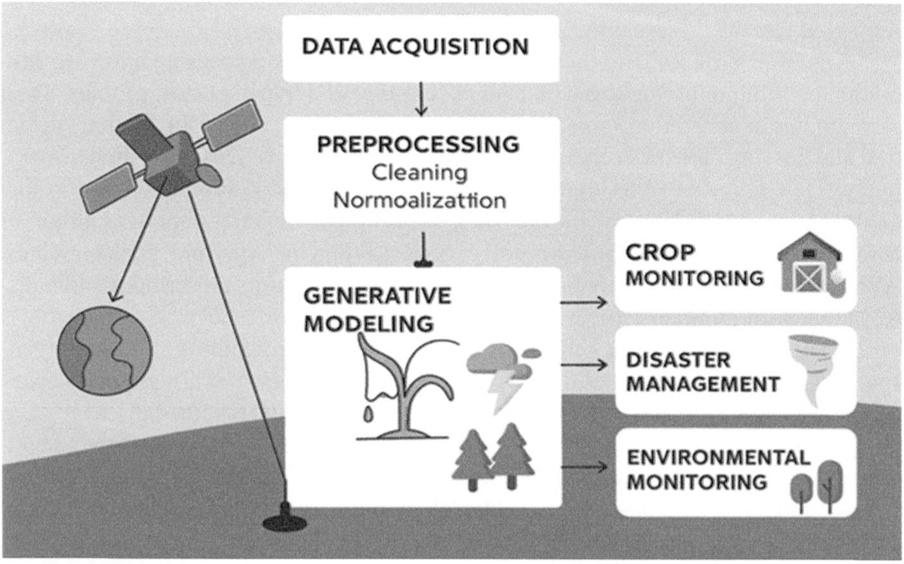

FIGURE 1.8 Generative AI pipeline for remote sensing image enhancement.

TABLE 1.4
Addressing Data Limitations in Remote Sensing Using Generative AI

Problem	Traditional Limitation	GenAI-Based Solution
Imbalanced datasets	Rare event underrepresentation	GANs synthesize minority class samples
Seasonal or temporal gaps	Limited satellite revisits	Diffusion models simulate missing seasonal frames
Labeling bottleneck	High manual annotation cost	LLMs generate pseudo-labels for training augmentation

1.5.2 Cloud Removal and Image De-occlusion

Clouds are among the most persistent challenges in optical satellite imagery. Traditional interpolation or fusion methods often fail when large cloud cover blocks land features. GenAI models like GANs and transformers are capable of learning underlying spatial patterns and hallucinating realistic details beneath the clouds (Li et al., 2023).

Figure 1.9 presents a before-and-after comparison of satellite imagery processed using a GAN-based model for cloud removal. The left panel shows a cloud-covered Sentinel-2 satellite image, while the right panel shows Pixels SpT-GAN re-creates, where sparse transformers are used to preserve ground features.

1.5.3 Super-Resolution Imaging

GenAI can upscale low-resolution satellite images, restoring finer spatial details that are otherwise unavailable due to sensor limitations.

Table 1.5 presents the performance of various image super-resolution methods using PSNR and SSIM as evaluation metrics. Bicubic upsampling shows the lowest scores, while diffusion-based EDiffSR achieves the highest, indicating superior

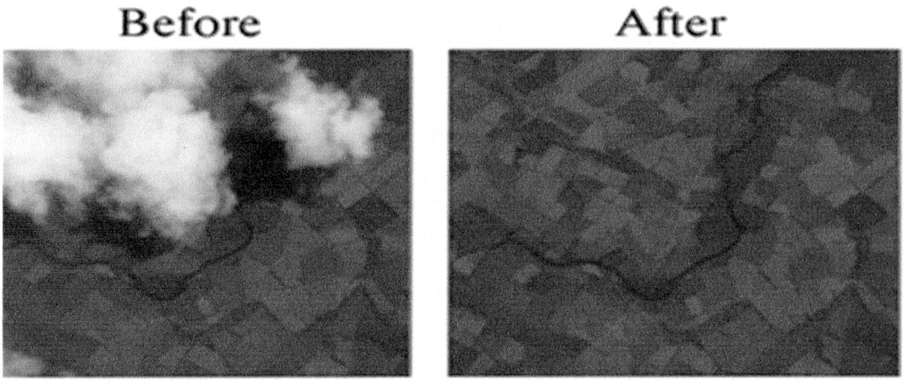

FIGURE 1.9 Cloud obstruction removal using GAN-based SpT-GAN architecture.

TABLE 1.5
Performance of Super-Resolution Models

Model	PSNR (dB)	SSIM Score
Bicubic Upsample	22.3	0.71
CNN-based SR	24.8	0.79
GAN-SR	26.1	0.83
EDiffSR (Diffusion)	28.4	0.88

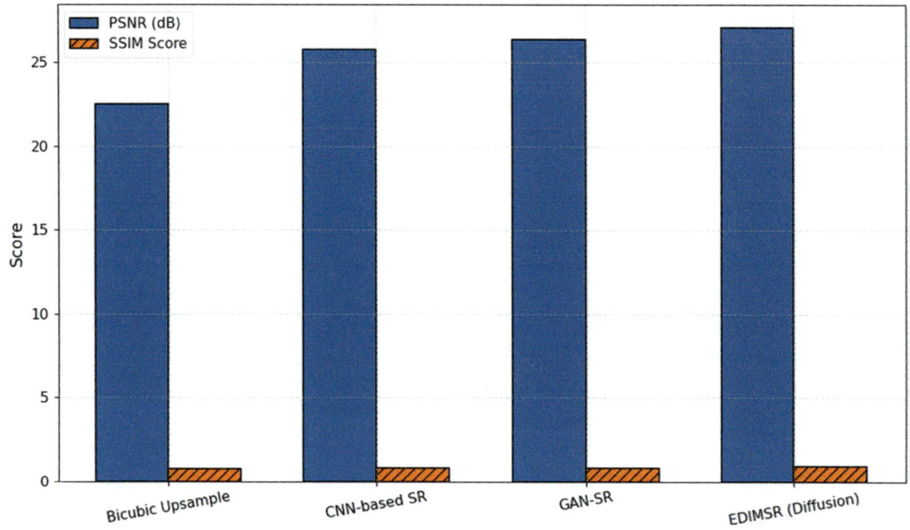

FIGURE 1.10 Comparison of PSNR and SSIM across super-resolution models.

image quality. The results highlight the effectiveness of advanced deep learning techniques over traditional approaches.

Figure 1.10 describes the comparison of models of super-resolution images in terms of clarity and structural similarity-Bicubic Up sample, CNN, GAN-SR, and EDiffSR.

1.5.4 SAR-to-Optical Image Translation

SAR images are often hard to interpret due to their grayscale, texture-heavy nature. Generative models like CycleGANs and Pix2Pix have been used to translate SAR images into optical-style imagery, aiding human and automated interpretation (Ledig et al., 2017).

Use Case: Disaster Zones Monitoring

- After a flood, SAR may capture water-logged areas even under cloud cover.
- GAN models can convert SAR images into an optical style for clear visual mapping, improving emergency response workflows.

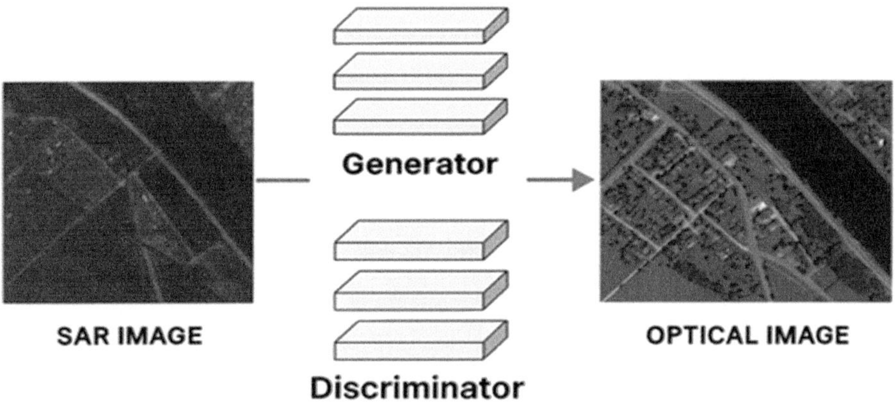

FIGURE 1.11 SAR-to-optical conversion using GAN.

Figure 1.11 illustrates the SAR-to-optical image translation process using a CycleGAN model. The flow shows how synthetic optical images are generated from radar-based SAR inputs, improving visual interpretability for remote sensing applications.

1.6 PRECISION AGRICULTURE WITH GENERATIVE AI

Precision agriculture refers to data-informed, data-centered method of farming that employs spatial, spectral, and environmental data to maximize crop sustainability and productivity. The integration of GenAI and remote sensing technologies offers farmers and agricultural researchers the best chance to understand the situation with crops in an unprecedented way, forecast yields much better than in the past, and make proper decisions concerning the required distribution of resources (Qiu et al., 2023). The synergy contributes to a decrease in waste, yield, and adaptation to climate change so that agriculture would be more resilient and compatible with the SDGs, including SDG 2 (Zero Hunger) and SDG 13 (Climate Action).

1.6.1 Enhancing Crop Health Monitoring

Remote sensing traditionally uses indices like NDVI (Normalized Difference Vegetation Index) to assess vegetation vigor. However, environmental factors such as cloud cover, atmospheric distortion, or sensor limitations often reduce the effectiveness of these indices.

Figure 1.12 illustrates a steady increase in NDVI values and corresponding maize yield from 2021 to 2023. This positive trend indicates improved vegetation health and productivity over the years. The parallel rise suggests a strong correlation between enhanced NDVI and agricultural output.

1.7 ENVIRONMENTAL MONITORING AND DISASTER RESPONSE

Environmental monitoring plays a critical role in understanding and managing the Earth's natural systems. From detecting deforestation and tracking glacier retreat to monitoring floods, wildfires, and droughts, remote sensing is essential. However,

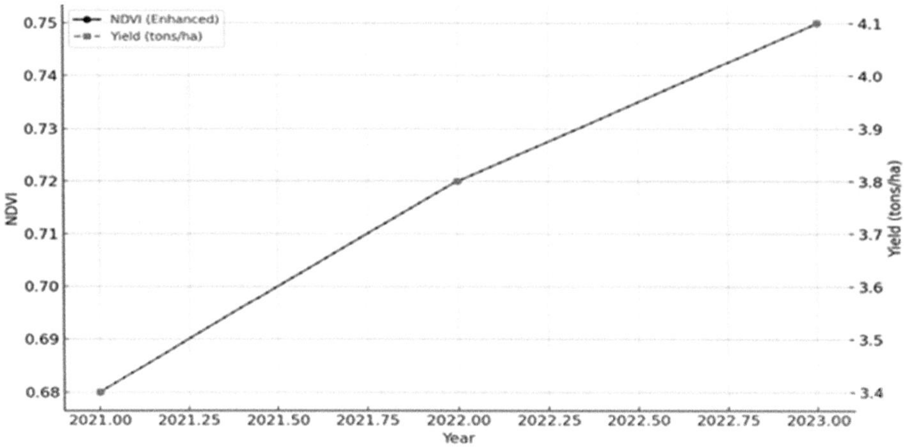

FIGURE 1.12 NDVI trend and yield variation for maize (2021–2023).

timely and accurate environmental data is often limited by sensor availability, resolution, and atmospheric conditions. Here, GenAI becomes a force multiplier, offering tools to simulate, reconstruct, and enhance environmental data for better situational awareness and response.

1.7.1 Change Detection and Land Use Monitoring

Change detection involves analyzing imagery from two or more time periods to identify environmental shifts like vegetation loss, urban sprawl, or glacier melt. The GenAI Meets SAR paper highlights the use of SAR-to-optical translation GANs to monitor terrain changes post-natural disasters like earthquakes and floods, improving the speed and accuracy of disaster mapping.

Figure 1.13 displays a side-by-side comparison for GAN-enhanced change detection using SAR imagery. The left image shows the original SAR image post-disaster, while the right highlights GAN-based model output of change detection in terrain shift detection.

1.7.2 Wildfire and Flood Mapping

During extreme weather events, traditional monitoring systems often fail due to cloud cover, low visibility, or missing temporal data. GenAI can fill gaps, simulate spread patterns, and support real-time emergency response (Li et al., 2024).

Case Highlight: Flood Simulation using Conditional GANs

- Input: Partially flooded SAR image
- Output: Simulated full-area flood map with temporal progression
- Benefit: Allows first responders to predict inundation zones ahead of time

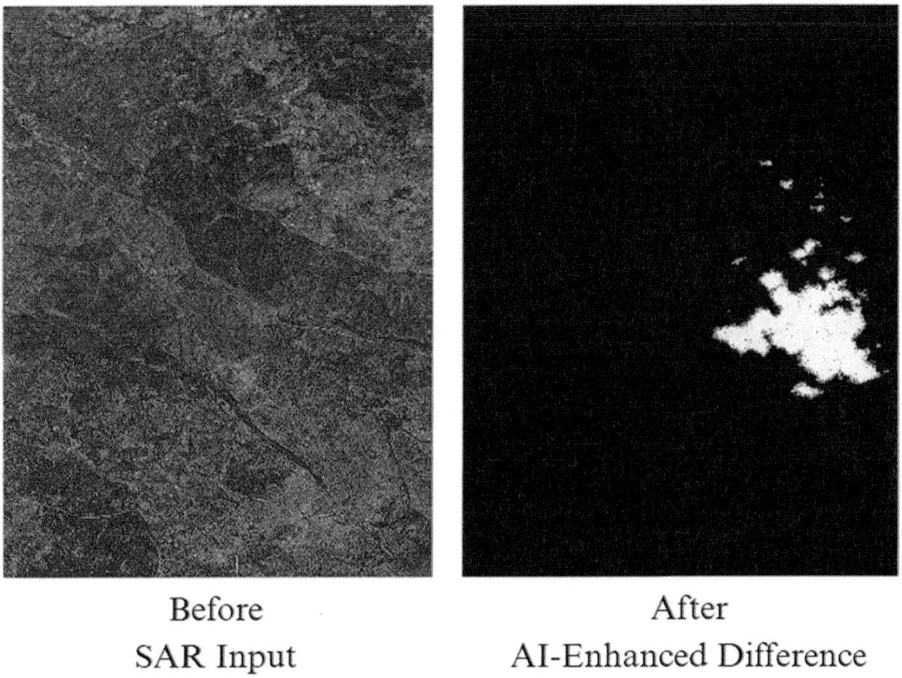

FIGURE 1.13 Change detection using SAR imagery and GAN enhancement.

1.7.3 Deforestation and Climate Monitoring

Monitoring subtle vegetation loss and forest degradation at a granular level is a persistent challenge in satellite monitoring. GenAI helps in generating high-resolution time-lapse maps, simulating future land cover change under different development or conservation scenarios, and enhancing the detection of illegal logging in cloud-prone areas (Sui et al., 2024).

1.7.4 Air and Water Quality Assessment

GenAI is also being applied to augment sensor networks for environmental health monitoring:

- **Air Quality**: GANs extrapolate from limited sensor stations to create full-city PM2.5 and NO2 maps.
- **Water Quality**: AI systems like Clean Water AI detect bacterial patterns using neural microscopy, aiding in SDG 6 (Clean Water and Sanitation).

Table 1.6 highlights how modern tools improve environmental monitoring. Air quality data from ground sensors is refined using interpolation techniques. Satellite images of chlorophyll are corrected for cloud blocking visibility. For water clarity, UAV images benefit from sharpness restoration methods.

TABLE 1.6
Environmental Indicators Improved via GenAI Techniques

Parameter	Sensor Type	GenAI Enhancement
PM2.5/NO2	Ground Sensors	GAN interpolation for full-grid visualization
Chlorophyll-a	Satellite (MODIS)	VAE fills gaps in cloud-covered zones
Turbidity Index	UAV/Drone Imagery	GAN-trained clarity restoration techniques

1.8 ADVANCING THE SUSTAINABLE DEVELOPMENT GOALS

The SDGs by the United Nations are an international list of issues that have to be tackled by the world in terms of issues such as hunger, health, environmental sustainability, and advancement in technology. The combination of GenAI and remote sensing technologies provides an effective framework to support advancements in most of these goals. Environmental data simulation, augmentation, and enhancement made possible by this approach help decision-making (Mirzaei et al., 2024) based on evidence, which benefits resource-limited or vulnerable areas.

Table 1.7 outlines how technology supports key SDGs. It enhances food security through better crop monitoring and disease alerts. Water quality and climate resilience benefit from advanced simulations and contamination detection. Urban planning and ecosystem protection are strengthened with accurate mapping and real-time insights.

1.9 TECHNICAL CHALLENGES AND RESEARCH FRONTIERS

The widespread implementation of advanced technology with remote sensing requires overcoming multiple substantial obstacles to attain widespread acceptance. Four main obstacles to medical research include technological constraints, restricted data availability, and concerns about ethics, with accurate interpretation needed in all cases. Researchers need to overcome these barriers in order to fully utilize GenAI for critical decision-making regarding agriculture and climate, as well as disaster response.

TABLE 1.7
Contribution of GenAI-Enhanced Remote Sensing to the SDGs

SDG	Target Domain	GenAI Role
SDG 2	Food Security	Crop yield prediction, disease detection
SDG 6	Clean Water Access	Turbidity modeling, synthetic water quality maps
SDG 9	Industry Innovation	Infrastructure mapping, model development
SDG 11	Urban Sustainability	Land use detection, unplanned settlement monitoring
SDG 13	Climate Action	Simulated climate scenarios, alert systems
SDG 15	Life on Land	Deforestation tracking, biodiversity habitat monitoring
SDG 17	Global Partnerships	Open-source models, federated learning across nations

1.9.1 Data Quality and Availability

Generative models need extensive datasets that contain both high-quality details and a wide variety of input information. However, in remote sensing, data can be incomplete, cloud-obstructed, or low-resolution. Labeled data is scarce, especially for rare events like pest outbreaks or floods. Privacy or geopolitical restrictions often limit data access.

The black-box nature of GANs and diffusion networks creates difficulties for users, particularly in government or environmental agencies, to trust their model outputs.

1.9.2 Computational Complexity and Cost

Models like diffusion transformers need expensive high-end GPUs alongside large memory capacity and lengthy training processes. Many developing areas, along with edge-based applications, which include drones and phones, experience performance limitations due to this bottleneck (Guo et al., 2023).

Table 1.8 highlights the computational resources required for different generative model architectures. It compares model size, estimated training time, and GPU needs, showing how complexity scales from GANs to transformers. The data illustrates the trade-off between model performance and resource demands.

Figure 1.14 compares GenAI model types by their complexity, training time, and GPU demands. Basic GANs and VAEs are lighter, needing fewer hours and GPUs, while diffusion and transformer models demand extensive time and hardware. Transformers, with up to 600M parameters, are the most resource-intensive. The graph reinforces the steep rise in compute as model size increases.

Emerging Solutions

- Model compression and pruning
- Distillation techniques
- Edge-compatible lightweight architectures

1.9.3 Ethical, Legal, and Social Implications

- **Fake Data Risk:** Synthetic images may be misused for misinformation.
- **Bias Amplification:** Poorly balanced training sets can propagate bias.
- **Data Ownership:** Who owns the generated content when training data is derived from multiple international satellites?

TABLE 1.8
Resource Requirements of Generative Models

Model Type	Parameters (M)	Training Time (hrs)	GPU Requirements
GAN (basic)	~25M	8–12 hrs	1× A100
VAE	~15M	5–10 hrs	1× V100
Diffusion	~75M	24–36 hrs	2–4× A100
Transformer	~300M–600M	72+ hrs	4–8× A100

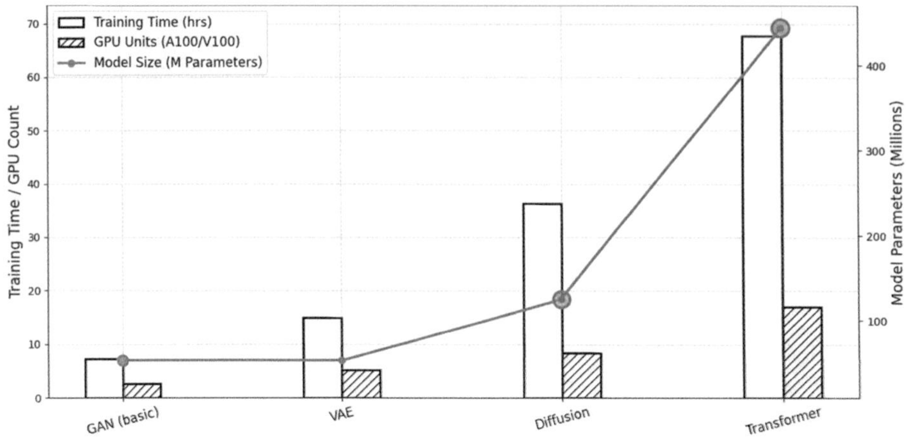

FIGURE 1.14 Training time and GPU requirements for GenAI Models (2023).

TABLE 1.9
Active Research Challenges and Directions in GenAI for Remote Sensing

Open Problem	Why It Matters	Suggested Research Direction
Explainable GenAI	Adoption in policy-sensitive environments	Saliency maps for diffusion/ transformers
Edge-compatible Models	Low-power, real-time deployment	Compressed GAN architectures
Few-shot Generation	Simulating rare events with minimal data	Meta-learning fused with VAEs
Multi-modal Data Fusion	Better accuracy from diverse sensors	Transformer-based cross-sensor fusion
Real-time Adaptation	Disaster detection and rapid response	Online learning and streaming GAN models

Table 1.9 outlines key research gaps in generative models used in geospatial contexts. Challenges like explainability, low-resource deployment, and cross-sensor fusion remain unresolved. Each problem is paired with promising research paths, such as pruned architectures or adaptive sampling. Addressing these issues is crucial for practical, real-world implementation.

1.10 CONCLUSION

GenAI has transformed the world of remote sensing with the ability to synthesize, enhance, and even simulate data on Earth observation. It has greatly enhanced the spatial resolution as well as filled in gaps in the data on the image to predict the setting of the environment. AI developments, like the GAN, the VAE, diffusion network, and transformer, allow industries such as precision agriculture, disaster response, and climate monitoring to obtain actionable intelligence. The innovations can help to directly boost the UN SDGs, such as Zero Hunger, Climate Action, and Clean Water, through AI-based yield predictions, disease detection in early stage, and efficient resource utilization. Generative models enable national agencies and farmers to measure the progress at different scales precisely and with speed. Nevertheless, the implementation of these tools does not come without its own difficulties, such as a small amount of training data, a considerable computational workload, and the dangers of transparency-impeding models, as well as the effect of governance that considers ethical implications. The solution to these threats lies in responsible research activities, open-source collaborations, and inclusive policies. Technical innovations are not enough, nor should they be, and must be characterized by transparency, trust, and equity in deployment plans. GenAI architectures that are domain-adaptive, transparent, and lightweight need to be developed to accommodate optimal performance even in an edge-type environment. As long as it innovates and continues to be ethically managed, GenAI presents an immense avenue to the democratizing and resiliency-enhancing access of environmental data. With a rising emergency in the environment, the AI-remote sensing interface is one of the solutions that can bring pivotal understanding, sustainability, and fair intervention to a smarter planet that is more proactive than ever before.

REFERENCES

Aleissaee, A. A., Kumar, A., Anwer, R. M., Khan, S., Cholakkal, H., Xia, G.-S., & Shahbaz Khan, F. (2022). Transformers in remote sensing: A survey. *IEEE Geoscience and Remote Sensing Magazine*, *10*(3), 31–58.

Goodfellow, I., Pouget-Abadie, J., Mirza, M., Xu, B., Warde-Farley, D., Ozair, S., Courville, A., & Bengio, Y. (2014). Generative adversarial nets. *Advances in Neural Information Processing Systems*, *27*. https://dl.acm.org/doi/10.5555/2969033.2969125

Guo, J., Lv, F., Shen, J., Liu, J., & Wang, M. (2023). An improved generative adversarial network for remote sensing image super resolution. *IET Image Processing*, *17*(6), 1852–1863. https://doi.org/10.1049/ipr2.12760

Jin, M., Wang, P., & Li, Y. (2024). HyA GAN: Attention driven cloud removal using channel and spatial attention in GAN. *International Journal of Remote Sensing*, *45*(6), 1755–1773. https://doi.org/10.1080/01431161.2024.2316673

Khanna, S., Liu, P., Zhou, L., Meng, C., Rombach, R., Burke, M., & Ermon, S. (2023). DiffusionSat: A generative foundation model for satellite imagery. arXiv. https://doi.org/10.48550/arXiv.2312.03606

Kong, Y., Liu, S., & Peng, X. (2022). Multi scale translation method from SAR to optical remote sensing images based on conditional generative adversarial network. *International Journal of Remote Sensing*, *43*(8), 2837–2860. https://doi.org/10.1080/01431161.2022.2072179

Längkvist, M., Karlsson, L., & Loutfi, A. (2014). A review of unsupervised feature learning and deep learning for time-series modeling. *Pattern Recognition Letters*, *42*, 11–24. https://doi.org/10.1016/j.patrec.2014.01.008

Ledig, C., Theis, L., Huszár, F., Caballero, J., Aitken, A. P., Tejani, A., … & Shi, W. (2017). Photo-realistic single image super-resolution using a generative adversarial network. *Proceedings of the IEEE Conference on Computer Vision and Pattern Recognition (CVPR)*, 105–114. https://doi.org/10.1109/CVPR.2017.19

Li, F., Yang, J., & Zhang, X. (2023). Cloud removal using SAR and optical images via attention mechanism–based GAN. *Remote Sensing of Environment*. Advance online publication.

Li, L., Ma, H., Zhang, X., Zhao, X., Lv, M., & Jia, Z. (2024). Synthetic aperture radar image change detection based on principal component analysis and two level clustering. *Remote Sensing*, *16*(11), 1861. https://doi.org/10.3390/rs16111861

Ma, W., Pan, Z., Yuan, F., & Lei, B. (2019). Super resolution of remote sensing images via a dense residual generative adversarial network. *Remote Sensing*, *11*(21), 2578. https://doi.org/10.3390/rs11212578

Mirzaei, A., Bagheri, H., & Khosravi, I. (2024). Enhancing crop classification accuracy by synthetic SAR optical data generation using deep learning. *Remote Sensing*, *16*(5), 1234.

Pang, B., Zhang, Y., & Pei, J. (2024). PRO SSRGAN: Stable super resolution generative adversarial network for Gaofen 5 remote sensing images. *International Journal of Remote Sensing*, *45*(3), 224–239. https://doi.org/10.1080/01431161.2024.2339205

Prasad, A., Mehta, N., Horak, M., & Bae, W. D. (2022). A two step machine learning approach for crop disease detection using GAN and UAV technology. *Remote Sensing*, *14*(19), 4765. https://doi.org/10.3390/rs14194765

Qiu, L., Zhang, X., Gu, C., & Zhu, S. (2023). A dual attentive generative adversarial network for remote sensing image change detection. arXiv. https://doi.org/10.48550/arXiv.2310.01876

Sui, J., Wu, Q., & Pun, M.-O. (2024). Denoising diffusion probabilistic model with adversarial learning for remote sensing super-resolution. *Remote Sensing*, *16*(7), 1219. https://doi.org/10.3390/rs16071219

Wang, H., Haeri Boroujeni, S. P., Chen, X., Bastola, A., Li, H., Zhu, W., & Razi, A. (2024). FLAME Diffuser: Wildfire image synthesis using mask-guided diffusion. arXiv. https://doi.org/10.48550/arXiv.2403.03463

Xiao, Y., Yuan, Q., Jiang, K., He, J., Jin, X., & Zhang, L. (2023). EDiffSR: An efficient diffusion probabilistic model for remote sensing image super-resolution. *Neural Computing and Applications*.

Zhu, X. X., Tuia, D., Mou, L., Xia, G.-S., Zhang, L., Xu, F., & Fraundorfer, F. (2017). Deep learning in remote sensing: A comprehensive review and list of resources. *IEEE Geoscience and Remote Sensing Magazine*, *5*(4), 8–36. https://doi.org/10.1109/MGRS.2017.2762307

2 Core Concepts of Understanding Generative AI Algorithms and Models

Vaibhavi Chavan and R. Ayswarya

2.1 INTRODUCTION

Generative Artificial Intelligence (GAI) represents one of the most revolutionary developments in modern artificial intelligence. Generative AI models stand apart from traditional system AI models due to their unique capability for generating genuinely new content from various mediums, including text, images, music, and video. The capability has transformed operations in multiple sectors, including natural language processing (NLP), artistic creation, medical research, investigations, and scientific simulation work. Generative models evolved through probabilistic mechanisms into deep learning systems (Kingma & Welling, 2014), resulting in a significant advancement in machine learning and transforming computer data generation and data interaction methods. Implementing and testing modern remote sensing technologies necessitates a comprehensive understanding of the ethical considerations inherent in AI-driven solutions. Additionally, contemporary digital data enables researchers to identify emerging technological patterns. This revolutionary AI content generation relies on fundamental models that enhance AI's capacity to produce novel content. Ian Goodfellow (2014) and his team developed Generative Adversarial Networks (GANs), a competitive neural network system that generates highly realistic outputs using two neural networks: a generator and a discriminator. VAEs utilize probabilistic encoding and decoding systems to learn from and generate data distributions that are complex. These models have numerous applications in various fields, including image generation, anomaly detection in industrial environments, and data augmentation for tasks such as facial recognition and other computer vision applications.

Transformer-based models emerged as a key advancement in generative AI development, completely transforming NLP operations. The widespread adoption of transformers was made possible by the "Attention Is All You Need" paper published in 2017 and resulted in the groundbreaking OpenAI Generative Pre-Trained Transformer (GPT) series, which now shows exceptional abilities in text generation, conversation modeling, and code synthesis. These models can now handle various domains through pre-training and fine-tuning processes that need minimal human

DOI: 10.1201/9781003616207-3

interaction. Diffusion models represent the current front-runner solution when compared to both GANs and VAEs. These model training methods perform two sequential steps: applying noise to samples and then removing it to produce high-quality outcomes. Stable and consistent image generation results are achieved through diffusion-based methods, which is why both DALL-E by OpenAI and Stable Diffusion from Stability AI utilize these approaches (Dhariwal & Nichol, 2021).

The advancement of GAI technology has improved to a significant degree, yet technical solutions remain incomplete. Generative AI faces key challenges, including technical issues like GAN mode collapse and transformer inefficiency, as well as ethical concerns related to AI-generated content. Separates for logical clarity. Scientific experts are developing hybrid models by merging beneficial elements from different generative approaches to address their limitations. AI systems require immediate attention to their ethical deployment practices, as well as the reduction of misinformation and bias during the implementation of AI. This provides extended explanations about all fundamental components that power generative AI algorithms and models in operation. The assessment examines generative models starting from their core operating principles, followed by a comparison of key attributes across models, which leads to insights on real-world utilization success (Vaswani et al., 2017). A basic understanding of the foundational principles of generative AI enables researchers, developers, and practitioners to identify the technology's performance boundaries and strengths, thereby advancing its future development and applications.

Figure 2.1 illustrates four key types of generative models employed in advanced technology. GANs are known for creating realistic images and enhancing their quality. Transformers help produce and summarize text in natural language. VAEs are used to detect irregular patterns and manage data in a smooth latent space. Diffusion models play a key role in enhancing medical imaging and reducing noise in visuals.

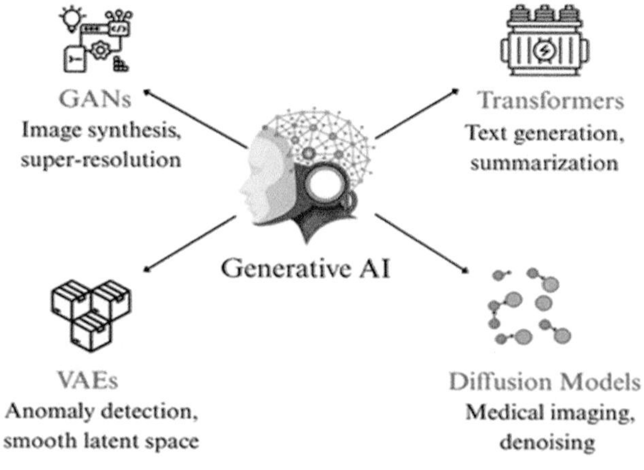

FIGURE 2.1 Key generative AI models—GANs, VAEs, transformers, and diffusion models—and their primary functions in image and text generation.

2.2 THE ESSENTIAL ARTIFICIAL INTELLIGENCE GENERATION MODELS WITH THEIR STRUCTURAL DESIGNS

The field of GAI has advanced significantly with the development of sophisticated models capable of producing photorealistic images, generating natural-sounding text, synthesizing audio, and delivering detailed scientific information. "These models operate on distinct principles, often integrating statistical methods with deep learning to generate data that mimics real distributions." Generative AI technologies have expanded through key models, including GANs, Variational Autoencoders (VAEs), transformer-based models, and diffusion models, collectively becoming leading forces in artificial intelligence. These different computational modeling techniques work toward the common purpose of creating superior artificial content, which can help numerous domains with their diverse requirements.

Through GANs introduced by Ian Goodfellow (2014), AI entered an innovative stage because the generative AI mechanism operates through an adversarial process consisting of a generator and a discriminator neural network that compete until optimized. Within GANs, the generator produces artificial data that resembles real-world patterns, as the discriminator determines whether the content is original or synthetic. Throughout the training process, the adversarial dynamic of generator and discriminator progressively enhances the generator's ability to produce increasingly realistic outputs with each iteration.

GANs have demonstrated success across various fields, including deepfake creation, image super-resolution, medical imaging, and art generation. However, their practical implementation often encounters several challenges. One common issue is that the generator tends to develop performance limitations over time, causing it to focus on generating narrow, repetitive outputs rather than producing diverse and varied results. GAN training becomes complicated because monitoring the generator-discriminator ratio remains essential to prevent the system from becoming unstable. Research on GAN training issues led to the development of three enhanced versions: Conditional GANs (cGANs) by Mirza & Osindero (2014), which produce controlled outcomes; StyleGAN, which enhances image synthesis; and Cycle GAN, used for image-to-image translation by Zhu, Park, Isola, and Efros (2017). The high computational demands of adversarial training for GANs pose a significant limitation on their practical application in real-time scenarios, despite producing outstanding visual outputs for images and videos.

VAEs achieve efficient data encoding and reconstruction by implementing probabilistic methods that utilize statistical inference techniques for data modeling. The traditional autoencoder executes direct data compression, whereas VAEs design an orderly latent space for generating smooth, continuous data variations. The controlled synthesis methodology can benefit from this function because medical imaging, anomaly detection, and structured data creation belong to this category.

The VAE architecture divides its operation into two separate parts: an encoder that transforms data into a probabilistic latent space and a decoder that uses the encoded information to reconstruct the data. Standard autoencoders differ from VAEs because the probability distribution approach, rather than concrete encoding, results in varied outputs that avoid direct copycat behavior on training information. VAEs deliver interpretable and stable results; however, their reconstructed outputs are often less detailed and appear blurrier compared to those generated by GANs. VAEs excel in controlling variations in data output, making them particularly suitable for tasks that require structured data representation rather than high-definition image generation (Brock, Donahue, & Simonyan, 2019).

Over the past couple of years, transformer models have brought substantial advancements to generative AI systems, enabling text generation in NLP applications. Transformers enabled self-attention processing to analyze complete sequence information as a whole rather than the sequential data processing performed by recurrent neural networks (RNNs) and long short-term memories (LSTMs). Transformers addressed long-range dependency issues by introducing modern design features that made them suitable for processing large-scale datasets. Transformers offer capabilities beyond text generation, incorporating multimodal operations with DALL-E to create pictures from written instructions. However, the resource-intensive nature of these generative models poses scalability issues, limits accessibility, and contributes to environmental concerns such as high energy consumption (Bengesi et al., 2024).

2.2.1 Generative AI Diffusion Models

Diffusion models have surpassed GANs in the field of image synthesis, emerging as a more effective solution for generating realistic visual content. These models operate through a two-step process: initially, they add noise to render images unrecognizable, and subsequently, they learn to reverse the noise, reconstructing clear and natural visuals. This systematic denoising approach yields more reliable and higher-quality results compared to GANs, demonstrating superior performance in generating high-fidelity images (Dhariwal & Nichol, 2021). Scientists have successfully applied diffusion models to create text-to-image generation systems, develop scientific simulations, and reconstruct medical images. The distinct advantage of diffusion models over GANs is their ability to deliver consistent and numerous reliable output results despite GAN limitations on training stability and mode collapse. The main drawback of using diffusion models is their slow processing speed, which requires multiple successive steps to create an image, thus outperforming other modeling approaches. Stable Diffusion and DALL-E tools showcase the high image quality and precision of diffusion models, which makes them dominate in AI creative content operations.

2.2.2 Convergence of Generative Models and Future Prospects

Scientists widely agree that combining different generative AI techniques, such as GANs, VAEs, transformers, and diffusion models, can lead to more powerful and effective solutions than using any single method alone. Contemporary research focuses on developing interpretable generative systems by combining the structured latent representations of VAEs, the adversarial training efficiency of GANs, the scalability of transformers, and the stability of diffusion models. The ethical pitfalls associated with generative models stem from the propagation of inaccuracies, inherent biases, and the unethical control of synthesized content. To safeguard these

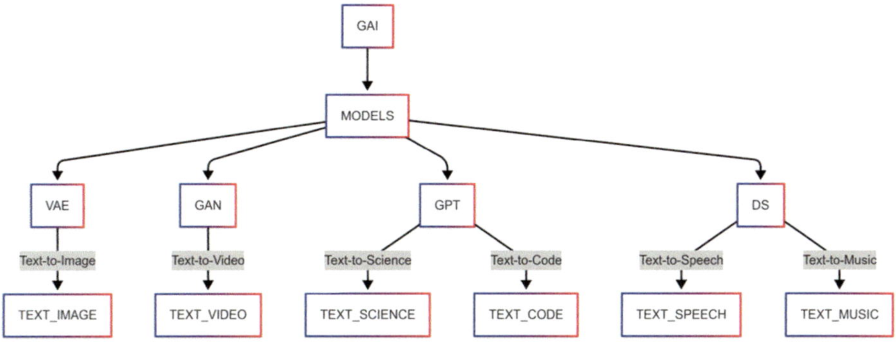

FIGURE 2.2 Generative AI models mapped to tasks: VAEs for images, GANs for videos, GPT for text/code, and DistilBERT for voice and music synthesis.

technologies for public welfare, the advancement of generative models must align technological advancement with ethical responsibility. Iterative tuning of model architectures, coupled with efforts to enhance usability, will improve overall performance while promoting solutions that prioritize user-oriented decision-making.

Figure 2.2 illustrates various generative AI models used for specific tasks. Specifically, VAE transforms text into images, whereas GAN converts text into videos. GPT is used to generate scientific content and to write programming code. DistilBERT (DS) produces voice and music creation based on written input. All these models are derived from a core generative framework to fulfill diverse creative tasks.

2.3 GAN-BASED TEXT-TO-IMAGE CONVERSION

GANs enabled new possibilities in translating linguistic descriptions into natural visual imagery. It generates high-quality images based exclusively on linguistic information, allowing users to visualize conceptual or nuanced perceptions through human language. In contrast, classical image generation approaches require text-to-image GANs to analyze semantic context and connections from the input text while simultaneously generating accurate graphical and pixel-level details. This dual-processing capability makes them particularly effective for applications requiring precise alignment between visual outputs and descriptive content, such as in design, content creation, and scientific visualization. The "Generator" network relies on text interpretation to produce images, while the "Discriminator" network checks both authenticity and fidelity of the generated images. Between competitive and iterative operations, the two networks generate successively refined images until they produce highly accurate and realistic outputs.

2.3.1 Operational Steps in Remote Sensing Applications

Specific and informative input plays a significant role in the final visual output. Hence, after an input definition has been made, the linguistic data is processed through NLP approaches. The preprocessing step enables NLP algorithms to read textual data and identifies the necessary semantic elements, such as geographic locations, geographic features, weather conditions, vegetation types, and elements of built infrastructure. Embeddings analytically create semantic vectors of the input text materials and convert descriptive language into numerical variables, enabling the computation of GAN.

The generator of the GAN uses semantic inputs to create the initial synthesized images, which are refined through consecutive iterations to obtain greater fidelity. This process is illustrated in the work of Reed et al. (2016). The GAN model has the discriminator evaluate the validity of each created image in relation to authentic remote sensing images that can be gathered using historical satellite or Unmanned Aerial Vehicles (UAV) observations. The discriminator is used to refine the outputs and becomes progressively closer to real ones over a series of repetitions, as it assists the generator by providing continuous feedback on the quality of the generated images to build synthetic satellite images that are increasingly similar to the actual images obtained through remote sensing. The process of GAN training is repeated until the synthetic results achieve a realistic accuracy, essentially eliminating the detectable differences between them and accurate remote sensing imagery. By the end of the process, the images produced are indescribably similar to that of authentic satellite imagery except that they faithfully reflect the meaning of the original text input.

2.3.2 The Role of NLP in the Generation of Image

NLP is an essential component in the frameworks of text-to-image generation, where GAN serves as the primary driver. It is capable of manipulating and processing the semantic meaning of textual input through translation into numerical values by employing linguistic algorithms, thus making it applicable in the process run by GAN. The primary process of converting verbal input into an image output is based on the text transfer format, which includes the following stages: token splitting, grammatical analysis, meaning-based encoding, and context-sensitive representation, all of which are conducted using transformer RNNs.

Language precision is also crucial, especially in the context of remote sensing activities, where the data describing the environment must be clearly and accurately expressed. Poor interpretation of natural language can be a significant issue for the accuracy and stability of the system, leading to the failure to generate images. To eliminate such mistakes, the organized NLP preprocessing techniques are adopted, and subsequently, accuracy rules are applied. The establishment of semantic fidelity prior to the fusion with the GAN model is necessary in such routines to keep the textual data in balance between the parity of representation. Interpreting complex textual input using high-quality NLP models is a giant leap toward the development of GAN-based remote sensing systems. Effective semantics at high levels essentially provide a good representation of complex text in a visual form, which enhances effectiveness and usability for an individual in reality.

2.3.3 Addressing Practical Remote Sensing Challenges

Remote sensing is also a data-dependent task, with several data-related issues, including unrecorded data, sensor-to-sensor variability, and signal degradation due to atmospheric influence, as well as site-specific limitations. GAN-based models overcome the problem of acquiring a dataset, as GANs can generate missing information based on synthetic images. Methods of silencing, such as these frameworks, introduce possible ways of sustaining data integrity. The reduced loss estimates and gradient management technique of Wasserstein GANs (WGANs) facilitate consistent training, preventing numerical instability and mode collapses, while also producing high-quality and consistent visuals in remote sensing (Arjovsky et al., 2017). GAN-generated images can reduce operational expenses and logistics complications in traditional remote sensing by supplementing real data. The ability to produce a visual product from raw data can help organizations streamline their environmental monitoring, emergency response, and urban planning tasks without the need for field surveys.

The next promising direction in the field of remote sensing is the further application of models based on GANs to enable text-to-image transformations and enhance performance. Such frameworks facilitate the creation of graphically legitimate outcomes while enhancing realistic flexibility, accuracy, and functionality. The impact of GANs on remote sensing is likely to grow rapidly due to the development of more effective denoising techniques and enhanced methods for generating semantic representations.

2.4 REMOTE ENVIRONMENTAL MONITORING USING GENERATIVE AI

The generative methodologies are revolutionizing remote environmental monitoring by enabling the generation of detailed and context-aware pictures from limited or incomplete information sources. Scientists apply state-of-the-art models, such as GANs, VAEs, and diffusion models, in their efforts to generate high-quality images of distant or obscure terrain, with minimal to no sensor information or satellite visibility. By enhancing the spatial and temporal accuracy of environmental data sets, these intelligent systems not only improve forecasting capabilities for events such as forest degradation, urban development, and changes associated with climate change but also enhance the understanding of these phenomena. As generative frameworks continue to evolve, they will become an increasingly important aspect in live environmental monitoring, emergency efforts, and sustainability planning for resources in the long run.

2.4.1 Advantages of GAN Models in Environmental Monitoring

Studies have demonstrated the growth of GAN in the field of remote sensing, which offers enhanced observation capabilities in ecological and environmental applications. With a highly realistic portrayal of the environment, such systems provide specialists with the visual insights essential for undertaking crucial planning and analysis (Zhu et al., 2022). Compared to conventional remote sensing activities, GAN-driven visuals significantly reduce the need for large sensor networks and eliminate the time lag associated with manual data collection. The transition facilitates increased productivity and workflow flexibility, enabling analysts to experiment with and interpret situations that are otherwise difficult to track.

GANs deliver innovative solutions for environmental observation by resolving long-standing obstacles, such as cloudy skies, insufficient sensor distribution, and large geographic expanses. They generate convincing datasets that compensate for missing or incomplete information, especially in regions where observations are sparse. Consequently, these systems enable environmental organizations to expand their capacity to observe and address challenges such as extreme weather, climate impact, and urban planning, even with limited resources.

Figure 2.3 outlines the common uses of remote sensing, which involves processing various input data sources to produce valid environmental data. The satellite images provide land cover maps, which utilize cloud data for fog detection. Crucial patterns are revealed through geospatial data analysis, making planning more effective. Blurred images at night are enhanced to reveal clear images through light activity. Such developments enhance better decision-making in ecological and city management systems.

2.4.2 Enhancement of Processing and Quality of Data

The implementation of GAN-based models in remote sensing is expected to result in a significant reduction in response time, thanks to improved data processing and optimized workflows. GANs provide precise and high-resolution images using adversarial learning techniques, which reduces the time required to analyze the data. Rapid data processing is a crucial factor in ecological emergencies, as in this case,

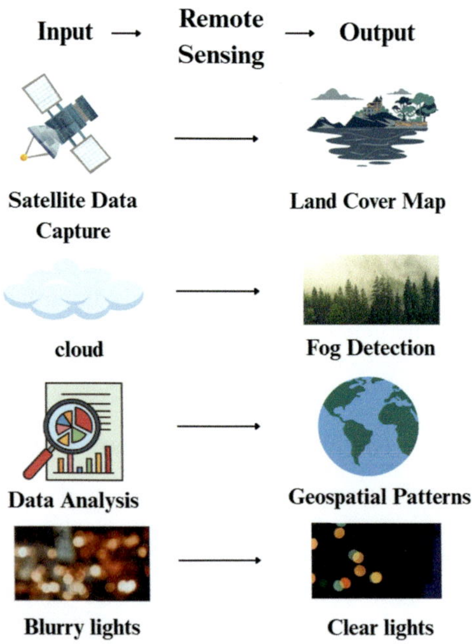

FIGURE 2.3 Common remote sensing applications, including land cover mapping, fog detection, night-time image enhancement, and geospatial pattern analysis.

prompt decisions are necessary to facilitate the mitigation and restoration processes. These systems accelerate ecological surveillance, as the input of simple data and descriptive information is translated into swift visual images. The instant creation of graphics enables first responders to locate and analyze areas of interest promptly, thereby increasing situational awareness during operational levels. The result is a more effective disaster management technique, where authorities and first responders are better equipped to operate quickly and allocate supplies to areas of concern. It enhances workflow efficiency and simultaneously strengthens the impact of disaster management work.

Modern environmental monitoring involves ensuring real-time observation, prompt data incorporation, and effective communication of information in the present day. GAN-based generative models generate new visuals based on real-world environmental conditions as defined by the requisite specifications. Real-time GAN applications can be utilized to survey disasters and monitor biodiversity, as well as deliver services to city developers and operators within the agribusiness system. The creation of real-time images on GAN models in the case of wildfires begins with sensor-based textual or visual reports that enable emergency control groups to monitor rapidly shifting fire lines and temperatures. The Fine-Grained Environmental Generated Samples (FEGS) images generated by the GAN models enhance real-time experiences for emergency responders, enabling them to make rapid and more informed decisions regarding evacuations and the effective utilization of their firefighting services (Bailey, Xu, & Nogueira, 2021).

GAN models in precision agriculture enable farmers to conduct regular assessments of crop health and productivity through the analysis of text data. Field conditions become immediately visible to farmers through visual interpretations of textual crop reports and sensor data, enabling them to promptly detect pest infestations, water stress, and nutrient deficiencies, and apply corresponding actions. Agricultural sustainability, along with improved productivity, benefits significantly from the real-time monitoring capabilities enabled by GAN models. The utilization of GAN-powered real-time surveillance enables municipal authorities, along with city planners, to analyze urban expansion patterns within metropolitan areas. Planners can prevent and resolve environmental challenges by generating real-time images from demographic data and zoning documents, leading to effective urban planning for sustainability purposes.

Figure 2.4 highlights the process of using data for environmental monitoring. It begins with collecting information, which is then used to create visual representations. They are helpful for immediate disaster action and agricultural enhancement. The knowledge acquired also guides sound urban planning. The combination of these measures enables sound choices that protect the environment.

2.4.3 Practical Case Studies

Several illustrative cases demonstrate the effective role of GAN systems in monitoring environmental factors. Remote sensing technology, when combined with GAN models, has proven helpful in monitoring the Amazon rainforest to detect forest felling and illegal logging. The GAN models enable the prevention of impediments such

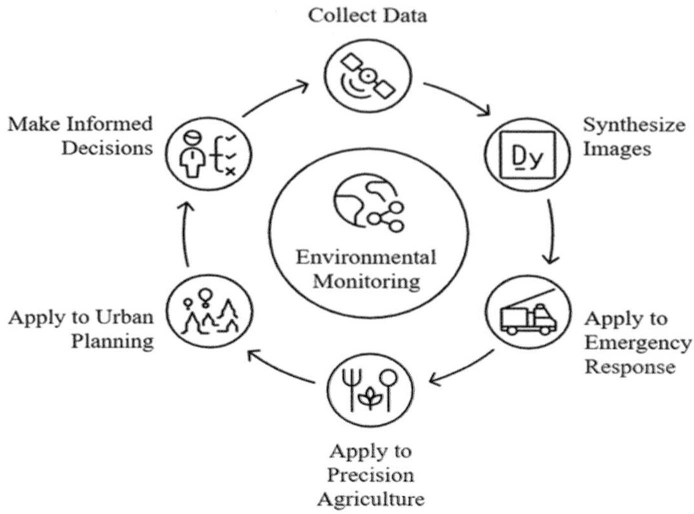

FIGURE 2.4 Workflow of environmental monitoring from data collection to visualization and decision-making for disaster response, agriculture, and urban planning.

as constant cloud cover and outdated satellite imagery, as it is possible to generate satellite images using ground-level textual inputs and coordinate systems (Silva Junior et al., 2020; Silva et al., 2024). This has also helped law enforcement agencies better detect illegal logging activities and preserve biodiversity and natural forests (Reiche et al., 2018).

Another important application of GAN technology is monitoring coastal erosion efficacy, where researchers can obtain detailed visual forecasts formed through the combination of sensor signals and weather information. These derived images allow for improving the comprehension of erosion patterns, which enables coastal administrators to establish targeted protection strategies (Almonacid-Caballer et al., 2016; Agarwal, Jahan, & Ghosh, 2021). GAN-generated images can also be helpful in strategic planning to protect critical infrastructure and natural biosystems along vulnerable coastal regions, making the actual use of GAN technology apparent (Luijendijk et al., 2018). Systems based on the GAN are gaining increasingly practical applications in various environmental monitoring activities. They encompass wildfire tracking and illegal forest clearing via multimodal multitasking learning, as well as the tracking of palm habitats using high-resolution satellite data, thereby enhancing coverage, detail, and responsiveness in ecology-related fields (Zhang et al., 2023; Pereira et al., 2023; Mahmud et al., 2024).

Technologies that have been widely applied in controlling flood disasters experienced in recent years are GAN-driven technologies. The emergence of GAN models enabled relevant environmental agencies to issue flood warnings that automatically identified submerged regions and predicted zones at risk. Emergency services could plan to respond to emergencies beforehand, reducing fatalities among humans as well as the economic losses in the event of disasters. The GAN-based models have led to significant advances in remote environmental monitoring, offering enhanced data enhancement capabilities and the ability to process data in real-time and at high speed, in addition to surveillance. The enhanced capabilities of the GAN model are linked to the effectiveness of environmental crisis response, sustainable urban development, and agriculture and conservation. With the evolution of technology, GAN models will also remain an essential tool in enhancing and refining methods of environmental monitoring worldwide, directly contributing to global sustainability goals.

2.5 PROCEDURES FOR GAN-BASED REMOTE SENSING

The effective combination of GANs with remote sensing may require a rational and properly organized implementation system. Proper data entry at the initial stage and the thorough analysis of performance at the final stage make a significant contribution to the consistency and success of this model in each step. This section provides an in-depth description of the most significant operations in GAN-based remote sensing, focusing on the operational flow required to achieve optimal performance with well-tuned models and integrated systems.

2.5.1 Data Collection and Method of Preparation

One of the essential requirements for utilizing GAN models in remote sensing applications is the careful collection and thorough preparation of the relevant data. The remote sensing application requires imagery that represents varying geographic areas, ecology, and weather conditions as compared to a general image processing application. Due to these various scenarios, GAN-based solutions primarily depend on their ability to obtain high-quality data that captures them. The primary sources of such information are satellite images, aerial photos, and images obtained by drones. Optical satellite sensors used in data collection are often combined with radar systems during data collection, as they maintain constant data collection regardless of cloud cover or low visibility. The active use of a set of historical data and current observational data enables GANs to perform better in their environmental modeling (Isola et al., 2017).

The acquisition of data is followed by preprocessing to clean the data, ensuring it is accurate and consistent for use in GAN implementation. Data cleansing begins by removing problematic images with errors and ensuring the consistency of common dimensional measurements and resolution specifications. The standardization of images originating in various sources is possible with the use of both image normalization strategies and histogram equalization coupled with color correction strategies. Critical data augmentation process prepares the datasets. Data augmentation techniques that involve the use of rotation functions, scaling, cropping, and contrast-adjusting techniques are ideally suited for use due to the scarcity of high-quality labeled images. Data augmentation techniques enhance the learning process of the model's training by introducing enriched training opportunities that provide the GAN with skills in detecting good environments, as noted by Radford et al. (2016).

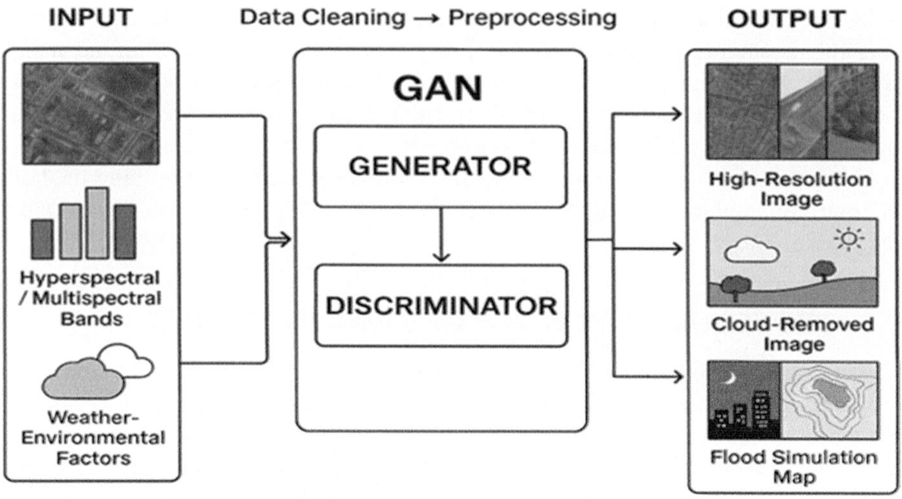

FIGURE 2.5 GAN-based framework for remote sensing image refinement, producing high-resolution outputs and simulation maps.

Figure 2.5 illustrates the process by which image data and environmental factors are combined to create useful outputs. It begins with inputs such as satellite imagery and spectral bands. After cleaning and preprocessing, the data is fed into a GAN model. The generator and discriminator collaborate to refine the results. The final outputs include clear, high-resolution images and simulations, such as flood maps.

2.5.2 Algorithm Development and Hyperparameter Tuning

The algorithm development process requires ongoing user input, as fine-tuning hyperparameters and refining the model involve several iterations. GAN model configuration begins with conventional parameters, which are derived from proven successful models and academic publications. The evaluation system begins with the first implementation phase to determine GAN performance in environmental image generation. GAN performance undergoes significant transformations because developers utilize gradient descent in combination with adaptive optimizers, such as Adam and RMSProp, to optimize their operations. The training of remote sensing models requires sequential adjustments to the learning rate and momentum parameters to stabilize operations and prevent training issues, such as mode collapse and gradient vanishing.

Multiple sophisticated tuning methods combine the evaluation of deep architectural depth with large convolutional filters, while also considering dropout rate values. The selected deep convolutional networks are an optimal choice because they provide efficient computational processing with superior performance outcomes. Weight decay, along with spectral normalization, is combined with other regularization methods in this model to prevent overfitting, enabling the model to operate effectively on both seen and new data. The implementation of vertical and horizontal configuration updates requires strict version release procedures as well as detailed documentation for configurations and result outputs. Experimental success, combined with systematic model development, becomes achievable when previous results serve as a basis for future refinements by development teams.

2.5.3 The Evaluation Metrics for Performance Analysis Consist of FID Together with Precision and Recall

Remote sensing must use precise GAN performance evaluation to achieve dependable and helpful synthetic imagery. Evaluating the performance of GANs requires metrics that can quantitatively assess both image quality and diversity, as visual inspection alone is insufficient for this purpose. One of the most commonly adopted performance assessment approaches utilizes the Fréchet Inception Distance (FID). The FID evaluation measures the similarity between synthetic imagery made by GAN and real photographs by extracting features using Inception v3. The statistical match between synthetic and native images improves as FID scores decrease, indicating superior performance. Remote sensing benefits from this metric, as environmental visuals contain multiple complex details that can result in poor decision outcomes or misinterpretations when precision is imprecise (Salimans et al., 2016).

The pair of metrics, precision and recall, operates in conjunction with FID when evaluating image generation quality. The precision evaluation method calculates the number of output images that experts identify as realistic in comparison to the real data samples, thereby quantifying the authenticity of the results. The recall metric evaluates the GAN output coverage against the real dataset diversity to determine whether the GAN can reproduce all natural environmental variations. The combination of these evaluation standards leads to thorough performance assessments. A GAN model can produce realistic images with high precision, though its limited recall indicates inadequate coverage of the diverse environmental data present in the original dataset. A low recall rate combined with low precision indicates the generation of many diverse results that lack realistic visual quality. When assessing a GAN model through both metrics, practitioners ensure the model operates as expected for real-world application needs.

Routine assessments during training enable parameter adjustments to produce a sophisticated, adaptable system that integrates information gathered from tests. The evaluation metrics serve as a system to help developers make quality improvements to their model, resulting in realistic synthetic imagery that accurately matches actual environmental elements. Creswell et al. (2018).

Figure 2.6 presents a step-by-step algorithm for implementing a GAN-based remote sensing model. It begins with data collection, followed by preprocessing, which is performed to enhance the quality of the data. The algorithm and model structure can be developed based on the cleaned data. Then, model hyperparameter tuning is performed as a process of refining the model, attempting to optimize the model configuration, and subsequently measuring model performance to confirm efficiency and accuracy.

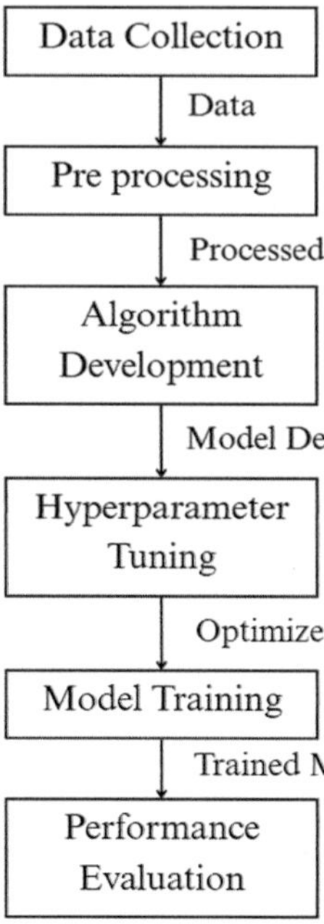

FIGURE 2.6 Step-by-step GAN implementation pipeline, including data preparation, model setup, tuning, training, and evaluation metrics.

2.6 SECURITY STANDARDS AND DATA MANAGEMENT PROTOCOLS

The use of GAI in environmental tracking and remote sensing requires professionally built security systems and robust data management, as evidenced by the extensive research base. credibility, privacy, and trustworthiness of synthetic data should be reflected in its synthetic properties, ensuring detailed and precise management through the generation of GAN results. It is essential to protect sensitive environmental information detected through remote sensing, and it is also vital to implement stricter access control, as the information requires a high level of confidentiality. It is crucial to ensure the safety of such data to maintain its validity, prevent abuse, and inform decision-making related to environmental policies using evidence.

2.6.1 Management of Synthetic Data

GAN synthetic datasets can be seen as a viable alternative to actual data, especially when data is costly to acquire and gather, when using the data requires a complex process, or when it is of a sensitive nature. Standardization of the synthetic data processes and proper governance are vital in maintaining data integrity and its value. The standard data management process begins by recording all details of the data used to generate it, including algorithms, hyperparameters, and simulation conditions. Correct documentation will enable the verification of data origin, ensure the possibility of repeating the experiments, and disclose all elements of the processes required to validate and the regulatory procedures.

The generated synthetic data must be subjected to numerous verification processes that should be carried out strictly. Validation aims to integrate the testing of realism with the verification that the generated images can accurately replicate natural environments, while maintaining the uncompromising performance of the metrics. Validation of synthetic data happens continuously through quantitative and qualitative performance assessments, which compare synthetic data to genuine observational datasets and require expert inspections to validate data accuracy. The rigorous validation procedures minimize the potential risk of releasing faulty environmental information, thereby enhancing credibility during decision-making processes. Organizations need to develop specific guidelines for synthetic data storage together with standards for its accessible use. The system requires defined data accessibility standards and proper utilization policies, together with specific user rights definitions. Since synthetic data contains information that replicates natural geographical and environmental patterns, organizations require strict protocols to determine who can access synthetic data and how it is distributed within operations.

2.6.2 Security and Privacy Protocols in Remote Sensing Applications, along with Strategies

Applications that utilize remote sensing devices require robust security protocols, as they handle sensitive geographic, ecological, and infrastructural information. The fundamental need arises from risks to data integrity, unauthorized access, and exploitation, emphasizing the importance of confidentiality and authentic data sources. Initially, organizations need to deploy security-enhanced data acquisition protocols that protect sensor data from satellites and drones during movement and when the data remains dormant. Data remains confidential throughout its entire existence, from collection to storage and use, thanks to the implementation of end-to-end encryption standards.

Organizations need to establish thorough identity management systems that accurately regulate effective access rights. Role-based access control (RBAC), combined with multi-factor authentication (MFA) and biometric security, enhances system protection by restricting access to authorized users. Continuous monitoring of access logs and real-time detection of unauthorized attempts and security breaches enable quick detection and response capabilities through proactive monitoring. Remote sensing data requires privacy protocols that meet both national and international regulatory standards for their procedures. Data protection guidelines must be documented in protocols with definitions of PII and procedures for handling geographically sensitive information. To observe essential ecological hotspots and strategic locations, as well as private residential and commercial areas, stakeholders should take the highest level of precautions. The evaluation process of data privacy methods must occur frequently to ensure alignment between technology and law, while adhering to public standards and legal updates.

Security issues arise for GAN-based remote sensing systems when they launch their operations, although synthetic images from GANS introduce misinformation risks and threaten the exposure of confidential data. Organizations need to establish preventive measures to protect against risks that may arise during the deployment of GAN-based remote sensing systems (Bengesi et al., 2024). Governance policies that leverage algorithms provide organizations with a practical approach to managing risks. The security and reliability maintenance of generative models depend on systematic assessments accomplished through regular auditing and performance tracking functions that exist within the framework structure. Through systematic model outcome analysis, organizations obtain the ability to detect unusual patterns that demand prompt remedial procedures. Testing needs to operate through multiple tests to identify problematic situations and operational weaknesses. Risk mitigation implementation largely depends on backup and redundancy procedures that organizations must perform. Business operations require active system backup practices to safeguard against disruptions because creating swift backups of authentic and synthetic data repositories remains a critical task for operational protection. Various authentication methods and fine-grained protocols within the organization prevent unnecessary access to sensitive data, thereby maintaining the integrity and security of critical information.

2.7 FUTURE PROSPECTS

Research groups accelerate the development of GAN technology through continuous improvements in computational capabilities, introducing innovative algorithms, and collaborating on interdisciplinary partnerships. Extended improvements to GAN architectures over the next few years will transform the way remote sensing, combined with environmental monitoring, functions. The future of GAN technology development focuses on expanding multimodal integration features as its primary advancement. Future generations of GANs will demonstrate enhanced performance by processing multiple simultaneous input formats, which include both written descriptions and audio signals, sensor readings, and real-time monitoring data. Through multimodal functionality, environmental researchers and professionals will acquire the ability to develop realistic visual representations of complex real-world conditions. Refined multimodal GAN systems will enhance the quality of interpretation and expand their application areas throughout environmental monitoring practices.

Additional breakthroughs will be made in achieving higher computational speeds as well as expanding scalability ranges. The current GAN models require substantial computing infrastructure because they necessitate large processing capacities, particularly when analyzing environmental datasets. Future GAN algorithms will prioritize lightweight designs for efficient real-time operations, enabling integration into drones, satellites, and mobile platforms. This will broaden access to sophisticated remote sensing for a broader range of stakeholders. The interactive GAN-based technology will evolve through systems that enable users to steer model output transformations during real-time interactions. The implementation of reinforcement learning functionalities in GANs would enable them to react more effectively to human-corrected input, thereby improving their performance and reducing operational complexity. Through interactive approaches, professionals would gain the power to direct models for producing specific visualizations that align with their present surveillance and analytical requirements.

2.7.1 Opportunities for Further Integration in Remote Sensing

GAN and GAI technologies offer broad opportunities for integration with existing remote sensing applications. GAN-generated synthetic imagery shines as an essential tool for environmental tracking, as scientists increasingly recognize the importance of sustainability. GAN technology creates valuable opportunities for improving predictions about natural disaster occurrences, including wildfires, floods, and hurricanes. Such predictive visual capabilities enable organizations to enhance their ability to prepare for disasters and allocate resources, resulting in decreased damage to both the environment and the economy.

The combination of GAN technology with urban planning tasks and sustainable development holds substantial potential applications. The integration of GAN technology enables urban planners and authorities to predict future urban development scenarios by combining text-based forecasts with population statistics. Simulation models facilitate the development of informed planning policies that align with environmentally friendly urban development and long-term sustainability in city infrastructure. The more widespread adoption of GAN technology in remote sensing approaches would enhance their agricultural applications. High-quality, real-time visual displays of agricultural conditions that detect pests and indicate soil degradation help precision farming evolve. By utilizing GAN-generated imagery derived from periodic textual reports and forecasts, farmers can identify crop stress, predict yields, and optimize their resource usage for improved food security and sustainable agricultural practices.

2.7.2 Research Priorities and Areas for Further Investigation

The future success of GAN and GAI technologies depends on concentrated research efforts that will expand their beneficial applications while identifying and addressing any limitations and ethical challenges. Researchers need to enhance users' understanding and ability to decipher the internal logic of algorithms. The realism of GAN-generated visuals diminishes visibility into the algorithms that produce them. Further research is needed to establish proper methods that demonstrate GANs through the step-by-step process they use for generating specific visual outputs. Developed explanations of synthetic imagery generation would enable users and regulators to comprehend the reasoning processes that lead to the creation of synthetic images, thereby enhancing trust and acceptance in sensitive areas, including environmental policy-making and resource management.

Future work should focus on two essential objectives: stabilizing GAN models while simultaneously reducing their computational requirements. The training of the GAN network requires a substantial computational capacity, which may become too expensive or environmentally unsustainable to maintain. Scientists should direct their research toward building advanced, designed architectures along with training approaches and combination models that deliver equivalent performance with decreased computational needs. GAN technology accessibility through research efforts will empower a broader range of users, making the technology practical in low-resource environments.

Research requires immediate attention to both ethical and biased aspects before proceeding. The use of GAN models that process biased datasets yields negative results in decision systems because incorrect information spreads discriminatory stereotypes, creating problems for both policy outcomes and performance. Scientists should establish systematic approaches to detect and quantify biases in GAN-generated synthetic data, along with techniques to decrease these biases. Comprehensive bias detection strategies, along with mitigation protocols, must be used to preserve environmental patterns. Research studies must begin to advance GAN's reliability level and output stability. GAN systems must produce real-looking environmental images in critical public safety operations and times that require disaster response intervention. Scientists need to develop appropriate stress-testing methodologies for GAN systems under various extreme operational conditions, ensuring the systems demonstrate stability across different real-world scenarios. A baseline challenge needs proper solutions because it demands effective handling of regulations and ethical questions involving GAN-produced data. Achieving ethical use of GAN systems requires research to identify privacy threats as well as authentic data validation methods and potential misuse scenarios for synthetic data. Research teams must create detailed ethical framework definitions and practical deployment guidelines for GAN systems since responsible remote sensing benefits require this framework.

The successful application of generative AI technologies for environmental monitoring requires addressing four core challenges: simplifying algorithm readability, reducing computing requirements, eliminating bias, and developing robustness frameworks and ethical governance frameworks. The establishment will produce enduring benefits by developing accountable technologies that serve the public good. The given information detailed the principles of GAI along with the actual GAN methodologies, in conjunction with various remote sensing applications. It analyzed GAN structures that convert written documents into realistic satellite images to determine the operational advantages of their use in monitoring activities and quicker image production. The results of environmental monitoring systems generate high-quality visual outputs based on the respective capabilities of NLP and GAN models, accommodating real-time observational data quite well.

The current chapter provides a detailed elaboration on the understanding of how to apply GAN, focusing on how they perform within a system of monitoring the environment. There was a need for a well-planned approach to carry out the operational plan, which might have involved intensive data collection activities, building advanced models, and training hyperparameters, ultimately leading to an overall performance evaluation scheme. To properly evaluate the GAN model, it was necessary to use the FID, as well as Precision and Recall, to analyze the results. Security procedures and processing of synthetic data were given a particular emphasis during the analysis. The management systems that can provide reliable protection of information require proper convergence of optimistic safety measures with complete privacy compliance, based on the guidelines cited in the section. The security of Generative AI systems, particularly in the context of environmental monitoring, involves algorithmic monitoring, a specifically designed security platform, as well as training programs and simulated responses, as discussed in the presentation. The uses of remote sensing were also discussed, including the effects of GANs on remote sensing equipment.

REFERENCES

Agarwal, V., Jahan, S., & Ghosh, S. (2021). Shoreline change detection using deep learning on Sentinel-2 imagery. *Remote Sensing Applications: Society and Environment*, 22, 100488. https://doi.org/10.1016/j.rsase.2021.100488

Almonacid-Caballer, J., Ruiz, L. A., Pardo-Pascual, J. E., & Palomar-Vázquez, J. (2016). Evaluation of beach nourishment erosion using UAV images. *Remote Sensing*, 8(10), 795. https://doi.org/10.3390/rs8100795

Arjovsky, M., Chintala, S., & Bottou, L. (2017). Wasserstein generative adversarial networks. *arXiv preprint*. https://doi.org/10.48550/arXiv.1701.07875

Bailey, A., Xu, C., & Nogueira, K. (2021). A survey on anomaly detection using deep learning in remote sensing. *Remote Sensing*, 13(22), 4615. https://doi.org/10.3390/rs13224615

Bengesi, S., El-Sayed, H., Sarker, M. K., Houkpati, Y., Irungu, J., & Oladunni, T. (2024). Advancements in generative AI: A comprehensive review of GANs, GPT, autoencoders, diffusion models, and transformers. *IEEE Access*, 12, 69812–69830. https://doi.org/10.1109/ACCESS.2024.3397775

Brock, A., Donahue, J., & Simonyan, K. (2019). Large-scale GAN training for high-fidelity natural image synthesis. In *Proceedings of the International Conference on Learning Representations (ICLR)*. https://openreview.net/forum?id=B1xsqj09Fm

Creswell, A., White, T., Dumoulin, V., Arulkumaran, K., Sengupta, B., & Bharath, A. A. (2018). Generative adversarial networks: An overview. *IEEE Signal Processing Magazine*, 35(1), 53–65. https://doi.org/10.1109/MSP.2017.2765202

Dhariwal, P., & Nichol, A. (2021). Diffusion models beat GANs on image synthesis. *arXiv preprint*. https://doi.org/10.48550/arXiv.2105.05233

Goodfellow, I., Pouget-Abadie, J., Mirza, M., Xu, B., Warde-Farley, D., Ozair, S., Courville, A., & Bengio, Y. (2014). Generative adversarial networks. *arXiv preprint*. https://doi.org/10.48550/arXiv.1406.2661

Isola, P., Zhu, J. Y., Zhou, T., & Efros, A. A. (2017). Image-to-image translation with conditional adversarial networks. In *Proceedings of the IEEE Conference on Computer Vision and Pattern Recognition (CVPR)* (pp. 1125–1134). https://doi.org/10.1109/CVPR.2017.632

Kingma, D. P., & Welling, M. (2014). Auto-encoding variational Bayes. *arXiv preprint*. https://doi.org/10.48550/arXiv.1312.6114

Luijendijk, A., Hagenaars, G., Ranasinghe, R., Baart, F., Donchyts, G., & Aarninkhof, S. (2018). The state of the world's beaches. *Scientific Reports*, 8(1), 6641. https://doi.org/10.1038/s41598-018-24630-6

Mahmud, K., Rahman, M., & Wang, Y. (2024). Monitoring mangrove ecosystems using deep learning and Sentinel-2 imagery. *arXiv preprint*. https://doi.org/10.48550/arXiv.2410.05443

Mirza, M., & Osindero, S. (2014). Conditional generative adversarial nets. *arXiv preprint*. https://doi.org/10.48550/arXiv.1411.1784

Pereira, L. F., Nogueira, K., & dos Santos, J. A. (2023). Time-series SAR-based deforestation detection using recurrent residual CNNs. *arXiv preprint*. https://doi.org/10.48550/arXiv.2310.05697

Radford, A., Metz, L., & Chintala, S. (2016). Unsupervised representation learning with deep convolutional generative adversarial networks. *arXiv preprint*. https://doi.org/10.48550/arXiv.1511.06434

Reed, S., Akata, Z., Yan, X., Logeswaran, L., Schiele, B., & Lee, H. (2016). Generative adversarial text-to-image synthesis. In *Proceedings of the 33rd International Conference on Machine Learning (ICML)* (pp. 1060–1069). PMLR. https://proceedings.mlr.press/v48/reed16.html

Reiche, J., Hamunyela, E., Verbesselt, J., Wielaard, N., & Herold, M. (2018). Improving near-real-time deforestation monitoring in tropical dry forests by combining dense Sentinel-1 time series with Landsat and ALOS-2 PALSAR-2. *Remote Sensing of Environment*, 204, 147–161. https://doi.org/10.1016/j.rse.2017.12.034

Salimans, T., Goodfellow, I., Zaremba, W., Cheung, V., Radford, A., & Chen, X. (2016). Improved techniques for training GANs. *arXiv preprint*. https://doi.org/10.48550/arXiv.1606.03498

Silva, J. R., Monteiro, A. M. V., & Souza, C. M. (2024). Mapping selective logging in the Brazilian Amazon using Sentinel-2 and deep learning approaches. *International Archives of the Photogrammetry, Remote Sensing and Spatial Information Sciences*, 48(3), 121–128. https://doi.org/10.5194/isprs-archives-XLVIII-3-2024-121-2024

Silva Junior, C. H. L., Anderson, L. O., Silva, A. L., Almeida, C. T., Rufin, P., & Cochrane, M. A. (2020). Persistent collapse of biomass in Amazonian forest edges following deforestation. *Science Advances*, 6(40), eaaz8360. https://doi.org/10.1126/sciadv.aaz8360

Vaswani, A., Shazeer, N., Parmar, N., Uszkoreit, J., Jones, L., Gomez, A. N., Kaiser, Ł., & Polosukhin, I. (2017). Attention is all you need. *arXiv preprint*. https://doi.org/10.48550/arXiv.1706.03762

Zhang, Y., Liu, J., & Gonzalez, R. (2023). Deep multimodal learning for rapid forest fire and deforestation detection. *arXiv preprint*. https://doi.org/10.48550/arXiv.2307.04916

Zhu, J., Wang, Y. Y., & Guo, Q. (2022). GAN-based remote sensing image generation: A comprehensive review. *Remote Sensing*, 14 (13), 3185. https://doi.org/10.3390/rs14133185

Zhu, J. Y., Park, T., Isola, P., & Efros, A. A. (2017). Unpaired image-to-image translation using cycle-consistent adversarial networks. In *Proceedings of the IEEE International Conference on Computer Vision (ICCV)* (pp. 2223–2232). https://doi.org/10.1109/ICCV.2017.244

3 The Use of Generative AI in Environmental Remote Sensing

Mohamed A.E. AbdelRahman

3.1 INTRODUCTION

Remote sensing has long been a fundamental tool for environmental monitoring, providing synoptic, multitemporal, and multispectral views required to gain insight into Earth's surface processes. Satellite-induced remote sensing platforms, including Landsat, Sentinel, and MODIS, have made the large-scale monitoring of land-use practices, soil health condition, vegetation productivity, and hydrological processes feasible (Chen et al., 2024). Despite such advances, environmental remote sensing still suffers persistent problems including the presence of missing or spoiled data, for example, due to cloud coverage, sensor failures, and inadequate temporal range. These are among the limiting factors of both completeness and temporal fidelity in providing reference datasets for accurate models and decision support (Rahimi & Jung, 2024).

AI has become a game changer in all science domains, including geospatial analysis and Earth observation lands; In AI, generative models have given a very interesting twist. Unlike conventional predictive or classification models which map inputs to fixed outputs through the solution of a training task generative AI can learn the underlying probability distribution of the data, making it possible to generate new realistic samples that have similar statistical characteristics as real-world observations (Liu et al., 2023). This data generation ability makes generative AI particularly applicable to the challenges raised in remote sensing, notably in image completions, synthesis of super-resolution composites, and dataset simulation of temporal trends.

The use of generative artificial intelligence in environmental remote sensing first drew attention when scientists began experimenting with Generative Adversarial Networks (GANs) to clean and enhance satellite images affected by clouds (Ajayi & Olaide, 2024; Wang et al., 2023). Over time, the field moved past these early experiments and started adopting more advanced techniques conditional GANs (cGANs), diffusion models, and transformer-based approaches. Each step brought something new, from creating realistic spectral data to generating large-scale climate simulations (Calleo et al., 2025). For example, diffusion models have been used to reconstruct cloud-covered regions in Sentinel-2 imagery, producing clear, cloud-free mosaics that maintain natural color and spectral relationships between bands

DOI: 10.1201/9781003616207-4

(Jing et al., 2023). Altogether, these developments have made remote sensing more consistent and trustworthy, particularly in places where cloudiness or missing data once made continuous environmental monitoring difficult.

To illustrate the differences between the main generative architectures and how each contributes to environmental applications, Table 3.1 summarizes their fundamental principles along with key examples of how they have been applied in remote sensing contexts.

Generative AI is not only a technology but also a paradigm shift in the sensing and processing of environmental data. Traditional remote sensing procedures focus on data interpretation, whereas generative AI provides the function of generating and enhancing data (Janga et al., 2023). This revolution transforms remote sensing into a more dynamic, responsive, and predictive science (Sharifi & Mahdipour, 2024). With generative synthesis, scientists are able to recreate missing measurements, enhance spatial resolution to detect finer environmental information, and even forecast future conditions under varied climatic or land-cover scenarios (Reichstein et al., 2019). To environmental managers, such is valuable particularly in developing countries and arid regions where field measurements are often lacking (Rolnick et al., 2022).

Open-access satellite imagery and powerful computing tools have really sped up the use of generative techniques. With platforms like Google Earth Engine and cloud-based AI services, researchers can now train huge models and combine data from different sources think optical, radar, and hyperspectral imagery (Gorelick et al., 2017; Janga et al., 2023). Generative AI is quickly becoming a go-to tool for building

TABLE 3.1
Generative AI Techniques and Their Applications

Generative AI Technique	Core Principle	Remote Sensing Application	Example Output
GANs (Generative Adversarial Networks)	Adversarial training between a generator and discriminator to create realistic data	Cloud removal, super-resolution of satellite images, data gap filling	Cleaned and enhanced satellite imagery
VAEs (Variational Autoencoders)	Learning latent representations for probabilistic data reconstruction	Data compression, spectral unmixing, hyperspectral image restoration	Reduced data size with preserved spectral fidelity
Diffusion Models	Iterative denoising and probabilistic image generation	Synthetic image generation, terrain reconstruction, data augmentation	High-fidelity terrain or land-cover simulations
Transformer-Based Generative Models	Attention-based contextual learning for sequence or spatial data	Multitemporal prediction, anomaly detection, image-to-image translation	Predictive environmental change patterns

Note: This table summarizes the principal generative AI models applied in environmental remote sensing, illustrating their mechanisms and outcomes across diverse applications.

robust, data-driven environmental intelligence systems. These systems help with early warnings, land-use planning, and managing resources in a sustainable way (Sharifi & Mahdipour, 2024). In this chapter, you'll get a clear look at how generative AI is changing environmental remote sensing. We start with the basics of what generative algorithms are and how they've developed then dive into real-world uses in monitoring the environment. That includes areas like soil, vegetation, and water. Later, we look at the software people are using, the technical and ethical challenges, and where things are heading as generative AI becomes part of sustainable Earth observation. At the end, I'll share some thoughts on how all this progress can deepen our scientific understanding and help shape better global environmental policies.

3.2 CONCEPTUAL BACKGROUND: GENERATIVE AI AND ITS EVOLUTION

3.2.1 The Foundations of Generative Modeling

Generative artificial intelligence is a group of machine learning algorithms that learn what actual data looks like and then use that knowledge to create new, realistic samples. Unlike simply finding the edge between different classes like discriminative models, generative models go into where exactly the data is being generated in the first place (Goodfellow et al., 2014). This is why they can accomplish more than sorting or tagging things they can rebuild data, fill in gaps, and recognize patterns no one has ever seen before.

The roots of generative modeling lie in unsupervised learning and probabilistic graphical models. The early models, like Restricted Boltzmann Machines and Deep Belief Networks, started this practice by learning hidden patterns in data (Hinton et al., 2006; Hinton & Salakhutdinov, 2006). But the real turning point was 2014 with Generative Adversarial Networks, or GANs. Creating data then became a game: one network tries to generate fake data, and the other tries to spot the fakes (Goodfellow et al., 2014). This adversarial setup sparked a wave of innovation in AI, and these models can now generate images that are almost indistinguishable from the real ones.

3.2.2 Key Generative Architectures

Several major architectures underpin current generative AI applications in remote sensing and environmental sciences:

- **Generative Adversarial Networks**
  GANs remain among the most common models for remote sensing due to the fact that they can generate high-fidelity spatial patterns. GANs have been applied to super-resolution enhancement, cloud removal, and realistic terrain synthesis (Benedetti et al., 2018). Extensions such as cGANs are able to make generation driven by auxiliary data, for example, temporal information or spectral bands, thus improving environmental consistency.
- **Variational Autoencoders**
  VAEs learn to encode input data into a latent space and then from the latent space to reconstruct the input. Though their outputs are smoother than those

of GANs, VAEs provide stable training and explicit probabilistic modeling and therefore are especially suited to reconstructing gaps in spectral data and compressing multispectral image data (Kingma & Welling, 2019).

- **Diffusion and Denoising Models**
New technologies such as Denoising Diffusion Probabilistic Models (DDPMs) and Score-Based Generative Models excelled the initial GANs in visual quality and stability (Ho et al., 2020). The models learn to reverse a diffusion process and iteratively transform random noise into structured data. They have been extremely beneficial in modeling rich data distributions for the needs of filling cloud-covered areas in Sentinel-2 images and synthesizing high-resolution temporal composites (Sui et al., 2024).
- **Transformer-based and Multimodal Models**
Building on the success of massive language models, transformer models are becoming increasingly employed in geospatial generative tasks. They have the ability to learn long-range dependencies in both spatial and temporal contexts, allowing for multimodal conditioning on text, elevation, or weather information (Ramesh et al., 2021). New "foundation models" for earth observation, learned from huge satellite archives, hold out the prospect of further merging generative and predictive capabilities (Reichstein et al., 2019).

3.2.3 Training Paradigms and Data Representation

Generative models are to a large extent founded on extensive, varied data sets that capture spatial, spectral, and temporal variation. Training usually reduces to minimizing a loss function that measures the discrepancy between synthetic and actual data. For GANs, this loss is adversarial; for VAEs, reconstruction loss augmented with a Kullback–Leibler divergence penalty; and for diffusion models, a denoising objective that aims to minimize clean reconstruction from noisy data.

Environmental remote sensing is an exceptional challenge for generative learning since the geospatial data is multi-modal in nature, combining optical, radar, thermal, and topography information. Recent work thus concentrates on data-fusion and domain-adaptation techniques to enable generative models to leverage cross-sensors and cross-temporal correlations. For instance, spatiotemporal GANs can combine low-resolution high-frequency MODIS observations with high-resolution low-frequency Landsat observations to synthesize imagery at high spatial resolution on a daily basis (Mu & Tian, 2022).

3.2.4 From Image Generation to Environmental Intelligence

In environmental applications, the generative paradigm is also employed for more than image synthesis purposes to reconstruct the environmental state. Through learning spatial relationships and physical patterns, generative models can interpolate missing data, predict environmental dynamics, or synthesize hypothetical conditions. Generative diffusion models, for example, have been employed to predict future states of floodplains under future rainfall conditions in order to allow for risk assessment and climate adaptation planning (Wang et al., 2024).

These applications represent a change of pace from the employment of AI as a secondary analysis tool to that of a co-creative partner to environmental modeling. Instead of interpreting satellite observations passively, generative AI actually generates realistic environmental situations consistent with physical principles and observed trends. When combined with domain-specific knowledge for the case of hydrological or erosion-soil models, these systems can produce physically constrained and interpretable output a requirement critical to decision-making in environmental management (Sun et al., 2025; Yazdani et al., 2025).

This blending of generative paradigms into environmental remote sensing applications is depicted in Figure 3.1, a conceptual model of data inputs, AI-driven synthesis operations, and application domains.

3.2.5 Evolution and Recent Milestones

The evolution of generative AI in environmental remote sensing tracks general trends in deep learning. Between 2016 and 2020, most research employed GAN-based architecture for cloud removal, pansharpening, and spectral reconstruction (Xu et al., 2021). Between 2021 and 2023, there was extensive application of diffusion models due to their training stability and the capability to model multi-scale structures. At the same time, foundation-scale models such as DiffusionSat and SatFlow began integrating gigantic databases of remote sensing imagery, enabling pre-trained generative models with the ability to generalize across geographic locations (Khanna et al., 2023).

In the future, generative AI can be expected to converge with physics-informed neural networks (PINNs) and Earth digital twins (EDTs) to provide dynamic and interactive environmental prediction platforms. This convergence is a transition from passive data reconstruction to continuous environmental simulation and positions

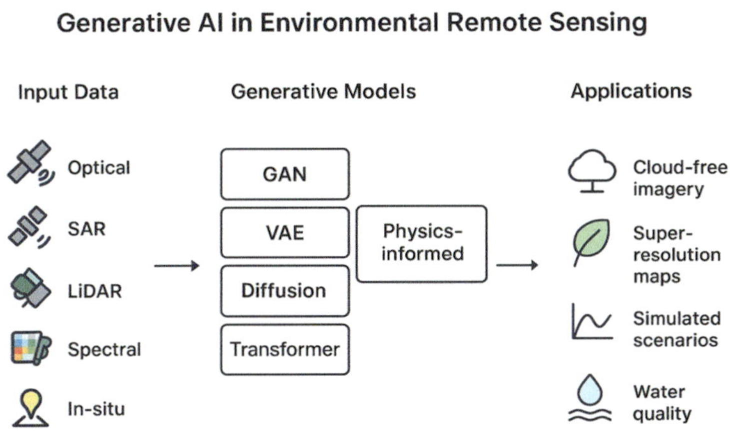

FIGURE 3.1 Conceptual framework illustrating the integration of generative AI in environmentally remote sensing linking data sources, model architectures, and environmental applications such as image synthesis, cloud removal, and digital twin modeling.

generative AI as more compatible with sustainable development goals and Earth system science (Karniadakis et al., 2021).

3.3 GENERATIVE AI FOR ENVIRONMENTAL REMOTE SENSING: APPLICATIONS AND CASE STUDIES

Generative AI has also demonstrated excellent performance under diverse environmental conditions, including soil mapping, vegetation interpretation, land-use/land-cover mapping, and water-quality assessment. By generating synthetic but physically realistic data, generative models can fill in missing records, enhance image resolution, and replicate environmental dynamics beyond the domain of routine observation systems (Janga et al., 2023; Whig et al., 2024).

The range of environmental application domains augmented by generative AI is succinctly provided in Table 3.2, indicating the limitations of traditional approaches and the success achieved with data-driven generative modeling.

3.3.1 Cloud Removal and Image Restoration

Cloud cover remains the largest problem in optical remote sensing, particularly for regions located in tropical and subtropical areas with persistent cloud cover, resulting in incomplete temporal observations. Traditional techniques temporal compositing, multi-sensor merging, and interpolation fail to provide a good reconstruction of spatial information in highly clouded areas. Generative AI has revolutionized this task through its ability to inpaint missing regions without sacrificing the spatial and spectral coherence of the scene.

TABLE 3.2
Environmental Applications Enhanced by Generative AI

Environmental Domain	Traditional Limitation	Generative AI Enhancement	Impact
Soil mapping	Sparse or outdated field data; limited spectral resolution	Synthetic spectral data generation; image super-resolution	Improved classification and mapping accuracy
Land-use/Land-cover change	Missing temporal observations; cloud contamination	Scene synthesis and temporal interpolation	Continuous, reliable change detection
Water-quality monitoring	Cloud-affected or low-frequency imagery	Cloud-gap filling and spectral reconstruction	Higher data availability and cleaner time series
Vegetation and carbon studies	Mixed spectral signals; inconsistent NDVI data	Spectral unmixing; synthetic NDVI generation	Better assessment of vegetation health and carbon dynamics

Note: This table summarizes how generative AI addresses key data and modeling limitations across major environmental monitoring themes, improving spatial, temporal, and spectral accuracy.

The cGAN process, as it is designed for cloud removal, is illustrated in Figure 3.2, showing the discriminator-generator interaction, the training objective, and the produced output cloud-free.

GANs and diffusion models are some of the top-performing architecture models for this type of task. cGAN-based models, for instance, were learned to de-cloud Sentinel-2 images from paired cloud-free references and generated high-quality reconstructions with natural patterns and texture of vegetation preserved (Zhao & Jia, 2023). More contemporary approaches combine Synthetic Aperture Radar (SAR) data, which can penetrate cloud cover, with optical imagery with a multimodal generative model to forecast the hidden surface reflectance (Zhan et al., 2023).

The most effective architectures for this application are GANs and diffusion models. cGAN-based models, for example, have been trained to restore natural images of cloudy Sentinel-2 images with paired cloud-free references, producing high-quality reconstructions with maintained natural texture and vegetation patterns. Figure 3.3 shows the Schematic architecture of a cGAN for cloud removal in optical satellite imagery.

New developments in Diffusion Probabilistic Models (DPMs) provide further improvements by progressively denoising cloud-occluded pixels using learned statistical priors and rendering physically plausible reconstruction even without temporal references (Wang et al., 2024). The models not only fill cloud gaps but also ensure spectral accuracy, which is important in vegetation and soil research a critical requirement in environmental monitoring tasks is also guaranteed (Yu et al., 2025).

New developments in DPMs have the advantages of iteratively denoising cloud-occluded pixels based on learned statistical priors to create physically realistic reconstructions even when temporal information is not available (Wang et al., 2024). DPMs not only fill cloud gaps but also preserve the spectral accuracy required in vegetation and soil studies essential requirements in environmental monitoring applications.

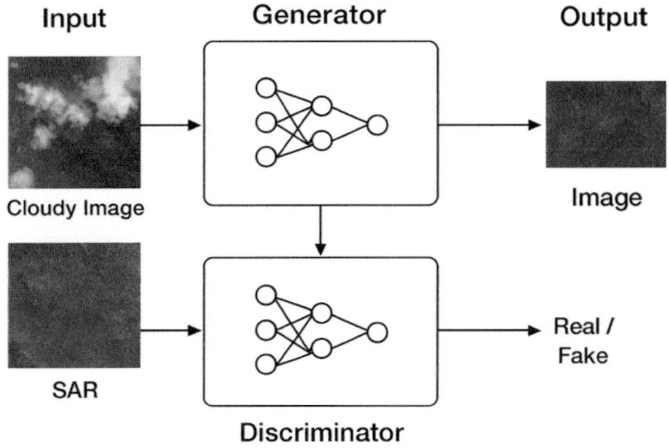

FIGURE 3.2 Architecture of a conditional GAN (cGAN) model for satellite cloud removal. The generator learns to translate cloudy images into cloud-free outputs, while the discriminator distinguishes real, clear-sky images from generated ones.

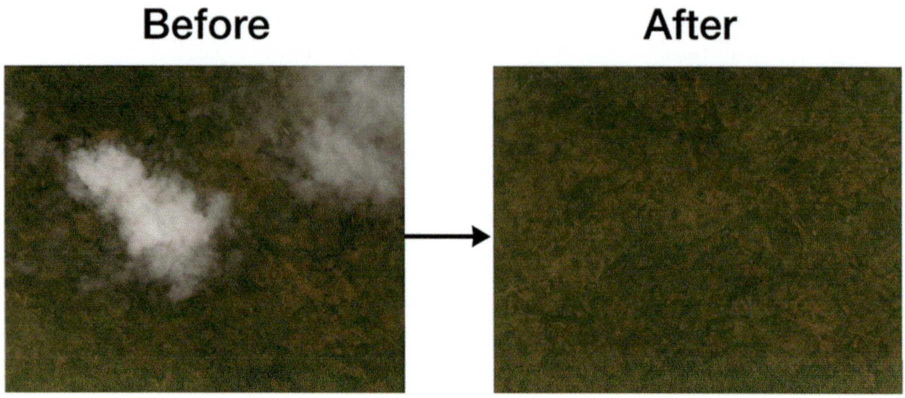

FIGURE 3.3 Schematic architecture of a conditional Generative Adversarial Network (cGAN) for cloud removal in optical satellite imagery. The generator learns to inpaint cloud-covered regions using contextual cues from the input, while the discriminator ensures that outputs are spectrally and spatially consistent with true cloud-free references.

Figure 3.4 shows an example of generative cloud removal using cGAN and diffusion models.

3.3.2 Data-Fusion and Super-Resolution Synthesis

Temporal versus spatial resolution compromises in satellite systems are time-honored constraints on environmental monitoring. Landsat offers high spatial resolution but low temporal frequency, and MODIS is available daily but at low spatial resolution. Data fusion compositing images, taking the best from every source is enabled by generative AI.

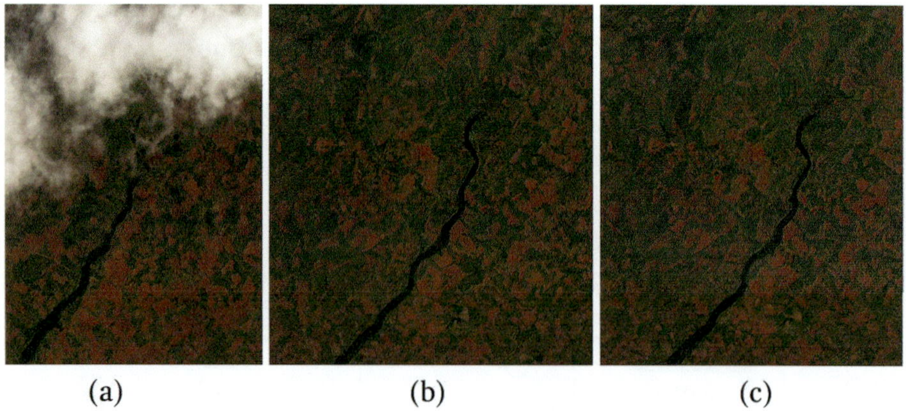

FIGURE 3.4 Example of generative cloud removal using cGAN and diffusion models. Image (a) shows the original cloud-covered Sentinel-2 scene, image (b) shows the generated cloud-free output, and image (c) panel presents the difference visualization highlighting reconstructed regions.

In a normal workflow, a VAE or GAN is trained to transform low-resolution MODIS inputs into high-resolution Landsat-similar outputs with co-registered samples. SRGAN (Super-Resolution GAN), for instance, has already been effectively utilized to generate 10-meter spatial resolution from 250-meter MODIS data for daily vegetation mapping (Sudhakara & Bhattacharjee, 2024). Similarly, diffusion-based super-resolution models capture finer spatial gradients in soil moisture and vegetation cover better than conventional interpolation methods (Ahmad et al., 2024).

Moreover, cross-sensor generative models allow for optical–SAR fusion through the complementary nature of reflectance and radar backscatter data. Fusion in this sense allows for trusted land-surface reconstructions under cloudy weather or night-time scenes, greatly improving environmental monitoring continuity (Ahmad et al., 2024).

Doublet low- and high-resolution training images can be used to train generative models to generate high-resolution fine-scale spatial features with spectral consistency. This can be utilized for the generation of daily high-resolution observations that are crucial for precision agriculture and monitoring the environment (Hong et al., 2024). Figure 3.5 shows a super-resolution example generated using a cGAN, illustrating enhanced spatial detail and spectral coherence compared to the original low-resolution image.

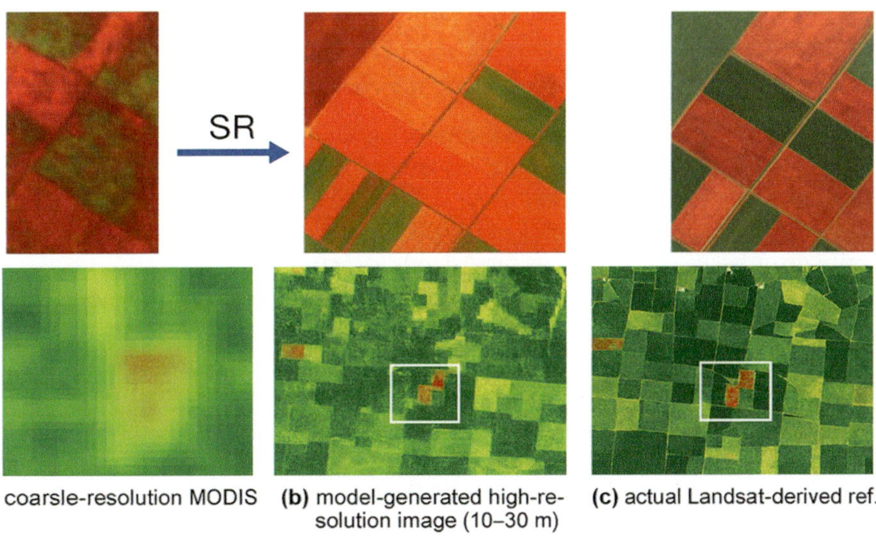

FIGURE 3.5 Super-resolution example generated using a conditional GAN (cGAN), illustrating enhanced spatial detail and spectral coherence compared to the original low-resolution image. (a) Coarse-resolution MODIS NDVI input (blurred, low spatial fidelity). (b) Model-generated high-resolution NDVI image (10–30 m), synthesized using SRGAN-style enhancement. (c) Actual Landsat-derived NDVI reference (high spatial fidelity). All panels use consistent NDVI color mapping (green = high vegetation, yellow = moderate, red = low), and a zoom box highlights a fine-scale feature recovered in panels (b) and (c).

3.3.3 Change Detection and Environmental Dynamics

Change detection the measurement of environmental change through the passage of time is critical for tracking land degradation, deforestation, and urban development. Pixel-by-pixel comparative techniques are fraught with radiometric inhomogeneities and misregistration artifacts. Generative AI offers simulation-based alternatives where a temporal state is generated from another, and differences between simulated and real imagery highlight areas of change.

For example, CycleGAN-informed designs can convert "pre-change" images into a simulated "post-change" setting, making differences directly attributable to environmental change (Zhou et al., 2024). It has worked very effectively in monitoring urban sprawl in fast-developing cities as well as in detecting shoreline changes in coastal ecosystems.

In crop monitoring, spatiotemporal generative models forecast vegetation indexes, such as NDVI or EVI, at a later time by learning from a previous time series (Gao et al., 2023). They are not only for early warning but also as sources of synthetic data to train drought and yield models in areas lacking data.

Emerging technologies of generative modeling allow for the construction of constantly updated digital replicas of Earth systems, so-called Earth Digital Twins. They combine satellite remote sensing, sensor networks, and forecast simulation to provide real-time environmental perception and scenario forecasting. Figure 3.6 shows a conceptual schematic of an AI-driven Earth Digital Twin framework integrating remote sensing inputs, generative synthesis layers, and real-time environmental feedback loops.

3.3.4 Environmental Anomaly Detection and Simulation

Anomaly detection in environmental remote sensing usually entails modeling the "normal" background environmental condition. Generative AI, insofar as it can model high-dimensional distributions, is well positioned to learn what normal spectral–spatial activity resembles and, by extension, identify deviations from it that signify disturbance events such as fires, floods, or infestations.

Autoencoder-type generative models have also been used to reconstruct baseline conditions from archives of satellite imagery, where reconstruction errors of large magnitude indicate anomalies (Hinton et al., 2006; Hinton & Salakhutdinov, 2006). For instance, a generative model trained on multi-year Landsat data over wetlands will identify sudden changes caused by water level loss or pollution incidents.

Furthermore, generative diffusion models were recently employed to simulate environmental processes such as rainfall-induced sediment transport and floodplain morphological evolution (Liu et al., 2024). By generating multiple realistic realizations under varying climatic inputs, the models help quantify uncertainty and facilitate risk-informed decision-making in land restoration and water management.

3.3.5 Earth Observation Enhancement and Digital Twins

Generative AI also plays a key role in the development of EDTs, real-time, data-synched digital twins of the planet that integrate real-time sensing and predictive

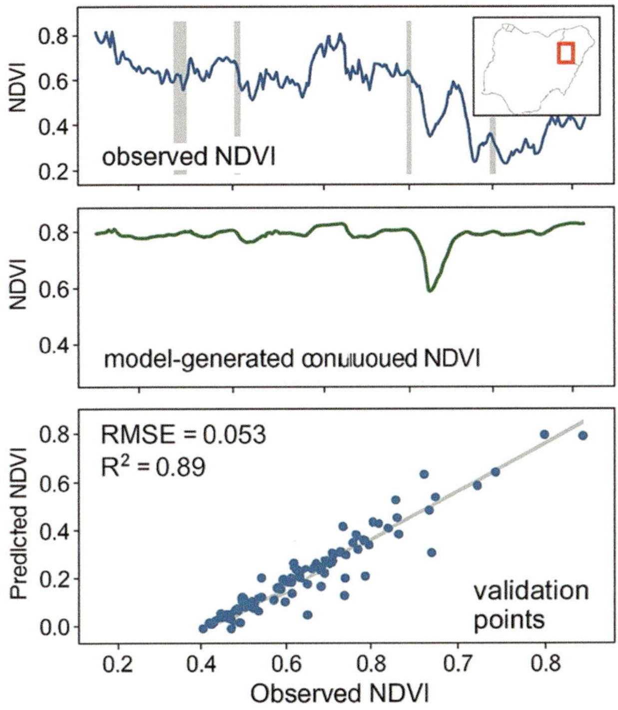

FIGURE 3.6 Conceptual schematic of an AI-driven Earth Digital Twin framework integrating remote sensing inputs, generative synthesis layers, and real-time environmental feedback loops.

simulation. With physics-informed generative models, for example, such systems can simulate environmental processes in continuous time and space.

For example, the generation of physically realistic soil moisture maps from hydrological principles through the intercombination of diffusion models and PINNs, completing observation gaps (Tee et al., 2024). Scenario simulation, restoration planning, and ecosystem forecasting become possible with the application of digital twins using generative AI.

They represent the extension of generative AI from image synthesis to complete environmental synthesis, whereby global dynamic systems of the world can be emulated, imaged, and manipulated according to hybrid data–model paradigms.

3.3.6 Representative Case Studies

- **Case 1 Tropical Agriculture Cloud Removal**
  Researchers applied conditionally GAN-trained Sent2S1 model to Sentinel-2 and Sentinel-1 pairs to recover 98% of optical tropical crop pixels obscured by clouds, improving yield estimation from NDVI by 15% (Zhou et al., 2024).

- **Case 2 Daily Crop Monitoring Super-Resolution**:
  A GAN–Diffusion hybrid generated 10 m NDVI maps every day to facilitate near-real-time crop stress tracking and precision irrigation scheduling improvement (Sudhakara & Bhattacharjee, 2024).
- **Case 3 Environmental Change Prediction**:
  A spatiotemporal VAE predicted areas of high soil erosion in Egypt's desert regions under different rainfall regimes to facilitate timely management intervention (Wang et al., 2025).

3.4 DISCUSSION

The aforementioned examples indicate that generative AI extends the analytical function of remote sensing from observation to simulation and creation. It intensifies the quality and quantity of environmental data, thereby enabling informed decision-making for ecosystem, water, and land management.

However, successful deployment requires vigorous validation, cross-sensor calibration, and physical interpretability. While GANs and diffusion models generate visually correct results, scientific validity, that is, radiometric integrity and physical plausibility is an area of research that is important to tackle.

3.5 CHALLENGES, LIMITATIONS, AND ETHICAL CONSIDERATIONS IN GENERATIVE AI FOR REMOTE SENSING

3.5.1 Data Quality, Bias, and Representativeness

The generative AI model's accuracy is inherently associated with the quality and diversity of training data. Remote sensing environmental data are spatially, temporally, and spectrally biased, especially toward intensively studied regions, such as Europe and North America, and still lack data in developing countries and desert regions (Reichstein et al., 2019). Since generative models learn from such unbalanced inputs, they could possibly increase geographic bias to produce fake images that misrepresented local land-cover patterns or soil reflectance characteristics.

Furthermore, satellite images are afflicted with radiometric noise, sensor drift, and atmospheric artifacts that can contaminate learned distributions. If left uncorrected, such imperfections are propagated via the generative process to generate fraudulent reconstructions. Such issues are optimally addressed via standard preprocessing pipelines, domain-adaptation techniques, and bias-sensitive model evaluation for the purpose of ensuring outputs that maintain both spatial coherence and environmental truth (Janga et al., 2023).

3.5.2 Physical Consistency and Scientific Validity

Unlike traditional remote sensing algorithms founded on physical radiative-transfer models, generative AI algorithms are mostly statistical correlation-based. This introduces a nagging problem with the physical interpretability and correctness of the produced results. For example, a GAN may accurately mimic the visual appearance of vegetation, but its synthetic reflectance values will violate known spectral relations or energy-balance constraints (Koldasbayeva et al., 2024).

To minimize such risks, hybrid models Physics-Informed Generative Models (PIGMs) have come into existence in recent times. PIGMs incorporate physical equations or conservation laws directly into the loss function, which forces the generated data to remain within physically realistic scales. This adds scientific validity with a loss in computational complexity and diminishes model flexibility. Developing effective hybrid algorithms that meet both realism and physical precision remains a primary research frontier of the field (Bastek et al., 2024).

3.5.3 Computational Cost and Scalability

Training of high-resolution generative models, such as diffusion networks or multimodal GANs, takes monstrous computational power and energy. As globally scaled Earth Observation datasets become increasingly reliant on environmental applications, the carbon cost of training on the latter presents itself as a paradox unto itself using power-greedy computation toward sustainable research ends (Iftikhar & Davy, 2024).

Scalable solutions require parameter-efficient architectures to be implemented, cloud-based training on renewable-energy servers, and federated learning strategies employing decentralized regional data without aggregating centrally. These solutions not only reduce computational costs but also enhance data privacy and universality among institutions.

3.5.4 Model Explainability and Transparency

Another significant generative AI limitation in environmental applications is that it is "black-box." Decision-makers and scientists normally require transparent models so that they can view how predictions are built, especially when the outputs inform policies or environmental interventions. Yet, generative networks hardly provide explicit reasoning paths from input features to outputs.

Strategies for obtaining explainable generative AI (XGAI) are visualizing latent space activations, quantitatively measuring feature sensitivity, and incorporating uncertainty estimation into outputs (Doshi-Velez & Kim, 2017). Such technologies play a crucial role in building user trust, especially in applications that have regulatory or risk-sensitive connotations, such as land degradation monitoring, flood-risk assessment, and climate modeling.

Ultimately, explainability necessitates both algorithmic architecture and interdisciplinary collaboration between AI engineers, environmental scientists, and policymakers.

3.5.5 Ethical, Legal, and Environmental Implications

The use of environmental remote sensing by generative AI also raises several ethical issues. Synthetic data generation may render it impossible to identify and differentiate between real and fake observations, and unless metadata are accurately annotated, this can be misused or misinterpreted. For instance, unannotated synthetic images might inadvertently be used in official environmental reports, undermining data authenticity.

Intellectual-property issues also come into play when training generative models on proprietary satellite datasets without explicit permission. Compliance with licensing terms and clear data provenance practices are therefore paramount (European Space Agency, 2024a).

From an environmental viewpoint, the energy expended to perform massive-scale AI training results in carbon emissions, and the necessity for green AI practices such as model efficiency, hardware efficiency, and carbon offset (Patterson et al., 2021).

Finally, the application of ethical AI needs to prioritize equity and access. Generative AI platforms need to reach all areas geographically evenly and not be concentrated in data-intensive economies, and promote open-source collaboration for the support of global environmental sustainability.

3.6 SUMMARY

Generative AI can revolutionize environmental remote sensing but will progress depending on the resolution of the interconnected problems of bias, physical realism, transparency, and sustainability. Creating physically rooted, interpretable, and morally governed models will ensure that generative AI enriches and does not distort our understanding of the Earth system.

3.7 FUTURE DIRECTIONS AND RESEARCH OUTLOOK

3.7.1 Evolution Toward Foundation and Multimodal Models

The trend for generative AI in environmental remote sensing will be toward foundation models high-capacity, pre-trained models that are able to learn universal representations from multimodal sources of data. Analogous to the type of models like GPT or Stable Diffusion in images and words, foundation models for Earth Observation would integrate optical, radar, LiDAR, hyperspectral, and climatic data streams into synergistic generative frameworks (Reichstein et al., 2019).

These systems will enable cross-domain synthesis, where the knowledge of one modality of sensing can generate complementary output in another (e.g., generating synthetic SAR data from optical imagery). This resulting fusion will enable unparalleled continuity in the observation of land-surface dynamics, vegetation health, and soil moisture, even during data gaps or sensor failures. These advancements will similarly demand careful calibration, open model governance, and energy-saving training processes in order to ensure both scientific reliability and sustainability.

3.7.2 Real-Time Generative Earth Observation

The next ten years should witness the emergence of real-time generative Earth Observation (GEO) systems. Powered by streaming satellite data and cloud computation, they will generate continuously updating Earth representations more or less dynamic digital twins of real environmental states and near-future forecasts (Pan et al., 2025).

These real-time models can potentially transform environmental decision-making. For instance, generative diffusion networks that are trained on multi-sensor time-series data can model the short-term evolution of drought, crop water stress, or flood hazard before they are detected by conventional observation intervals. Matching such predictive generative systems with Geographic Information Systems (GIS) would enable adaptive land and water management, whereby policy co-evolves with real-time model outputs.

3.7.3 Integration with Physical Modeling and Climate Simulation

Increased research will increase the complementarity of data-driven AI and process-driven physical models. AI–physics integration is a rapidly growing class that combines the generalization capability of machine learning with physical interpretability. Physics-Guided Diffusion Models (PGDMs), for example, can be used to limit generative output to obey hydrological or radiative-balance laws so that physically interpretable simulation of soil moisture, evapotranspiration, and carbon fluxes is feasible (Xi et al., 2025).

Such convergence will be key to the creation of Earth System Digital Twins with the ability to simulate environmental conditions as well as model human–ecosystem interactions across different climate regimes. A paradigm shift from static observation to dynamic, intervention-possible modeling a milestone on the way to global environmental resilience.

This merging shall be central in creating Earth System Digital Twins that are not only able to replicate environmental states but also able to depict human–ecosystem interaction under different states of the climate. Figure 3.7 shows conceptual schematic of an Earth Digital Twin framework integrating generative AI, physical modeling, and real-time environmental data streams for dynamic climate and ecosystem simulation.

3.7.4 Edge AI and Democratization of Generative Tools

Another promising area is edge computing and low-power deployment of AI. As sensor networks increase, there will be an increasing demand for generative models to operate on-site, such as on unmanned aerial vehicles (UAVs) or field monitoring stations. Edge-based generative AI can produce or enhance imagery in the field, enabling real-time environmental awareness without relying on high-bandwidth data transmission (Koubaa et al., 2023).

Concurrently, open-source generative platforms will be democratized by efforts from space agencies and research consortia, allowing top-level models to become accessible to low-resource institutions as well as developing nations. This democratization process ensures that the value of generative AI is made available beyond the research laboratory to improve local-scale sustainability practice in forestry, agriculture, and water management.

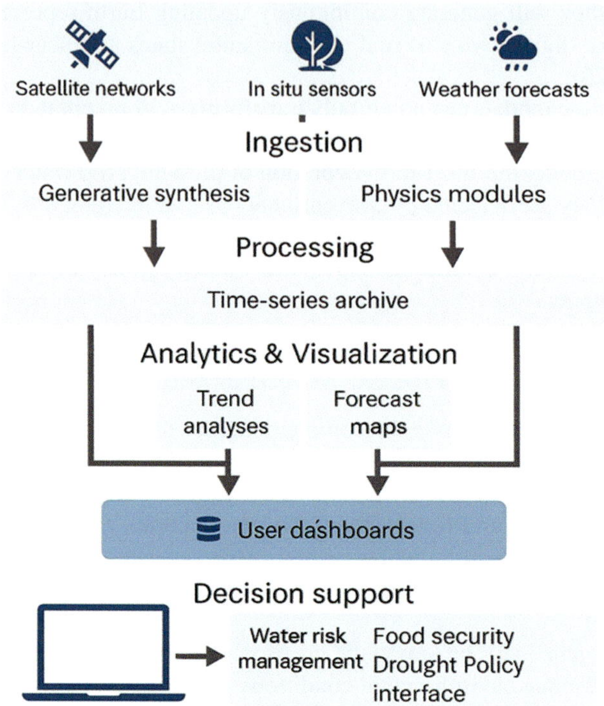

FIGURE 3.7 Conceptual schematic of an Earth Digital Twin framework integrating generative AI, physical modeling, and real-time environmental data streams for dynamic climate and ecosystem simulation.

3.7.5 Policy, Governance, and Ethical AI Futures

In order to achieve the maximum potential of generative AI responsibly, policy frameworks and ethics guidelines must evolve alongside technology developments. International institutions such as the Group on Earth Observations (GEO) and European Space Agency (ESA) have initiated the process of developing standards for labeling, verification, and documentation of artificial environmental data (European Space Agency, 2024b).

Future governance systems should focus on the following:

- Transparency: Demand clear metadata for every hybrid and synthetic product.
- Responsibility: Developing reproducibility standards and traceable model documentation.
- Justice: Enabling access to AI instruments and open environmental data for all on equal terms.
- Sustainability: Promoting carbon-free model training and energy-efficient architectures.

These policies will ensure that generative AI contributes to the global environmental agenda in a positive way enhancing, rather than replacing, human scientific intuition.

The innovations presented in the preceding subsections highlight the dynamic development of generative AI in environmental remote sensing. Combined, these horizons indicate a developing science moving from model-centric experimentation toward holistic, scalable, and ethically governed systems. Table 3.3 consolidates an integrated overview of the emerging directions in research, attendant innovations, and their anticipated contribution to the future of environmental monitoring and decision-making.

3.7.6 Summary Outlook

Generative AI transforms remote sensing from a passive observation science to an active simulation and synthesis science. Its capability to fill data holes, upscale resolution, and fake future environmental conditions has earth-shattering applications in ecosystem, water, and land management.

But turning this vision into reality relies on bridging technical and ethical thresholds at the same time to render models precise, interpretable, inclusive, and sustainable. With growing interdisciplinarity among AI researchers, Earth system researchers, and policymakers, generative AI will be a cornerstone for building resilient societies and sustainable landscapes for many generations to come.

To briefly summarize the listed outcomes in this chapter, Table 3.4 presents the most significant applications, technical shortcomings, and strategic research

TABLE 3.3
Emerging Frontiers and Research Priorities in Generative AI for Environmental Remote Sensing

Frontier Area	Core Concept/ Innovation	Expected Impact on Remote Sensing	Key References/ Examples
Foundation & Multimodal Models	Unified generative architectures integrating optical, radar, LiDAR, and climate data	Comprehensive Earth representations; improved cross-sensor synthesis	Reichstein et al. (2019)
Real-Time Generative Earth Observation	Streaming satellite data coupled with generative diffusion frameworks	Near–real-time environmental monitoring and forecasting	Pan et al. (2025)
AI–Physics Integration	Physics-guided diffusion and hybrid models enforcing physical constraints	Physically consistent simulations of soil, water, and carbon cycles	Koubaa et al. (2023)
Edge & On-Site AI	Low-power generative models operating on UAVs or IoT sensors	Rapid in-field image enhancement and synthesis; reduced data latency	Koubaa et al. (2023)
Ethical & Governance Frameworks	Policies for transparency, accountability, and sustainability in synthetic data	Trustworthy AI adoption in global environmental monitoring	European Space Agency (2024a,b)

Note: The table synthesizes the key research frontiers shaping the next decade of generative AI-driven Earth observation, linking each innovation to its operational and societal implications.

TABLE 3.4
Summary of Generative AI Applications, Challenges, and Future Recommendations for Environmental Remote Sensing

Application Domain	Key Achievements and Benefits	Major Challenges	Future Recommendations
Cloud Removal and Image Restoration	GANs and diffusion models reconstruct cloud-obstructed scenes with high spectral and spatial consistency, enabling continuous time-series analysis.	Visual accuracy may not always align with physical validity; limited generalization across sensors and regions.	Integrate physics-informed loss functions and multi-sensor calibration to preserve radiometric fidelity.
Data Fusion and Super Resolution	Generative frameworks combine optical, SAR, and multispectral data to enhance spatial and temporal resolution beyond single-sensor limits.	Risk of spectral distortion and unrealistic textural detail; high computational demand.	Develop energy-efficient fusion networks and benchmark against physically based upscaling techniques.
Change Detection and Predictive Simulation	CycleGANs and spatiotemporal models enable realistic prediction of environmental changes and future states for early warning and planning.	Difficulty in differentiating true environmental changes from model artifacts.	Employ hybrid AI–physics models with uncertainty quantification and validation using in-situ data.
Environmental Anomaly Detection	Autoencoders and diffusion models detect unusual events (floods, fires, pollution) by modeling baseline conditions and identifying deviations.	False positives in heterogeneous landscapes; limited interpretability for decision-makers.	Enhance explainability through saliency mapping and contextual integration with GIS layers.
Digital Twins and Real-Time Earth Simulation	Generative AI supports creation of EDT for dynamic visualization, forecasting, and environmental scenario testing.	Data governance and computational scalability remain key limitations.	Adopt modular architectures and open-source standards to ensure transparency, interoperability, and ethical use.
Sustainability and Ethical Governance	Generative AI promotes data accessibility and resilience but also raises concerns about energy use, fairness, and transparency.	High training costs, carbon footprint, and unequal access to AI infrastructure.	Encourage carbon-neutral AI pipelines, equitable data sharing, and inclusive capacity building in developing regions.

recommendations for further advancement of generative AI in environmentally remote sensing.

The future of remote sensing generative AI is therefore not merely technological but also philosophical and cooperative.

3.8 CONCLUSION

The emergence of Generative AI is a revolutionary leap in environmentally remote sensing. Breaking away from conventional analytical frameworks, generative algorithms such as GANs, VAEs, and DPMs have transformed the way we monitor, reconstruct, and simulate dynamic Earth systems. By examining the complex statistical relationships within multimodal data, such models are able to generate new, realistic, and scientifically meaningful environmental data well beyond the capabilities of sensors alone.

In practical applications, the use of generative AI has addressed all of the long-standing challenges in remote sensing, including cloud obstruction, absence of temporal completeness, and resolution limitations of sensors. With intelligent inpainting, super-resolution, and data-fusion techniques, generative models are able to reconstruct and merge satellite observations with excellent spatial and spectral precision. These advancements have enriched the accuracy of indicators derived from them such as vegetation indices, soil moisture maps, and land-surface temperature retrieval thus contributing to the improved reliability of environmental surveys at local, regional, and global scales.

The growing application of generative AI in environmental remote sensing has brought about a series of significant developments across different fields of use. At the same time, it has also revealed a variety of scientific and ethical questions to be solved in order to progress responsibly. Table 3.4 offers an overview of the main contributions, limitations, and forward-looking recommendations made throughout this chapter, forming a short summary of the current state of affairs on generative AI for environmental monitoring.

Additionally, generative AI has opened new horizons for predictive environmental modeling. Using time-series learning and spatiotemporal synthesis, models can emulate future landscape states, forecast environmental hazards, and support proactive management strategies. For instance, the generation of synthetic drought and flood sequences enables policymakers to analyze prospective impacts and develop adaptation prior to their occurrences. This foresight function redefines remote sensing as a reverse-looking observational mechanism to a forward-looking system for environmental sustainability and resilience.

However, with these come profound scientific, ethical, and operational issues. One of the most immediate is to give physical credibility to generated data. While generative outcomes may appear visually appealing, they need to uphold physical laws of energy balance, reflectance behavior, and environmental causality. Hybrid physics-informed generative approaches offer a strong pathway forward, balancing deep learning flexibility and scientific interpretability.

No less timely are ideas related to transparency, equity, and sustainability. Generative models of AI depend on massive computational power, often many times, over producing sizeable carbon profiles. While the world's scientific community continues to utilize these methods for research into sustainability, model development must itself be an eco-friendly process. Utilization of energy-efficient hardware architectures, optimal training pipeline efficiencies, and carbon offsetting measures are ethical responsibilities in keeping with the very goals of environmental science.

Also, data justice and global access remain priority concerns. The benefit of generative AI must be greater than that of technologically advanced nations. Democratization of rights-monetization-free open-source frameworks, equitable access to good-quality training data sets, and regional capacity development are critical steps toward a globally inclusive environmental AI framework. By making it possible for institutions in the developing world those in Africa, the Middle East, and South Asia to use and adapt generative AI platforms, the globe can assist in developing more equitable and fairer environmental monitoring systems.

The future of generative AI in remote sensing is therefore not only technical but also philosophical and social. The Earth Digital Twin concept, enabled by multimodal generative paradigms, is the synoptic convergence of physics, data, and computation. These digital replicas of Earth will enable real-time visualization, simulation, and planning for climate adaptation, land restoration, and water resource management.

This vision is possible only with continued interdisciplinary collaboration unifying geosciences, artificial intelligence (AI), ethics, and policy scientists collectively. It is possible only through such collaborations that generative AI can translate from the status of a technological advancement to a tool for sustainable development change. The next decade will dictate how responsibly and effectively humankind leverages generative intelligence to perpetuate the world's resources and maintain the tenuous equilibrium of its ecosystems.

In short, generative AI presents both unheralded opportunity and sacred responsibility. It enables us to generate new knowledge of the environment, but it requires us to live with integrity, be transparent, and be good stewards of the environment. If informed by scientific rigor and ethical vision, generative AI will be a foundation of the next generation of Earth observation one that not only charts the planet as it exists today but also imagines the planet as it might be: restored, resilient, and sustainable.

The future trajectory of generative AI in remote sensing is therefore not only technological but also philosophical and interdisciplinary. EDTs vision, supported by multimodal generative paradigms, is the confluence of data, physics, and computation (Figure 3.8).

DECLARATIONS

Author Contributions: The author solely conceived, structured, and wrote the chapter "The Use of Generative AI in Environmental Remote Sensing." The author reviewed the current literature, synthesized findings, and prepared all textual, graphical, and tabular content.

Funding: No specific grant from any public, commercial, or not-for-profit funding agency was received for this work.

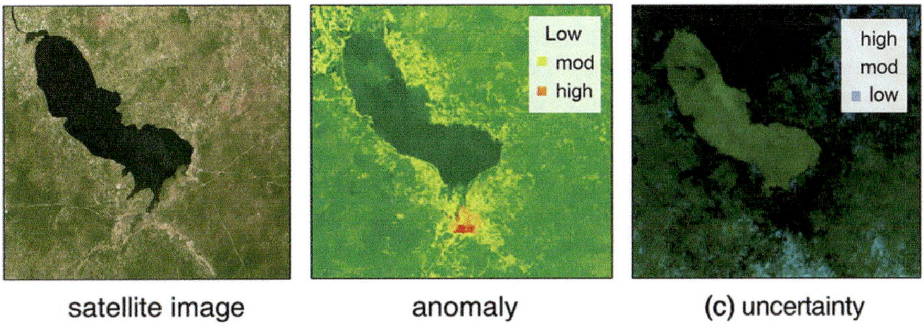

Anomaly detection with a generative autoencoder: (a) satellite inage; (b) anomaly map highlighting deviations from baseline; (c) model uncertainty/confidence map used to reduce false positives.

FIGURE 3.8 Integrated framework summarizing the role of generative AI in environmental remote sensing, connecting data sources, model architectures, key applications, and ethical governance within a sustainable development context.

Data Availability: This chapter does not include any original datasets. All data supporting the findings are derived from publicly available literature and cited accordingly.

Ethical Approval and Consent to Participate: Not applicable. This study does not involve human participants, animals, or data collected from individuals.

Consent for Publication: Not applicable. No personal or identifiable information is included in this work.

Competing Interests: The author declares no competing financial or non-financial interests relevant to the content of this chapter.

ARTIFICIAL INTELLIGENCE (AI) USE DECLARATION

AI tools were used solely to support language editing, formatting consistency, and citation organization under the full supervision and intellectual control of the author.

All scientific interpretation, conceptual framing, and analytical content were entirely created by the author.

No AI-generated text or image was used without human verification or critical revision, and all sources of information have been independently validated and cited.

REFERENCES

Ahmad, M. N., Shao, Z., Javed, A., Ahmad, I., Islam, F., Skilodimou, H. D., & Bathrellos, G. D. (2024). Optical–SAR data fusion based on simple layer stacking and the XGBoost algorithm to extract urban impervious surfaces. *Remote Sensing*, *16*(5), 873. https://doi.org/10.3390/rs16050873

Ajayi, O. G., & Olaide, D. M. (2024). Assessment of generative adversarial networks for cloud occlusion removal on remotely sensed images. *Arabian Journal of Geosciences*, *17*, 141. https://doi.org/10.1007/s12517-024-11939-y

Bastek, J.-H., Sun, W., & Kochmann, D. M. (2024). Physics-informed diffusion models. *arXiv preprint arXiv:2403.14404*. https://arxiv.org/abs/2403.14404

Benedetti, R., Ienco, D., Gaetano, R., & Del Frate, F. (2018). SAR to optical image synthesis for cloud removal with generative adversarial networks. *ISPRS Annals of the Photogrammetry, Remote Sensing and Spatial Information Sciences, IV-1*, 5–12. https://doi.org/10.5194/isprs-annals-IV-1-5-2018

Calleo, Y., Pilla, F., & Di Zio, S. (2025). Generative pre-trained transformers for climate scenarios: A statistical coefficient for future policy development. *Quality & Quantity*. https://doi.org/10.1007/s11135-025-02423-0

Chen, S., Wang, J., Liu, Q., Liang, X., Liu, R., Qin, P., Yuan, J., Wei, J., Yuan, S., Huang, H., & Gong, P. (2024). Global 30 m seamless data cube (2000–2022) of land surface reflectance generated from Landsat 5, 7, 8, and 9 and MODIS Terra constellations. *Earth System Science Data, 16*(11), 5449–5475. https://doi.org/10.5194/essd-16-5449-2024

Doshi-Velez, F., & Kim, B. (2017). Towards a rigorous science of interpretable machine learning. *arXiv preprint arXiv:1702.08608*. https://arxiv.org/abs/1702.08608

European Space Agency. (2024a). Disruptive Satcom Systems design with digital generative design. https://connectivity.esa.int/news/european-space-agency-kicks-groundbreaking-generative-ai-project-satellite-communication-design

European Space Agency. (2024b). ESA space environment report 2024. https://www.esa.int/Space_Safety/Space_Debris/ESA_Space_Environment_Report_2024

Gao, P., Du, W., Lei, Q., Li, J., Zhang, S., & Li, N. (2023). NDVI forecasting model based on the combination of time series decomposition and CNN–LSTM. *Water Resources Management, 37*, 1481–1497. https://doi.org/10.1007/s11269-022-03419-3

Goodfellow, I., Pouget-Abadie, J., Mirza, M., Xu, B., Warde-Farley, D., Ozair, S., Courville, A., & Bengio, Y. (2014). Generative adversarial nets. *Advances in Neural Information Processing Systems, 27*, 2672–2680. https://doi.org/10.48550/arXiv.1406.2661

Gorelick, N., Hancher, M., Dixon, M., Ilyushchenko, S., Thau, D., & Moore, R. (2017). Google Earth Engine: Planetary-scale geospatial analysis for everyone. *Remote Sensing of Environment, 202*, 18–27. https://doi.org/10.1016/j.rse.2017.06.031

Hinton, G. E., Osindero, S., & Teh, Y. W. (2006). A fast learning algorithm for deep belief nets. *Neural Computation, 18*(7), 1527–1554. https://doi.org/10.1162/neco.2006.18.7.1527

Hinton, G. E., & Salakhutdinov, R. R. (2006). Reducing the dimensionality of data with neural networks. *Science, 313*(5786), 504–507. https://doi.org/10.1126/science.1127647

Ho, J., Jain, A., & Abbeel, P. (2020). Denoising diffusion probabilistic models. *arXiv preprint arXiv:2006.11239*. https://doi.org/10.48550/arXiv.2006.11239

Hong, D., Li, C., Zhang, B., Yokoya, N., Benediktsson, J. A., & Chanussot, J. (2024). Multimodal artificial intelligence foundation models: Unleashing the power of remote sensing big data in Earth observation. *The Innovation Geoscience, 2*(1), 100055. https://doi.org/10.59717/j.xinn-geo.2024.100055

Iftikhar, S., & Davy, S. (2024). Reducing carbon footprint in AI: A framework for sustainable training of large language models. In Arai, K. (eds) *Proceedings of the Future Technologies Conference (FTC) 2024, Volume 1. FTC 2024. Lecture Notes in Networks and Systems*, vol. 1154. Springer, Cham. https://doi.org/10.1007/978-3-031-73110-5_22

Janga, B., Asamani, G. P., Sun, Z., & Cristea, N. (2023). A review of practical AI for remote sensing in Earth sciences. *Remote Sensing, 15*(16), 4112. https://doi.org/10.3390/rs15164112

Jing, R., Duan, F., Lu, F., Zhang, M., & Zhao, W. (2023). Denoising diffusion probabilistic feature-based network for cloud removal in Sentinel-2 imagery. *Remote Sensing, 15*(9), 2217. https://doi.org/10.3390/rs15092217

Karniadakis, G. E., Kevrekidis, I. G., Lu, L., Perdikaris, P., Wang, S., & Yang, L. (2021). Physics-informed machine learning. *Nature Reviews Physics, 3*(6), 422–440. https://doi.org/10.1038/s42254-021-00314-5

Khanna, S., Liu, P., Zhou, L., Meng, C., Rombach, R., Burke, M., Lobell, D., & Ermon, S. (2023). DiffusionSat: A generative foundation model for satellite imagery. *arXiv preprint arXiv:2312.03606*. https://arxiv.org/abs/2312.03606

Kingma, D. P., & Welling, M. (2019). An introduction to variational autoencoders. *Foundations and Trends® in Machine Learning, 12*(4), 307–392. https://doi.org/10.1561/2200000056

Koldasbayeva, D., Tregubova, P., Gasanov, M. et al. (2024). Challenges in data-driven geospatial modeling for environmental research and practice. *Nature Communications, 15*, 10700. https://doi.org/10.1038/s41467-024-55240-8

Koubaa, A., Ammar, A., Abdelkader, M., Alhabashi, Y., & Ghouti, L. (2023). AERO: AI-enabled remote sensing observation with onboard edge computing in UAVs. *Remote Sensing, 15*(7), 1873. https://doi.org/10.3390/rs15071873

Liu, S., Li, K., Liu, X., & Yin, Z. (2023). Geospatial AI in Earth observation, remote sensing, and GIScience. *Applied Sciences, 13*(22), 12203. https://doi.org/10.3390/app132212203

Liu, Y., Yue, J., Xia, S., Ghamisi, P., Xie, W., & Fang, L. (2024). Diffusion models meet remote sensing: Principles, methods, and perspectives. *arXiv preprint arXiv:2404.08926*. https://arxiv.org/abs/2404.08926

Mu, P., & Tian, F. (2022). Spatiotemporal fusion of multi-temporal MODIS and Landsat-8/9 imagery for enhanced daily 30 m NDVI reconstruction: A case study of the Shiyang River Basin cropland. *Remote Sensing, 17*(9), 1510. https://doi.org/10.3390/rs17091510

Pan, J., Lei, S., Fu, Y., Li, J., Liu, Y., Sun, Y., He, X., Peng, L., Huang, X., & Zhao, B. (2025). EarthSynth: Generating informative earth observation with diffusion models. *arXiv preprint arXiv:2505.12108*. https://arxiv.org/abs/2505.12108

Patterson, D., Gonzalez, J., Le, Q., Liang, C., Munguia, L. M., Rothchild, D., … & Dean, J. (2021). Carbon emissions and large neural network training. *arXiv preprint arXiv:2104.10350*. https://arxiv.org/abs/2104.10350

Rahimi, E., & Jung, C. (2024). Evaluating the applicability of Landsat 8 data for global time series analysis. *Frontiers in Remote Sensing, 5*, Article 1492534. https://doi.org/10.3389/frsen.2024.1492534

Ramesh, A., Pavlov, M., Goh, G., Gray, S., Voss, C., Radford, A., Chen, M., & Sutskever, I. (2021). Zero-shot text-to-image generation. *arXiv preprint arXiv:2102.12092*. Introduced DALL·E, a transformer-based model capable of multimodal generation. https://doi.org/10.48550/arXiv.2102.12092

Reichstein, M., Camps-Valls, G., Stevens, B. et al. (2019). Deep learning and process understanding for data-driven Earth system science. *Nature* 566, 195–204. https://doi.org/10.1038/s41586-019-0912-1

Rolnick, D., Donti, P. L., Kaack, L. H., Kochanski, K., Lacoste, A., Sankaran, K., … & Bengio, Y. (2022). Tackling climate change with machine learning. *ACM Computing Surveys, 55*(2), 1–96. This survey emphasizes the role of generative AI in improving environmental monitoring in regions with limited field data. https://doi.org/10.1145/3485128

Sharifi, A., & Mahdipour, H. (2024). Utilizing geospatial artificial intelligence for remote sensing applications. *Environmental Earth Sciences, 83*, 274. https://doi.org/10.1007/s12665-024-11584-4

Sudhakara, B., & Bhattacharjee, S. (2024). Prediction of high-resolution soil moisture using multi-source data and machine learning. In *Distributed Computing and Intelligent Technology* (pp. 282–292). Springer. https://doi.org/10.1007/978-3-031-50583-6_19

Sui, J., Ma, Y., Yang, W., Zhang, X., Pun, M.-O., & Liu, J. (2024). Diffusion enhancement for cloud removal in ultra-resolution remote sensing imagery. *arXiv preprint arXiv:2401.15105*. https://arxiv.org/abs/2401.15105

Sun, C., Fan, X., Lu, X., Zhou, L., Zhao, J., Dong, Y., & Chen, Z. (2025). Feature constraints map generation models integrating generative adversarial and diffusion denoising. *Remote Sensing, 17*(15), 2683. https://doi.org/10.3390/rs17152683

Tee, J. T. J., Zhang, K., Yoon, H. S., Gowda, D. N., Kim, C., & Yoo, C. D. (2024). Physics informed distillation for diffusion models. *arXiv preprint arXiv:2411.08378*. https://arxiv.org/abs/2411.08378

Wang, G., Hu, H., Islam, A.R.M.T. et al. (2025). Spatiotemporal rainfall concentration and erosivity in a tropical monsoon country. *Theoretical and Applied Climatology* 156, 251. https://doi.org/10.1007/s00704-025-05479-y

Wang, X., Sun, L., Chehri, A., & Song, Y. (2023). A review of GAN-based super-resolution reconstruction for optical remote sensing images. *Remote Sensing*, *15*(20), 5062. https://doi.org/10.3390/rs15205062

Wang, Y., Zhang, H., Liu, Q., & Chen, F. (2024). Probabilistic diffusion models advance extreme flood forecasting. *Geophysical Research Letters*, *51*(4), e2025GL115705. https://doi.org/10.1029/2025GL115705

Whig, P., Bhatnagar Bhatia, A., Nadikatu, R. R., Alkali, Y., & Sharma, P. (2024). GIS and remote sensing application for vegetation mapping. In *Geo-Environmental Hazards Using AI-enabled Geospatial Techniques and Earth Observation Systems* (pp. 17–39). Springer. https://doi.org/10.1007/978-3-031-53763-9_2

Xi, Y., Zhuang, Q., & Liu, X. (2025). A hybrid physics-guided deep learning modeling framework for predicting surface soil moisture. *Journal of Geophysical Research: Machine Learning and Computation*, 2(3). https://doi.org/10.1029/2025jh000682

Xu, M., Wang, Z., Zhu, J., Jia, X., & Jia, S. (2021). *Multi-attention generative adversarial network for remote sensing image super-resolution*. arXiv. https://doi.org/10.48550/arXiv.2107.06536

Yazdani, S., Singh, A., Saxena, N., Wang, Z., Palikhe, A., Pan, D., Pal, U., Yang, J., & Zhang, W. (2025). Generative AI in depth: A survey of recent advances, model variants, and real-world applications. *Journal of Big Data*, *12*(230). https://doi.org/10.1186/s40537-025-01247-x

Yu, Z., Idris, M. Y. I., & Wang, P. (2025). When cloud removal meets diffusion model in remote sensing. *arXiv preprint arXiv:2504.14785*. Introduces DC4CR, a multimodal diffusion framework for cloud removal without temporal references. https://arxiv.org/abs/2504.14785

Zhan, T., Bian, J., Yang, J., Dang, Q., & Zhang, E. (2023). Improved conditional generative adversarial networks for SAR-to-optical image translation. In *Pattern Recognition and Computer Vision* (pp. 279–291). Springer. Demonstrates multimodal SAR-optical fusion using cGANs for surface reflectance reconstruction. https://doi.org/10.1007/978-981-99-8462-6_23

Zhao, X., & Jia, K. (2023). Cloud removal in remote sensing using sequential-based diffusion models. *Remote Sensing*, *15*(11), 2861. Applies cGANs to Sentinel-2 imagery for cloud removal using paired references. https://doi.org/10.3390/rs15112861

Zhou, Y., Wang, H., Yao, G., Liu, M., & Xu, Q. (2024). A novel remote sensing landslide semantic segmentation method: Using CycleGAN-based change detection algorithms. In *Engineering Geology for a Habitable Earth: IAEG XIV Congress 2023 Proceedings*, Chengdu, China (pp. 39–50). Springer. https://doi.org/10.1007/978-981-99-9069-6_3

4 Vision-Language Models in Remote Sensing

Balancing Geospatial Intelligence with Ethical and Responsible AI Practices

Thota Sivasankar and Varun Narayan Mishra

4.1 INTRODUCTION

Over the past few years, Vision-Language Models (VLMs) have drawn attention for combining computer vision and natural language processing to handle multimodal data effectively. These models are designed to understand and reason across visual and textual inputs, allowing for tasks such as image captioning, visual question answering, and semantic search to be conducted with a level of abstraction and flexibility greater than traditional single-modality systems. Despite notable success in natural imagery domains, the adoption of VLMs for handling geospatial data remains relatively underexplored, though the potential for advancing spatial reasoning and semantic interpretation is significant. Geospatial data, which represents objects, events, and phenomena on the Earth's surface, has been proliferating by at least 20% every year (Lee & Kang, 2015). Among the various sources, remote sensing data captured by Earth observation satellites and drones is a significant contribution to geospatial data. This data requires human expertise to retrieve meaningful information about the objects, events, and phenomena on the Earth's surface. Adopting VLMs for such data would revolutionize the geospatial sector by enabling language-guided analysis. These models could enable users to provide natural language queries about geospatial data, generate contextualized descriptions of land cover, or assist in time-sensitive tasks such as disaster response and monitoring.

This vision is supported by several foundational models, which have evolved significantly over recent years. Particularly, Contrastive Language-Image Pre-training (CLIP) (Radford et al., 2021) and A Large-scale ImaGe and Noisy-Text Embedding (ALIGN) (Jia et al., 2021) posed contrastive learning over image-text pairs at scale and showed how such learning could combine image and text information into shared representations. These models do not require explicit labels; instead, they learn from

DOI: 10.1201/9781003616207-5

the rich data of images and their corresponding captions present in vast datasets. More recent developments like BLIP-2 (Li et al., 2023a) and Flamingo (Alayrac et al., 2022) have added generative functions and enhanced adaptability to a few-shot learning paradigm, increasing the applicability of VLMs across multiple domains. With respect to geospatial Artificial Intelligence (AI), the use of these architectures is beginning to be implemented. The development of VLMs applied to remote sensing images has been made through works such as RSCLIP (Li et al., 2023b) and SkyEyeGPT (Zhan et al., 2025). These attempts emphasize both the potential challenges and the promise of integrating VLM into geospatial workflows, particularly where shifts in scale, sensor type (such as optical versus Synthetic Aperture Radar (SAR), and meaning overlap create constant challenges. Another considerable challenge is the lack of image-text pairs explicitly designed for Earth observation because this gap hinders the optimal use of the models in specialized workflows.

With that said, solving these issues will necessitate the integration of new architectural frameworks along with novel approaches to dataset creation, pretraining alignment, and evaluation. However, the benefits remain persuasive: VLMs provide a versatile and potentially low-supervision framework through which users can access and interact with geospatial information. This functionality may alleviate annotation efforts while enhancing interpretability and accessibility. This chapter provides a systematic and comprehensive study of VLMs about geospatial intelligence. It starts with basic definitions and moves through a review of model architecture training strategies and their geospatial customizations. It also focuses on additional applications, such as advanced environmental monitoring, interpretation of land use, and urban change detection. It provides an evaluation of current limitations with key focus areas for future research. This chapter aims to equip researchers and practitioners who wish to explore the recent developments in VLMs by reviewing existing literature from a geospatial perspective.

4.2 FOUNDATIONS OF VISION-LANGUAGE MODELS

Before diving into the geospatial applications of VLMs, it is necessary to explore the core functioning of these models. At a glance, these may appear to be advanced models that associate images with words, and in a sense, that is accurate. However, the process is more involved. The idea is to develop a system that can take a visual input, like a satellite image, and respond with something meaningful in natural language, like a description, an answer, or even an interpretation. To make this work, researchers have explored a few different ways of connecting images with text inside a model.

Many early systems fall into what are called dual-encoder architectures. In these, the image goes into one side of the model and the text into the other. Both are processed independently, and the goal is to project their outputs into a common embedding space where semantically related image-text pairs are positioned near each other. If the image is of a road and the sentence says, "a winding mountain road," the system should align them. CLIP and ALIGN are examples of this type. These models are surprisingly good at retrieval tasks, like matching a caption to an image, without needing a ton of hand-labeled training data (Li et al., 2024). They are not great when more nuanced interactions are needed.

Another branch includes what are called fusion encoders. These are more sophisticated. Instead of keeping the image and text separate for most of the process, these models blend them early. That allows joint processing to capture cross-modal relationships. If the system needs to understand where a school building is in an aerial photo relative to a park, a fusion encoder has the tools to do that. Vision-and-Language BERT (ViLBERT) and Universal Image-Text Representation Learning (UNITER) are known for this design. Of course, with added complexity comes more demand for data and computational power (Tan & Bansal, 2019; Lu et al., 2019).

There is also a third family: autoregressive models. These are built to generate language, not just pick from options. They take an image (and maybe a prompt), and respond with a sentence, one word at a time. It is a bit like how humans write or speak. Models like Lava and Idefics2 fit this mold. These are especially useful when the goal is to produce free-form captions or answer open-ended questions (Laurençon et al., 2024; Liu et al., 2023).

These models are unified by a shared mechanism known as the modality bridge, which enables the integration of visual and textual features. In simple systems, it might be a projection layer. In more advanced ones, it could be attention modules, or something called a Querying Transformer (Q-Former). The point is to create a place where the image and the language start influencing one another.

Another subtle but important factor is the choice of base encoders, the neural networks that handle the image and text inputs separately before they are fused. Better encoders often lead to better overall performance, even if the fusion part of the model stays the same. It is why using stronger vision backbones like Slip or language models like Mistral often pays off (Zhai et al., 2023; Jiang et al., 2023).

To bring clarity to the distinctions, Table 4.1 summarizes these three architecture types, along with their representative models and use cases.

From a geospatial perspective, however, some issues stand out. The first is data scarcity. Remote sensing images usually do not come with textual labels, which limits the usefulness of models trained on natural image-caption pairs. Although

TABLE 4.1
Common VLM Architectures and Their Characteristics

Model Type	Example Models	Training Strategy	Typical Use Cases	References
Dual-Encoder	CLIP, ALIGN	Contrastive	Retrieval, zero-shot classification	Radford et al. (2021); Jia et al. (2021)
Fusion-Encoder	UNITER, ViLBERT	Co-attention	Scene understanding, captioning	Chen et al. (2020); Lu et al. (2019)
Autoregressive	Idefics2, Flamingo	Instruction-tuning	VQA, captioning, chat-based analysis	Laurençon et al. (2024); Alayrac et al. (2022)

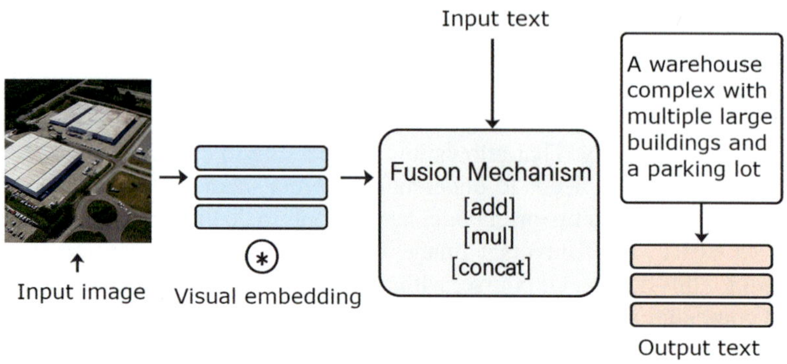

FIGURE 4.1 Generalized VLM pipeline.

datasets like Remote Sensing Image Captioning Dataset (RSICD) and DIOR-RSVG exist, they are small in scale and limited in diversity. To compensate, researchers have begun generating pseudo-captions using large language models, but this raises concerns about accuracy and bias (Tao et al., 2025).

Another challenge involves how images are tokenized. Initial approaches divided images into regular grids, which are not always optimal. Newer methods, such as Perceiver Resampled (Jaegle et al., 2021), aim to represent visual content in a more compact and task-sensitive manner, allowing for faster inference without significant losses in descriptive fidelity.

To help conceptualize the VLM structure, Figure 4.1 outlines a general pipeline: image and text inputs are encoded, aligned, and passed through a reasoning module to generate an output in natural language. With the structural understanding in place, we can now turn to how these systems are being adapted for geospatial tasks, where they promise to add a new layer of interpretability to complex Earth observation data.

4.3 VISION-LANGUAGE MODELS FOR GEOSPATIAL INTELLIGENCE

Applying VLMs to remote sensing is not just about repurposing models trained on web images. It represents a more foundational shift, one that redefines the principles underlying the interpretation of satellite data. Historically, this has meant classification, pixel labels, thematic maps, or maybe binary change detection outputs. These are precise, yes, but also rigid. Moreover, they do not scale well to questions of "what is happening here?" or "how has this landscape evolved?" VLMs change that conversation. Take image captioning, for instance. Instead of assigning the label "urban" or "water," models are now beginning to describe scenes in a way that resembles how analysts talk about them. "A dense residential cluster with visible road networks," or "open fields with scattered tree cover," yields more descriptive and analytically relevant information than conventional labels. That is the kind of thing Li et al. (2024) were exploring when they introduced the RSICap and RSGPT tools that try to bring structure and readability to satellite imagery captions. What is really interesting, though, is how VLMs let us search satellite data with natural language, not just by file name or metadata, but by meaning. SkyEyeGPT (Zhan et al., 2025) is one example. You can ask it questions like "where has construction expanded near rivers since 2015?" and it attempts to return a meaningful visual answer. This represents an approach that reproduces human spatial reasoning in interpreting and responding to queries.

Classification is another area where VLMs are making significant changes. Instead of retraining a model every time we want to add a new category, say, "solar installations" or "temporary shelters," prompt-based systems let you define that category in language. Remote CLIP (Liu et al., 2024) has shown this quite clearly. You write a description, the model uses it, and suddenly, classification becomes adaptable, not static. Moreover, in disaster settings, where quick insights are critical, these models can answer real-world questions like "Is this road blocked?" or "Are tents visible here?" Chappuis et al. (2022) and Bazi et al. (2022) both worked on this, combining conventional image encoders with transformers like Bidirectional Encoder Representations from Transformers (BERT). Instead of relying on map overlays or manual review, responders could get immediate feedback, which matters in the first 24 hours after a flood or earthquake. There is also emerging work on semantic change detection, which we find particularly exciting. Rather than flagging a change pixel by pixel, models like RSGPT can describe what has changed, "forest cleared along the hillside," or "new roads appearing west of the settlement." That is significantly more informative than conventional pixel-level change indicators.

Text-based image retrieval is another significant shift. You are not writing SQL queries or tagging layers. You ask: "coastal village with boats anchored nearby" or "valleys with visible terracing." AMFMN (Yuan et al., 2022) and VQA-TextRS (Al Rahhal et al., 2022) are examples of this. It is not perfect, but it significantly reduces the need for manually coding feature extractors or hand-labeling samples. The core opportunities for VLMs in geospatial intelligence are summarized in Table 4.2. All of this, from captioning to retrieval, hints at something broader: a change in how we treat satellite imagery. We are moving away from fixed, rule-bound systems to something more interactive, more language-driven. That does not mean the hard

TABLE 4.2
Core Opportunities for VLMs in Geospatial Intelligence

Application Area	Example Method/Model	Reference
Image Captioning	RSGPT, RSICap	Li et al. (2024)
Semantic Querying	SkyEyeGPT	Zhan et al. (2025)
Prompt-Based Scene Classification	RemoteCLIP	Liu et al. (2024)
Visual Question Answering	VQA-TextRS, BERT+ViT	Al Rahhal et al. (2022), Chappuis et al. (2022)
Semantic Change Detection	RSGPT	Li et al. (2024)
Text-based Image Retrieval	AMFMN, TextRS	Yuan et al. (2022)

science disappears. We still need spectral indices, model validation, and careful interpretation. However, the interface is changing. The way we interact with data is changing. Moreover, like all new tools, VLMs still have work ahead. They make mistakes, and some applications are far from field-ready. However, if you are thinking long-term about how remote sensing becomes more inclusive, more intuitive, and more explainable, this shift toward vision-language reasoning is one to watch.

4.4 APPLICATIONS AND USE CASES

While the use of VLMs in remote sensing is still relatively new, some recent studies and early deployments are showing that these models may contribute more significantly to the field than initially anticipated. Right now, most of what we see are demonstrations, proofs of concept rather than polished products, but even these early trials suggest that the approach is worth taking seriously. What follows are a few examples from the literature that reflect how these models are already being applied to real problems.

4.4.1 Disaster Damage Visualization and Geo-Localization

One of the more compelling examples comes from Sogi et al. (2024), who built an interactive retrieval system designed to identify and geolocate ground-level images from disaster events like flooding, earthquakes, hurricanes, fire, and landslides. Their framework utilized BLIP-2 to enable users to query by either image or text, retrieving relevant visuals from a large dataset. In cases where photos did not have GPS metadata, the model was still able to estimate location with surprising accuracy. In flood contexts, they achieved 94.1% Recall@K. The study result suggests that missing geolocation metadata may no longer critically hinder crisis mapping workflows.

Another group, Schmidt et al. (2025), tested four models—BLIP-2, OpenFlamingo, LLaVA1.6, and GPT-4o—on a similar task, this time using Flickr images from flood zones. GPT-4o demonstrated the highest spatial accuracy, with a median localization error of approximately 89 km, while OpenFlamingo identified the most significant number of usable images overall. One important thing they pointed out: most models worked best in Central Europe, suggesting a heavy training bias toward regions with more data coverage.

4.4.2 Monitoring Informal Settlements and Urban Expansion

When it comes to monitoring unplanned urban growth, traditional classification systems can only go so far. They are good at saying "this is a built-up area," but not much else. Recent VLMs like RSCLIP and RSGPT have added a new layer of description. Instead of tagging a region as just "urban," they can describe it as "scattered informal dwellings near high-density housing," or "recent construction visible in floodplain zone." According to Li et al. (2024), this level of semantic precision could be instrumental in planning or for early warning tools aimed at identifying at-risk communities.

4.4.3 Farmland Segmentation with Language-Guided Inputs

Agricultural monitoring has always been tricky, especially in areas with mixed land use. Wu et al. (2025) worked on a model called FSVLM, which used natural language to guide farmland segmentation. Instead of labeling images using predefined classes, users could give the model a phrase like "large rectangular plots with visible irrigation lines," and the system would find matching patterns in the imagery. When they tested the approach across four provinces in China, the results showed measurable improvements in both intersection-over-union and Dice scores.

4.4.4 Visual Reasoning for Environmental Change

Most satellite-based change detection systems focus on pixel shifts or class transitions. Tao et al. (2025) proposed a different approach: asking the model to "describe" the change. For example, if an area had lost vegetation over five years, the model might say, "shrubland replaced by bare soil, with linear features suggesting road construction." That is a lot more useful to a policymaker than a binary change mask. It is beneficial in long-term monitoring where trends matter more than momentary anomalies.

4.4.5 Visual Audit of Infrastructure Development

One of the persistent problems in rural development is that infrastructure often gets built before it gets recorded. Roads, bridges, or informal power lines can be visible in imagery long before they appear in official maps. Some researchers are now exploring whether VLMs can bridge that gap by letting analysts enter descriptions like "new road segment across river bend," and retrieving relevant image patches. These methods are still in prototype stages, but the potential use for NGOs or regional planners is obvious.

4.4.6 Geo-Semantic Climate Analysis

Climate monitoring, especially for glacial retreat or snowpack variation, usually requires complex remote sensing models. However, what if you could ask a system, "Show me where snow coverage has declined since 2015," and it returns both a map and a plain-language explanation? That is the vision behind some of the newer text-query systems trained for climate-linked tasks. Schmidt et al. (2025) noted, though, that while results are promising in data-rich regions, generalizing performance globally remains a significant hurdle.

What stands out across these examples is not just the technical novelty but the shift in how users can interact with satellite data. VLMs make it possible to frame questions in plain language and receive responses that are intelligible, flexible, and sometimes even intuitive. That does not mean they are perfect. There are still questions around accuracy, regional bias, and how reliable these systems are under complex scenarios. However, for now, they seem to offer a genuinely new way of thinking about geospatial workflows, and that alone makes them worth exploring.

4.5 CHALLENGES AND RESEARCH GAPS

The deployment of VLMs in geospatial contexts, although promising, faces several key challenges. These obstacles span across the availability of data, model architecture limitations, lack of explainability, and the absence of robust evaluation frameworks tailored to remote sensing. The following subsections elaborate on these issues and draw from recent literature to identify where the gaps lie.

4.5.1 Domain Misalignment and Dataset Scarcity

Most widely used VLMs are pretrained on image-text datasets derived from natural scenes, such as LAION-400M or COCO Captions. These corpora do not represent the structure, scale, and content typical of remote sensing data. Satellite imagery differs from natural scenes in key ways, such as features, top-down views, distinct spectral characteristics, and geographically specific features not typically found in ground-level photographs. The shortage of large, well-annotated, and geographically diverse datasets remains a significant bottleneck. Efforts like RSICD and DIOR-RSVG provide initial steps, but they lack the volume and richness needed for robust training (Li et al., 2024; Tao et al., 2025).

4.5.2 Spatial and Semantic Complexity

Geospatial imagery requires interpretation across multiple spatial scales. A single image may include fine-grained features, such as small agricultural plots, as well as broader land cover patterns. Models are expected to reason about both localized and contextual elements within the same scene. This requirement highlights constraints in the current design of VLMs that were designed primarily for object-centric natural images (Wu et al., 2025).

4.5.3 Explainability and Interpretability

The lack of transparency in model outputs is a pressing concern, especially in applications involving disaster response, climate assessment, or urban development. Stakeholders require not only predictions but also understandable justifications for those outputs. Although visual attention mechanisms and saliency maps offer partial insights, current implementations are limited in their applicability to geospatial domains. There is a growing need for explainable artificial intelligence (XAI) approaches adapted for geospatial VLMs (Sogi et al., 2024).

4.5.4 Geographic and Linguistic Bias

Biases in training data manifest spatially and linguistically. Schmidt et al. (2025) found that geocoding performance in flood-related imagery was more accurate in regions with high image density in training datasets, notably in Europe. This geographic bias compromises model reliability in less-represented regions such as Africa or Southeast Asia. Moreover, most VLMs are English-centric, excluding users who operate in local languages or dialects common in field operations and community engagement.

4.5.5 Model Size and Computational Burden

The complexity of VLMs like GPT-4V and Flamingo requires significant computational infrastructure. This presents challenges for real-time monitoring systems or resource-constrained environments, particularly in field deployments. While recent innovations like FSVLM (Wu et al., 2025) indicate progress in creating lightweight architectures, balancing performance with deployment feasibility remains an open issue.

4.5.6 Lack of Evaluation Frameworks

There is no universally accepted benchmark to evaluate VLMs in geospatial contexts. Metrics such as Bilingual Evaluation Understudy (BLEU) or Consensus-based Image Description Evaluation (CIDEr), adopted from natural image captioning, do not adequately capture the spatial or contextual accuracy required in Earth observation. As noted by Zhan et al. (2025), the development of remote sensing-specific benchmarks for tasks like visual question answering, change detection, and captioning is essential to advance reproducibility and trust.

4.6 FUTURE OUTLOOK

VLMs are increasingly finding their way into discussions on the future of geospatial analytics. While enthusiasm around their capabilities continues to grow, the reality is that their practical use in Earth observation workflows is still in its infancy. A lot of what we are seeing so far are proof-of-concept studies, often built in controlled environments. For these models to transition into operational tools that researchers, agencies, or practitioners can rely on, several things need to be systematically addressed. To begin with, there is a critical demand for datasets aligned with geospatial imagery. The general-purpose image-text datasets used to train current models, often derived from online photos, do not reflect the spatial and spectral complexity found in satellite imagery. What is missing are large-scale, geographically diverse datasets where captions or labels reflect actual land cover variations, regional patterns, and even socio-environmental context.

There is also the question of efficiency. Most VLMs today are built to run on robust cloud infrastructure. That is fine in research labs, but it is less feasible in field settings or under budget constraints. If we want these tools to work on the ground, say, in local government offices or environmental NGOs, then we will need ways to fine-tune and run them without needing massive Graphics Processing Unit (GPU) clusters. Lighter adaptation methods, like prompt tuning or adapter layers, could help bridge that gap, but more experimentation is needed to see how well these work in practice. Another issue that keeps coming up is explainability. It is not enough for a model to generate a caption or identify a region of interest. People using these tools need to know why the system made that call. Did it focus on a particular texture? A pattern in the landscape? Was it confused by something that looked similar?

Techniques like visual grounding and heatmaps offer partial insight, but they are still underused in this context. For geospatial applications where decisions can affect land rights, emergency responses, or conservation funding, interpretability becomes critical.

If these models are going to see wider use, they also need to fit into the platforms people already use. Tools like Google Earth Engine, QGIS, or SNAP are already part of the remote sensing ecosystem. Rather than developing entirely new systems around VLMs, a more effective strategy is to integrate them into existing platforms commonly used in remote sensing workflows. That kind of integration through APIs, plugins, or cloud connectors would allow researchers and analysts to test and adopt VLMs without leaving their established tools behind. Language is another underappreciated barrier. Most models assume that users will query in English, which leaves out a large portion of potential users. In many countries, field teams or analysts work primarily in their local languages. Making VLMs multilingual is not just a matter of fairness, which is essential for practical usability in regions where Earth observation could have the most impact. Finally, and perhaps most critically, there is the question of governance. If these models are going to inform decisions about infrastructure, disaster response, or land classification, then we need to think carefully about how they are audited. What safeguards are in place to detect bias or errors? Who is accountable when the output is wrong or worse, misused? The geospatial community has a chance to get ahead of these issues now, before these tools become ubiquitous.

4.7 CONCLUSION

As AI continues to transform how data is analyzed and interpreted, VLMs are beginning to change how satellite imagery is interpreted and used. These models move us beyond conventional routines of classifying pixels or detecting changes through rigid algorithms. Instead, they allow for a more descriptive, context-aware interaction with geospatial content, an advancement that addresses a long-standing gap in conventional remote sensing workflows. Over the course of this chapter, we have traced the development of VLMs from foundational architectures to their experimental use in various geospatial domains. From generating natural language descriptions of disaster-hit areas to supporting text-based retrieval of agricultural patterns, these models show clear potential. What they offer, above and beyond traditional methods, is not just accuracy, but interpretability. They produce outputs in natural language, which enhances interpretability and makes them especially relevant for tasks where effective communication is critical.

That said, we are not yet at a point where these tools can be widely deployed without concern. The challenges are substantial and multifaceted. There is a distinct lack of training data that reflects the kind of imagery and context typical in geospatial analysis. Most existing models are still grounded in internet photos and English-language corpora, which introduces both regional and semantic biases. Moreover, their computational demands remain high, often beyond the reach of institutions working in low-resource settings. Evaluation standards, too, are still evolving; there is no consensus on how to judge these systems fairly in spatial contexts. Looking ahead, it is clear that technical innovation will need to be matched by thoughtful design. Building datasets that reflect local realities, reducing dependency on massive computing infrastructure, and integrating these systems into platforms already used by analysts are the practical steps that will make VLMs more than just prototypes. At the same time, we will need interdisciplinary collaboration to ensure these tools are not only practical but also ethical and inclusive. Ultimately, what makes VLMs promising is not just their ability to process visual and linguistic inputs but their potential to reshape how we ask questions of the Earth and how we explain what we see.

REFERENCES

Al Rahhal, M. M., Bazi, Y., Alsaleh, S. O., Al-Razgan, M., Mekhalfi, M. L., Al Zuair, M., & Alajlan, N. (2022). Open-ended remote sensing visual question answering with transformers. *International Journal of Remote Sensing*, *43*(18), 6809–6823. https://doi.org/10.1080/01431161.2022.2145583

Alayrac, J. B., Donahue, J., Luc, P., Miech, A., Barr, I., Hasson, Y., ... & Simonyan, K. (2022). Flamingo: a visual language model for few-shot learning. *Advances in Neural Information Processing Systems*, *35*, 23716–23736.

Bazi, Y., Al Rahhal, M. M., Mekhalfi, M. L., Al Zuair, M. A., & Melgani, F. (2022). Bi-modal transformer-based approach for visual question answering in remote sensing imagery. *IEEE Transactions on Geoscience and Remote Sensing*, *60*, 1–11. https://doi.org/10.1109/TGRS.2022.3192460

Chappuis, C., Mendez, V., Walt, E., Lobry, S., Le Saux, B., & Tuia, D. (2022, July). Language transformers for remote sensing visual question answering. In *IGARSS 2022-2022 IEEE International Geoscience and Remote Sensing Symposium* (pp. 4855–4858). IEEE. https://doi.org/10.1109/IGARSS46834.2022.9884036

Chen, Y. C., Li, L., Yu, L., El Kholy, A., Ahmed, F., Gan, Z., ... & Liu, J. (2020, August). Uniter: Universal image-text representation learning. In *The European Conference on Computer Vision* (pp. 104–120). Cham: Springer International Publishing. https://doi.org/10.1007/978-3-030-58577-8_7

Jaegle, A., Gimeno, F., Brock, A., Vinyals, O., Zisserman, A., & Carreira, J. (2021, July). Perceiver: General perception with iterative attention. In *International Conference on Machine Learning* (pp. 4651–4664). PMLR.

Jia, C., Yang, Y., Xia, Y., Chen, Y. T., Parekh, Z., Pham, H., ... & Duerig, T. (2021, July). Scaling up visual and vision-language representation learning with noisy text supervision. In *International Conference on Machine Learning* (pp. 4904–4916). PMLR.

Jiang, A. Q., Sablayrolles, A., Mensch, A., Bamford, C., Chaplot, D. S., Casas, D., ... & Sayed, W. E. (2023). Mistral 7B. *arXiv preprint arXiv:2310.06825*. https://doi.org/10.48550/arXiv.2310.06825

Laurençon, H., Tronchon, L., Cord, M., & Sanh, V. (2024). What matters when building vision-language models?. *Advances in Neural Information Processing Systems*, *37*, 87874–87907.

Lee, J. G., & Kang, M. (2015). Geospatial big data: challenges and opportunities. *Big Data Research*, *2*(2), 74–81. https://doi.org/10.1016/j.bdr.2015.01.003

Li, J., Li, D., Savarese, S., & Hoi, S. (2023b). Blip-2: Bootstrapping language-image pre-training with frozen image encoders and large language models. In *International Conference on Machine Learning* (pp. 19730–19742). PMLR.

Li, X., Wen, C., Hu, Y., Yuan, Z., & Zhu, X. X. (2024). Vision-language models in remote sensing: Current progress and future trends. *IEEE Geoscience and Remote Sensing Magazine*, *12*(2), 32–66. https://doi.org/10.1109/MGRS.2024.3383473

Li, X., Wen, C., Hu, Y., & Zhou, N. (2023a). RS-CLIP: Zero shot remote sensing scene classification via contrastive vision-language supervision. *International Journal of Applied Earth Observation and Geoinformation*, *124*, 103497. https://doi.org/10.1016/j.jag.2023.103497

Liu, F., Chen, D., Guan, Z., Zhou, X., Zhu, J., Ye, Q., … & Zhou, J. (2024). Remoteclip: A vision language foundation model for remote sensing. *IEEE Transactions on Geoscience and Remote Sensing*, *62*, 1–16. https://doi.org/10.1109/TGRS.2024.3390838

Liu, H., Li, C., Wu, Q., & Lee, Y. J. (2023). Visual instruction tuning. *Advances in Neural Information Processing Systems*, *36*, 34892–34916.

Lu, J., Batra, D., Parikh, D., & Lee, S. (2019). Vilbert: Pretraining task-agnostic visiolinguistic representations for vision-and-language tasks. *Advances in Neural Information Processing Systems*, *32*, 1–11.

Radford, A., Kim, J. W., Hallacy, C., Ramesh, A., Goh, G., Agarwal, S., … & Sutskever, I. (2021, July). Learning transferable visual models from natural language supervision. In *International Conference on Machine Learning* (pp. 8748–8763). PMLR.

Schmidt, S., Díaz Fragachan, E., Arifi, D., Hanny, D., & Resch, B. (2025). Assessing the spatial accuracy of geocoding flood-related imagery using Vision Language Models. *Spatial Information Research*, *33*(2), 1–17. https://doi.org/10.1007/s41324-025-00609-0

Sogi, N., Shibata, T., Terao, M., Senzaki, K., Tani, M., & Rodrigues, R. (2024, July). Disaster damage visualization by VLM-based interactive image retrieval and cross-view image geo-localization. In *IGARSS 2024-2024 IEEE International Geoscience and Remote Sensing Symposium* (pp. 1746–1749). IEEE. https://doi.org/10.1109/IGARSS53475.2024.10640928

Tan, H., & Bansal, M. (2019). LXMERT: Learning cross-modality encoder representations from transformers. *arXiv preprint arXiv:1908.07490*. https://doi.org/10.48550/arXiv.1908.07490

Tao, L., Zhang, H., Jing, H., Liu, Y., Yan, D., Wei, G., & Xue, X. (2025). Advancements in vision–language models for remote sensing: datasets, capabilities, and enhancement techniques. *Remote Sensing*, *17*(1), 162. https://doi.org/10.3390/rs17010162

Wu, H., Du, Z., Zhong, D., Wang, Y., & Tao, C. (2025). FSVLM: A vision-language model for remote sensing farmland segmentation. *IEEE Transactions on Geoscience and Remote Sensing*, *63*, 4402813. https://doi.org/10.1109/TGRS.2025.3532960

Yuan, Z., Zhang, W., Fu, K., Li, X., Deng, C., Wang, H., & Sun, X. (2022). Exploring a fine-grained multiscale method for cross-modal remote sensing image retrieval. *IEEE Transactions on Geoscience and Remote Sensing*, *60*, 4404119. https://doi.org/10.1109/TGRS.2021.3078451

Zhai, X., Mustafa, B., Kolesnikov, A., & Beyer, L. (2023, October). Sigmoid loss for language image pre-training. In *2023 IEEE/CVF International Conference on Computer Vision (ICCV)* (pp. 11941–11952). IEEE. https://doi.org/10.1109/ICCV51070.2023.01100

Zhan, Y., Xiong, Z., & Yuan, Y. (2025). Skyeyegpt: Unifying remote sensing vision-language tasks via instruction tuning with large language model. *ISPRS Journal of Photogrammetry and Remote Sensing*, *221*, 64–77. https://doi.org/10.1016/j.isprsjprs.2025.01.020

5 Advanced Large Language Models for Satellite Image Processing

Ezil Sam Leni A, Revathi T, Adrian David Cheok, and Shalen S

5.1 INTRODUCTION

The field of Artificial Intelligence (AI) has recently witnessed a shift from task-specific and narrow models to generalized and multimodal systems that can understand and reason across various modalities. In this category, vision-language models (VLMs) (Li et al., 2019) stand out in that they can effectively handle visual and textual inputs in a unified paradigm. These models integrate computer vision and natural language processing (NLP), enabling tasks such as image captioning, visual question answering (VQA) (Zhang et al., 2024), and cross-modal retrieval. Given the rapid growth of visual data volume in images daily, particularly in the Remote Sensing (RS) domain, the incorporation of VLMs provides a more scalable and intelligent solution than manual annotation and standalone visual processing approaches.

Recent years have seen significant advancements in VLMs, enhancing multimodal learning capabilities. Base models such as VisualBERT (Li et al., 2019) and ViLBERT (Lu et al., 2019) adapted transformer architectures to model simultaneously both images and text, and contrastive models such as contrastive language–image pre-training (CLIP) (Radford et al., 2021) and SigLIP (Zhai et al., 2023) have presented successful representations that can be learned efficiently under Internet scale noise. In the RS field, models such as EarthGPT (Zhang et al., 2024), NOTC (Luo et al., 2025), and others have been designed to handle multi-sensor image processing that includes optical, synthetic aperture radar (SAR), and infrared imagery, for semantic understanding of satellite scenes. Such models are not only capable of classifying land use categories and disaster-struck areas but also generate natural language descriptions, and answer user queries, which introduces explainability and accessibility into RS analytics.

Nevertheless, the use of VLMs for RS poses many difficulties. The domain gap between natural images, where most VLMs are pretrained, and RS images is one of

DOI: 10.1201/9781003616207-6

the major open problems in VLM for RS data due to the significant differences in texture, resolution, orientation, and spectral signatures. RS data are limited, and annotation of the multimodal corpus is not available to fine-tune the large models. In addition, existing VLMs have limitations in conducting complex visual reasoning, detecting novel object classes that go beyond predefined vocabularies, and interpreting time-series data, for example, essential for applications such as urban change detection and flood assessment. The limitation of the output interpretability, the computational expense, the lack of generalization to out-of-distribution (OOD), and others are also highly limiting their applicability in operational RS workflows.

This chapter addresses these gaps by providing a systematic review of VLMs for RS. The aims of this study are as follows: (a) to summarize a historical and architectural perspective of VLMs with a special emphasis on their application to RS data; (b) to assess their performance on fundamental tasks such as object detection, scene classification, image captioning, VQA, visual grounding, and change detection; (c) to analyze the current challenges such as the size limitations of datasets, domain shifts, and complexities due to fusion; and (d) to outline potential research directions of lightweight architectures, interpretable models, and domain-specific training strategies toward sustainable and deployable VLMs in RS.

5.2 EVOLUTION OF VISION-LANGUAGE MODELS

VLM has seen dramatic advances, from simple multimodal fusion techniques to complex and scalable architectures, to tackle large-scale vision-language problems. Understanding this evolution is key to recognizing how VLMs have been adapted for RS applications. Adaptations from NLP almost entirely drove initial VLMs. Groundbreaking applications such as VisualBERT (Li et al., 2019) and ViLBERT (Lu et al., 2019) transferred the transformer architecture, which was originally proposed for texts (BERT) (Devlin et al., 2019), for joint image-text modeling. A comparison of the state-of-the-art VLMs is summarized in Table 5.1 with respect to their architecture, training, and applicability to RS.

5.2.1 Early Developments

Li et al. (2019) presented an entirely new model, VisualBERT, that worked on image parts and text at the same time, rather than just combining them separately. It adopted a single transformer encoder for multimodal fusion. ViLBERT proposed a two-stream model; initially, different transformer streams processed image and text input (Lu et al., 2019) before merging cross-modal attention layers. These early models used masked language modeling and image-text matching objectives to learn joint embeddings and to achieve state-of-the-art performance in VQA, image captioning, and retrieval tasks. However, most of these methods are based on fixed pretrained visual features and are weak in end-to-end learning. They also had poor scalability and fine-grained reasoning abilities, leading to a recent trend of contrastive and generative methods.

TABLE 5.1
Comparison of VLMs for Remote Sensing Tasks

Model	Architecture Type	Training Approach	Remote Sensing Adaptation	Key Strengths	Reference
CLIP[a]	Vision Transformer + Text Encoder	Contrastive pretraining	No; general-domain only	Zero-shot transfer, scalability	Radford et al. (2021)
BLIP-2[b]	Dual-encoder + LLM bridge	Bootstrapped language-image pretraining	Needs domain tuning for RS	Modular design, instruction tuning	Li et al. (2023)
LLaVA[c]	ViT + LLaMA (frozen) + Adapter	Instruction tuning	Adapted for RS QA and image captioning	Multimodal chat capability, lightweight tuning	Liu et al. (2023)
EarthGPT	ViT + CNN + LLM with fusion	Multitask, instruction-tuned	Designed specifically for RS	Multi-sensor fusion (optical, SAR, IR), VQA, captioning	Zhang et al. (2024)
NOTC[d]	Novel Object Transformer + CLIP	Captioning for novel object classes	Extended for unseen events in RS	Few-shot captioning for unseen objects	Luo et al. (2025)
Flamingo	Perceiver Resampler + Frozen LLM	Few-shot visual instruction tuning	Applicable to RS after fine-tuning	Few-shot learning, multilingual image understanding	Alayrac et al. (2022)

[a] CLIP—Contrastive language-image pretraining
[b] BLIP-2—Bootstrapping language-image pretraining
[c] LLaVA—Large language and vision assistant
[d] NOTC—Novel object transformers with CLIP

5.2.2 Contrastive Learning with the VLMs

The key breakthrough comes with contrastive learning frameworks for vision-language pretraining, most notably in CLIP (Radford et al., 2021). CLIP jointly trains a visual encoder (e.g., ResNet, ViT) and a text encoder with a contrastive loss using 400 million image-text pairs scraped from the web. The model is trained to maximize the similarity of the matching image-text pairs and minimize the similarity for mismatched image-text pairs. There are several advantages of adopting contrastive learning, including its scalability. It is possible to train with large, noisy datasets without requiring manual annotations. In a common embedding space, image and text representations can be directly compared by cosine similarity. This contrastive framework also inspired us to move away from VLMs designed for transfer learning to models that could learn from few or zero samples, which is particularly appealing for RS applications where annotated data are typically limited. After CLIP, SigLIP model investigated contrastive pretraining and made the multimodal alignment process more efficient.

5.2.3 Generative Vision and Language Models

While contrastive models aligned modalities well, they were less well-suited to generation tasks like captioning or open-ended QA. Therefore, generative VLMs were developed. Contrastive Captioner (CoCa) (Yu et al., 2022) is an image–text foundation model that combines contrastive and captioning objectives in an encoder-decoder model to support retrieval, classification, and high-quality captioning with a single pretrained model. The model managed to produce captions by maintaining representation alignment quite well. CM3leon and Chameleon focus on multimodal conditional generation and can support tasks such as captioning, visual reasoning, and image generation from text. These generative models leveraged masked language modeling, causal language modeling, and decoder-only transformer architectures to predict sequences conditioned on visual inputs. They were able to describe images and fill in gaps and even imagine unseen scenarios—capabilities required for more sophisticated RS tasks such as disaster evaluation and urban change detection.

5.2.4 Pretrained Backbones and Parameter-Efficient Models

Another prominent line of work has been to repurpose highly semantically precise pretrained large language models (LLMs) like generative pretrained transformer (GPT), large language model Meta AI (LLaMA) and plug them into visual encoders like MiniGPT-4 (Zhu et al., 2023), which connected a vision transformer (ViT) visual encoder to a frozen LLaMA model through a basic linear projection. It made vision-to-text generation possible with only a few examples. Large language and vision assistant (LLaVA) (Liu et al., 2023) employed a lightweight adapter to transfer the vision features into LLM's hidden space, adapting to multimodal conversations while only using a small number of fine-tuned parameters. This parameter-efficient fine-tuning significantly reduced the computation necessary for training strong multimodal models and led to the rise of instruction-tuning, training multimodal models to understand human-like instruction provided across tasks. This modular design is particularly well-suited for RS. Low-data regimes enable training of strong models even when RS-specific training data are limited. It can accommodate new modalities (e.g., SAR, infrared) by updating only a few adapter layers but not retraining the entire model.

5.2.5 Earth GPT (Zhang et al., 2024): A Universal MLLM for RS

Earth GPT is explicitly tailored for multi-sensory images: optical, SAR, and infrared images are used.

Multi-tasks: Scene classification, object detection, VQA, captioning, and visual grounding. Open-set reasoning: Dealing with new objects and classes, and unknown objects and classes. Visual-Enhanced Perception: Utilizes a large-scale architecture of CNNs and Vision Transformers to encode both local details and global context.

Cross-modal alignment: It enforces the vision-language fusion by aligning representations in training.

Cooperative Instruction Tuning: Finetunes the model for a variety of RS tasks with a shared instruction-based interface.

EarthGPT, importantly, builds upon MMRS-1M—a dataset of over 1 million image-text pairs from 34 diverse RS datasets—and sets a new state of the art for the task of multi-sensory multimodal learning in RS. Evaluation results revealed that EarthGPT outperformed traditional expert models as well as multimodal large language models (MLLMs) in a variety of RS tasks, demonstrating the importance of task-oriented VLM designs.

5.3 REMOTE SENSING AND VISION-LANGUAGE MODELS

RS has become an important method for Earth's surface observation, surveillance, and analysis. High-resolution satellite imagery, sUAS, and unmanned aerial vehicle (UAV) imagery enable the collection of vast amounts of data in a variety of locations, including agriculture, environmental management, urban planning, and disaster management. The complexity of RS data requires advanced tools to extract actionable insights efficiently. VLMs provide a possible solution by combining visual perception with human natural language understanding. They are also used for auto tags/keywords and to train models for automatic scene classification, object detection, change detection, image captioning, VQA, etc. VLMs generate natural language descriptions and respond to textual queries, enhancing object recognition for geographic and RS analysis.

5.3.1 Significance of Remote Sensing

A key issue in solving the significant global challenges is to use RS technologies for environmental monitoring. As satellite images are used to detect and monitor deforestation, glacial retreat, desertification, and biodiversity loss, they are critical to combating climate change and protecting ecosystems (Ankayarkanni & Leni, 2016).

Disasters: Quick mapping of disaster-hit areas caused by floods, earthquakes, hurricanes, and wildfires helps in making informed emergency responses and distributing resources (Usha Nandini & Leni, 2015).

Urban Planning and Infrastructure: RS applications include monitoring urban sprawl, traffic networks, urban green cover, and land use change—all are vital for sustainable urban planning (Cheng, Han, Lu, & Guo, 2017).

Agriculture and Food Security: Monitoring crop health, irrigation needs, and soil quality using multispectral images aids in food security by alerting to timely decisions (Yuh et al., 2017).

Defense and Security: RS is essential for surveillance, border monitoring, and strategic intelligence (Li et al., 2019).

Automating the interpretation of RS data with VLMs enables stakeholders to gain sharper, faster, and scalable insights for timely and evidence-based decisions. A language-only GPT-4 was developed by Liu et al. (2023) for multimodal language-image instruction and is named LLaVA, a large multimodal model connecting a vision encoder and LLM for vision and language experiments.

5.3.2 Multimodal Remote Sensing Analysis

RS data itself is multimodal. Optical imagery captures data in the visible and near-infrared spectra, which makes it useful for vegetation analysis, urban mapping, and environmental monitoring. SAR-active microwave sensing, which can be seen through both clouds and the dark, is suitable for all-weather day-night observation of surface structures. Thermal Imaging detects emitted infrared radiation (which gives information about temperature) and is valuable for observing water bodies, urban heat islands, and the progression of wildfires. Hyperspectral Imagery offers hundreds of adjacent spectral bands that support detailed analysis of material composition and land cover.

Multimodal fusion is crucial for combining the complementary advantages of different sensors. VLMs are well-suited for this setting because they learn cross-modal relationships between different modalities and text. Recent architectures exploit cross-attention modules and hierarchical fusion layers to incorporate multimodal features into structured representations that capture spatial, spectral, and contextual information of interest. Efficient multimodal fusion can benefit several tasks, including cross-modal retrieval.

i. Searching images using textual queries and retrieving relevant text descriptions from visual inputs, commonly referred to as cross-modal retrieval, as well as generating natural language descriptions for visual data (Roy et al., 2022).
ii. Change detection in time: Identifying, localizing, and characterizing semantic changes between multi-temporal images acquired at different time intervals. Solving visually complex IQ test problems often requires a holistic understanding across visual and text modalities.

Therefore, VLM-based multimodal analysis can revolutionize RS as a flexible and rich analytical tool, enhancing its impact across scientific research, environmental monitoring, and societal applications.

5.3.3 Types of Remote Sensing Images

RS collects information about the Earth's surface from sensors on satellites, aircraft, or ground-based stations. The RS images can be commonly divided into the following three types because of the distinction of the interaction between the sensors and the electromagnetic waves.

- **Synthetic Aperture Radar (SAR):** SAR (Moreira, Prats-Iraola, Younis, Krieger, Hajnsek, & Papathanassiou, 2013) is an active RS technique. The SAR sends microwave signals toward the Earth and records the reflected (backscattered) signal from the land. Since it relies on its power source, SAR can function under all-weather conditions: day and night, and in rain, fog, and smoke, and can penetrate the vegetation canopy. It has the range: microwave (usually L, C, X bands). These images are used for surface deformation (urban monitoring and disaster evaluation [flood extent], soil moisture, and sea ice mapping). It also functions in the dark and under cloud cover.
- **Optical Remote Sensing:** Optical RS is a passive technique that measures the reflected sunlight in the visible, near-infrared (NIR), and shortwave infrared (SWIR) spectral ranges. It takes pictures based on how different substances absorb and reflect different wavelengths of sunlight. It has a spectral range of visible (400–700 nm), NIR (700–1300 nm), and SWIR (1300–2500 nm). These images are used for land cover classification, vegetation monitoring, urban expansion, and natural resource mapping. The limitation is that it does not work on cloudy days or after dark.
- **Infrared Remote Sensing:** The infrared data consists of NIR and thermal infrared (TIR) images. The range NIR (700–1300 nm) is helpful to demonstrate the strong reflectance of healthy plants in vegetation research. Thermal IR (8–14 μm) measures emitted energy, not reflected sunlight, and helps to detect surface temperature differences such as heat islands, soil moisture, or wildfires. Infrared RS is particularly useful for vegetation health analysis, heat mapping in urban areas, and water stress and soil moisture determination.

5.3.4 Datasets for Remote Sensing

RS datasets play a vital role in the intelligent interpretation of insights for satellite images and are shown in Table 5.2. RS images are mainly used in object detection, scene classification, and segmentation. The recent evolution in computer vision and LLMs paved the way for the evolution of various datasets for image captioning, VQA, visual grounding, and region-level captioning.

TABLE 5.2
Datasets for Remote Sensing

Activity	Dataset	No. of Instances
Object Detection	DOTA (Xia et al., 2018)	1,63,486
	DIOR (Li, Wan, Cheng, Meng, & Han, 2020)	58,200
Object Classification	NWPU-RESISC45 (Cheng, Han, Lu, & Guo, 2017)	6,300
	UCMerced-Landuse (Yang & Newsam, 2010)	420
Image Captioning	RSICD (Lu, Wang, Zheng, & Li, 2017).	24,333
	UCM-Captions (Qu, Li, Tao, & Lu, 2016)	9,986
	RSVQA-LR (Lobry et al., 2020)	67,228
Visual Grounding	DIOR-RSVG (Zhan, Xiong, & Yuan, 2023).	30,820
Region Level Captioning	DIOR-RSVG (Zhan, Xiong, & Yuan, 2023).	30,820
Change Detection	LEVIR-CC (Liu, Zhao, Chen, Zou, & Shi, 2022)	10,077 (Unchanged image pairs—5,039/changed image pairs 5,038)

5.4 REMOTE SENSING APPLICATIONS USING VISION-LANGUAGE MODELS

VLMs have made tremendous progress on interpretation challenges in RS. VLMs, from the combination of visual and textual modalities, can better understand and reason in high-dimensional RS images. The following subsections discuss some of the most promising application areas that VLMs are revolutionizing in RS analysis. Key RS tasks and their best-suited VLMs are summarized in Table 5.3.

5.4.1 Object Detection

In RS, object detection is the task of detecting and localizing several objects of interest in an image. Different from natural scene images, objects in RS imagery tend to be small, clustered, and arbitrarily oriented, as well as recorded with various spectral bands and resolutions. The traditional object detection methods in RS based on the horizontal bounding boxes (HBB) were unable to identify tilted and elongated objects accurately. Nowadays, oriented bounding boxes (OBB) are utilized to accurately bound the object boundaries, particularly in aerial and SAR images (Moreira et al., 2013). The reasons VLMs improve object detection in RS are as follows:

- Incorporating class cues in the form of text prompts during training and inference
- Multimodal pretraining for strong generalization across diverse object categories

TABLE 5.3
Application Domains Versus Suitable VLMs for Remote Sensing

Remote Sensing Task	VLM(s) Used	Model Usage	Reference
Object Detection	EarthGPT, CLIP, LLaVA	Handles multi-class, rotated, and small objects	Zhang, Cai, Zhang, Zhuang, and Mao (2024); Radford et al. (2021); Liu et al. (2023)
Scene Classification	BLIP-2, CLIP, EarthGPT	Semantic scene labeling with open-set adaptability	Li et al. (2023); Radford et al. (2021); Zhang, Cai, Zhang, Zhuang, and Mao (2024)
Image Captioning	EarthGPT, NOTC, Flamingo	Generates dense, context-rich RS-specific captions	Zhang, Cai, Zhang, Zhuang, and Mao (2024); Luo et al. (2025); Alayrac et al. (2022)
Visual Question Answering (VQA)	LLaVA, EarthGPT	Supports natural language query answering on satellite images	Liu, Li, Wu, and Lee (2023); Zhang, Cai, Zhang, Zhuang, and Mao (2024)
Visual Grounding	CLIP, DIOR-RSVG with VLMs	Links language phrases to regions in multispectral imagery	Radford et al. (2021); Zhan, Xiong, and Yuan (2023)
Change Detection	EarthGPT, Siamese VLMs	Detects spatial and temporal differences; generates explanations	Zhang, Cai, Zhang, Zhuang, and Mao (2024); Liu, Zhao, Chen, Zou, and Shi (2022)

- Facilitating few-shot/zero-shot object detection with the alignment between object features and text descriptions, and requiring little annotated data

In addition, hierarchical transformers and cross-attention-based modules allow VLM to learn spatial hierarchies and contextual relationships between objects, thus dramatically improving the detection performance at different scales (Carion, Massa, Synnaeve, Usunier, Kirillov, & Zagoruyko, 2020). The advantages include efficient detection of rotated dense objects, reduced parameters through LoRA (Hu et al., 2022), and improved compatibility with various resolutions. The VLM-based object detection in RS generally consists of two stages, as shown in Figure 5.1. In the first stage, the original RS image is fed into a VLM encoder extended with the LoRA modules at the self-attention layers, allowing fewer parameters to be fine-tuned. This

FIGURE 5.1 A flowchart of an object detection process starting from an input RS image to detected objects.

step produces dense spatial feature maps that are read by a region proposal network (RPN) to propose object initial candidates. These proposals are further tuned by cross-attention layers that integrate visual and semantic features, ultimately producing HBB for object localization.

The second stage further improves this detection and improves the performance of the OBB model, which is crucial in RS scenarios with rotated or elongated objects. In this stage, the base layers of the VLM encoder are frozen, and LoRA is only applied to cross-attention and feedforward layers. New regression heads for OBB are presented, and the model is then further trained using rotated bounding box loss, for example, L1 loss, for better localization accuracy for objects in different orientations.

5.4.2 Scene Classification

RS scene classification is one of the tasks of labeling an entire satellite/aerial image into semantic classes like "urban," "forest," "agricultural land," or "water body." Previous RS scene classifiers were dominantly based on hand-crafted features or convolutional backbones (Cheng et al., 2017). However, such models often found it challenging to deal with the intra-class variation (different urban types) as well as the in-class similarity (farmland vs grassland) of RS data. VLMs significantly improve by matching global image representations with descriptive texts for robust cross-domain categorization (Radford et al., 2021) and applying masked modeling and cross-modal attention to attend to the semantically relevant region of the image, etc. Facilitating open-set recognition by synthesizing labels for previously unseen scenes using learned semantic attributes. When trained with textural context, VLMs also help the scene classification model to be more interpretable compared to visual-only models. The advantages are improved open-set classification performance, resilience to inter-class similarity, and support for incremental addition of new categories.

The RS scene classification based on VLMs consists of two important steps. First, the RS image is fed into the VLM encoder that is integrated with LoRA modules inside the self-attention layers (Hu et al., 2022) to ensure that the VLM can be efficiently trained. This stage is responsible for capturing the global semantic representation from the image (Zhang et al., 2024). These visual representations are further combined with textual embeddings of scene categories for the model to learn from joint representations. A contrastive learning approach is applied to adapt to two modalities (i.e., visual and textual), effectively comparing different types of scenes by semantic meaning. In the second part, the problem becomes that of open-set scene recognition. Here, the first few layers of VLM are frozen, and alignment heads are fine-tuned to accommodate new, unseen scene categories (Li et al., 2023; Zhang et al., 2024). Instruction tuning units are introduced to help improve the model's ability to generalize to new and temporal changes in land use types. Finally, the model is trained using a multi-loss approach, which includes classification loss and alignment loss, making it possible to classify scenes effectively even in complex and dynamic RS.

5.4.3 Image Captioning

Image captioning (IC) in the field of RS refers to the technique of generating textual descriptions in the natural language of RS images and can be used to interpret

RS images to end-users (e.g., urban planners, environmentalists, and policymakers). Specific challenges of video RS captioning are multi-scale modelling, multi-object and multi-class interactions, and accurate and diverse descriptions while maintaining context coherence.

To tackle these challenges, VLMs adopt multimodal encoders and decoder-based architectures that enable cross-modal attentional grounding between visual regions and textual descriptions. Such models support instruction-tuned captioning, allowing automatic caption styles to adapt to user inputs, and leverage large-scale pretraining on remote sensing image–caption datasets to improve linguistic fluency and semantic relevance (Shen et al., 2022). State-of-the-art captioning models in the RS domain can produce detailed reports, such as "A dense urban area with substantial road network expansion next to agricultural fields."

A two-stage structured method for image captioning of RS imagery with VLMs is used. In the first step, spatial features are mined from the input RS images by using a VLM enhanced with the LoRA-based intelligent attention mechanisms. These visual features are concatenated with prompt-based textual embeddings, which offer contextual signals for the generation of captions. The stacked embeddings are fed to an LLM decoder, which produces the initial descriptive captions for the photography, essentially translating the visual semantics into natural language. The second phase consists of developing detailed and domain-specific versions of the initial captions. It requires unfreezing the first-stage feature extractor and fine-tuning it on specialized RS datasets. To increase the model's generalization ability for various RS situations, the mixture-of-experts (MoE) architecture is used. For efficiency, LoRA modules are used during this fine-tuning process, as they can minimize the computations of parameter updates. The result is a collection of diverse, high-quality captions that more accurately reflect the spectral characteristics relevant to RS and that perform better in few-shot and low-resource settings. Such a design not only guarantees that the captions are fluent and relevant but also computationally efficient.

5.4.4 Visual Question Answering

VQA for RS works aims to answer open-ended, multiple-choice, or yes/no questions about satellite images. Examples: "Is there any evidence of flooding in this area?" How many bridges can be found in the picture? In contrast to conventional VQA on natural images, RS-VQA requires multi-sensory data reasoning and understanding specialized terminology in the domain (Lobry et al., 2020).

VLMs enhance RS-VQA by applying domain-adaptive pretraining approaches to match visual features with RS-specific linguistic queries (Du et al., 2022) and adopting instruction tuning to generalize to OOD question types—multi-step reasoning between the spatial hierarchies of images. Recent works show that VLM-based RS-VQA models improve performance against traditional visual-language baselines in terms of accuracy and relevance. The advantages include improved comprehension of complex spatial queries, and it has been designed to work well for both compositional and straightforward questions. A two-stage structured approach for VQA is shown in Figure 5.2 in RS with VLMs. The initial step involves performing feature extraction on the input RS images within a VLM that utilizes a LoRA-based intelligent attention mechanism to obtain both spatial and semantic features. These features

FIGURE 5.2 A flowchart of a reasoning process, starting with an Input RS Image, followed by a VLM Encoder + LoRA, Cross Attention, Reasoning Module, and finally producing an Answer Output.

are concatenated with prompt-to-textual embeddings generated based on the user-provided question. These multimodal embeddings are then combined and passed through a cross-modal reasoning module, which enables it to attend to the most question-relevant areas of an image where visual regions are aligned with appropriate linguistic cues. This stage involves locating the question and retrieving visual evidence of context based on spatial regions.

In the second stage, the aligned multimodal characteristics are fed to a language model LM decoder fine-tuned for RS-specific VQA duties. The decoder reasons over the visual-text embeddings incrementally, achieving this by using a hierarchical attention that combines both global context and local object details. For better generalization of unseen question types/OOD queries, instruction tuning and mixture-of-experts (MoE) architectures are employed. LoRA modules are used during fine-tuning to increase efficiency, reducing computation without sacrificing gains. Natural language answers are generated by extracting relevant tuples and augmenting RS domain-specific knowledge to answer different types of questions (e.g., spatial counting, land use classification, or disaster assessment) very accurately in their contexts.

5.4.5 Visual Grounding

Visual grounding is the process of associating textual phrases with the regions in an image. There are various critical applications of visual grounding in RS, including, but not limited to object localization, image–text retrieval, change interpretation, and semantic scene understanding. Disaster descriptions are linked to affected areas, thereby connecting urban planning annotations to specific city blocks. RS-specific visual grounding is challenged by the absence of pixel-wise annotations and high similarity of regions (Lan et al., 2024). VLMs contribute to visual grounding in RS by incorporating other (query-based) region proposal algorithms and utilizing multimodal pretraining to enhance the robustness of the model for ambiguities, and using a hierarchical attention mechanism to refine the dynamic relationship between global and local features (Zhan et al., 2022). By operating in this way, VLMs support precise groundedness even for abstract or underspecified textual descriptions. The advantages are grounding correctly under the ambiguous description and generalizing to abnormal phrase(s) and class(es).

Visual grounding in RS connects visual regions with corresponding textual phrases using a two-stage approach. At the first step, region-level embeddings are produced using the RS image of an agent and a VLM. At the same time, textual phrases are encoded with LLM forces matching their semantic meaning. These visual and textual cues are aligned via cross-modal transformer layers, and masked prediction is used to correlate correct image regions with phrases. For efficient training, LoRA modules are used in the grounding heads for parameter-efficient fine-tuning (Hu et al., 2022). The second phase is to improve the accuracy of this visual-textual alignment. The models are fine-tuned on supervised phrase-region matching, which establishes the direct correspondence between annotated text and image regions. For tasks with insufficient annotated data, the model's grounding ability can be preserved by incorporating self-supervised learning objectives. This two-stage method allows for robust visual grounding with ambiguous or underspecified textual inputs, such as those likely to be found in real-world RS applications.

5.4.6 Change Detection

Change detection concerns the recognition of alteration in multi-temporal images of the same part of the world's surface. Key challenges include correcting for misalignment from orbit dynamics or sensor differences and recording finer semantic variation at the level of pixels, rather than differences (Peng, Zhang, Guan, & Li, 2019; Daudt, Le Saux, & Boulch, 2018).

The features of VLMs enable extended change detection by aligning multi-temporal input through joint feature extraction modules. This approach outputs natural language descriptions of detected changes, such as "relevant new construction observed in the northern sector," while leveraging cross-modal contrastive learning tasks to effectively separate salient change signals from irrelevant noise. With the integration of spatial, temporal, and semantic data, VLMs present a new realm of scalable, explainable change detection. The advantages are supporting binarized and semantic change detection, readability through text results, and strong-to-small minor RS data changes.

5.5 RESEARCH CHALLENGES

Although VLMs have achieved remarkable performance in the analysis and interpretation of RS data, there still exist several challenges that hinder the broad uptake and practical applications in real-world scenarios.

5.5.1 Domain Gap and Dataset Scarcity

The vast majority of VLMs are pretrained on natural images (e.g., MS COCO or web crawls). In comparison, RS images have the following very distinct natures: (i) they are so huge; for example, there are millions of Half Earth images; (ii) they vary widely in spatial scale, resolution, and scene complexity; and (iii) most of them are of low spatial resolution or transformed from images with low resolution. Scale (tiny objects), viewpoint (top-down imagery), multi-sensor, and multi-band images are the spectral signatures. The discrepancies in domains of natural imagery and RS data make powerful VLMs underperform with little fine-tuning (Xia, Hu, Hu, & Shi, 2017). Additionally, the availability of well-annotated large-scale multimodal datasets for RS is much lower relative to the natural images' datasets, which makes it difficult to train or fine-tune effectively VLMs for the RS-related applications.

5.5.2 Multimodal Fusion Complexity

The problem of effectively combining heterogeneous multimodalities, such as optical, SAR, infrared, and hyperspectral data, is still open. Challenges include dealing with different spatial resolutions among sensors, matching features with very different spectral properties, and addressing sensor-specific distortion, noise, and artifacts. Procedures that allow for deep interactions between modalities are less developed, as existing cross-modal fusion methods, for example, concatenation and shallow attention, are not deep enough to model high-order interactions across modalities (Hong et al., 2020).

5.5.3 Open-Set and Out-of-Distribution Generalization

Because RS data is dynamic, new object classes, land use patterns, or changes in objects from season to season or due to human activities may occur, but they have never been seen before. Traditional closed-set VLMs that have been trained on a fixed set of object categories are unable to deal with novel object detection, open-vocabulary grounding, and zero-shot scene description. It is a significant research direction to develop VLMs that can have generalization capability to OOD samples without performance degradation.

5.5.4 Computational Costs and Efficiency

The training and application of VLMs for RS, multi-sensory, and high-resolution input signal are recognized to be computationally intensive due to GPU memory requirements, training time, and inference latency. There is an urgent demand to study light and efficient VLMs with other training methods, such as adapter tuning, knowledge distillation, and quantization (Hu et al., 2022).

5.5.5 Interpretability and Explainability

Explanation and trust are fundamental in the context of RS applications, such as in disaster response and military surveillance, where the stakes are high. Most existing VLMs are very opaque in that they reveal very little about why the object was an outlier, what these visual elements represent (the set of captioned phrases), and the relation between attention maps and textual queries. Successive generations of models need to include intrinsically interpretable components or explanation tools (Doshi-Velez & Kim, 2017).

5.6 CONCLUSION AND FUTURE ENHANCEMENTS

The incorporation of VLMs in RS pipelines has significantly advanced multimodal understanding. It has provided stronger tools for the resolution of challenging tasks such as object detection, scene classification, image captioning, question answering, and change detection. Such models, which incorporate both deep visual representations and linguistic context, can intuitively and effectively interpret Earth observation data. Their generalizability across modalities enables applications in situation-critical decision-making, such as disaster management, environmental monitoring, and urban planning.

In the future, VLMs are expected to be improved to create a larger impact on RS. In the future, more attention should be paid to creating domain-specific pretraining datasets, considering various geography, sensors, and environmental characteristics, so as not to rely more on natural image corpora. Advances in multimodal fusion will be instrumental in better leveraging diverse sources of input, including SAR, optical, and hyperspectral imagery. In addition, if VLMs could be fine-tuned for user-specific tasks and open-vocabulary settings, they would be more widely applicable and usable in human-AI interaction. As deployment rises in importance, light VLMs that can be executed on the edge, such as those implemented via LoRa and model quantization, will allow real-time processing in low-resource environments. Finally, making these models transparent and trustworthy with interpretable components built in will be critical for responsible and safe deployment in operational systems. Given this progress, VLMs are expected to play a central role in advancing RS intelligence.

REFERENCES

Alayrac, J. B., Donahue, J., Luc, P., Miech, A., Barr, I., Hasson, Y., Lenc, K., Mensch, A., Millican, K., Reynolds, M., Rutherford, E., Borgeaud, S., Anderson, K., Andreassen, A., Auten, A., Bradbury, J., Brown, T., Cai, T., Clark, A., … Zisserman, A. (2022). Flamingo: A visual language model for few-shot learning. *Advances in Neural Information Processing Systems*, 35, 23716–23736. https://doi.org/10.48550/arXiv.2204.14198

Ankayarkanni, B., & Leni, A. E. S. (2016). GABC-based neuro-fuzzy classifier with multi-kernel segmentation for satellite image classification. *Journal of Biomedical Research: Computational Life Science and Smarter Technological Advancement*, Special Issue, S158–S165.

Carion, N., Massa, F., Synnaeve, G., Usunier, N., Kirillov, A., & Zagoruyko, S. (2020). End-to-end object detection with transformers. *European Conference on Computer Vision*. https://doi.org/10.1007/978-3-030-58452-8_13

Cheng, G., Han, J., Lu, X., & Guo, L. (2017). Remote sensing image scene classification: Benchmark and state of the art. *Proceedings of the IEEE*, 105(10), 1865–1883. https://doi.org/10.1109/JPROC.2017.2675998

Daudt, R. C., Le Saux, B., & Boulch, A. (2018). Fully convolutional Siamese networks for change detection. *2018 25th IEEE International Conference on Image Processing (ICIP)*, 4063–4067. https://doi.org/10.1109/ICIP.2018.8451652

Devlin, J., Chang, M. W., Lee, K., & Toutanova, K. (2019). BERT: Pre-training of deep bidirectional transformers for language understanding. *Proceedings of the 2019 Conference of the North American Chapter of the Association for Computational Linguistics: Human Language Technologies (NAACL-HLT 2019)*, Volume 1 (Long and Short Papers), 4171–4186. https://doi.org/10.48550/arXiv.1810.04805

Doshi-Velez, F., & Kim, B. (2017). Towards a rigorous science of interpretable machine learning. *arXiv preprint arXiv:1702.08608*. https://doi.org/10.48550/arXiv.1702.08608

Du, Y., Wei, F., Zhang, Z., Shi, M., Gao, Y., & Li, G. (2022). Learning to prompt for open-vocabulary object detection with vision-language model. *IEEE/CVF Conference on Computer Vision and Pattern Recognition (CVPR)*, 14064–14073. https://doi.org/10.1109/CVPR52688.2022.01369

Hong, D., Gao, L., Yokoya, N., Yao, J., Chanussot, J., Du, Q., & Zhang, B. (2020). More diverse means better: Multimodal deep learning meets remote-sensing imagery classification. *IEEE Transactions on Geoscience and Remote Sensing*, 59(5), 4340-4354. https://doi.org/10.48550/arXiv.2008.05457.

Hu, E. J., Shen, Y., Wallis, P., Allen-Zhu, Z., Li, Y., Wang, L., Wang, W., & Chen, W. (2022). LoRA: Low-rank adaptation of large language models. *arXiv preprint arXiv:2106.09685.* https://doi.org/10.48550/arXiv.2106.09685

Lan, M., Rong, F., Jiao, H., Gao, Z., & Zhang, L. (2024). Language query-based transformer with multiscale cross-modal alignment for visual grounding on remote sensing images. *IEEE Transactions on Geoscience and Remote Sensing*, 62, Article 5626513. https://doi.org/10.1109/TGRS.2024.3407598.

Li, J., Li, D., Xiong, C., & Hoi, S. C. H. (2023). *BLIP-2: Bootstrapping language–image pretraining with frozen image encoders and large language models. arXiv preprint arXiv:2301.12597.* https://doi.org/10.48550/arXiv.2301.12597

Li, K., Wan, G., Cheng, G., Meng, L., & Han, J. (2020). Object detection in optical remote sensing images: A survey and a new benchmark. *ISPRS Journal of Photogrammetry and Remote Sensing*, 159, 296–307. https://doi.org/10.1016/j.isprsjprs.2019.11.023

Li, L., Su, W., Shen, X., Li, Z., Cao, Y., Gan, Z., Cheng, Y., & Liu, J. (2019). VisualBERT: A simple and performant baseline for vision and language. *arXiv preprint arXiv:1908.03557.* https://doi.org/10.48550/arXiv.1908.03557

Liu, Q., Zhao, L., Chen, Y., Zou, L., & Shi, Z. (2022). A large-scale dataset for change detection in remote sensing images. *IEEE Transactions on Geoscience and Remote Sensing*, 60, 1–15.

Liu, H., Li, C., Wu, Q., & Lee, Y. J. (2023). Visual instruction tuning. *Advances in neural information processing systems*, 36, 34892-34916. https://doi.org/10.48550/arXiv.2304.08485

Lobry, S., Marcos, D., Murray, J., & Tuia, D. (2020). RSVQA: Visual question answering for remote sensing data. *IEEE Transactions on Geoscience and Remote Sensing*, 58(12), 8555–8566. https://ieeexplore.ieee.org/document/9088993

Lu, J., Batra, D., Parikh, D., & Lee, S. (2019). ViLBERT: Pretraining task-agnostic visiolinguistic representations for vision-and-language tasks. *Advances in Neural Information Processing Systems*, 32. https://doi.org/10.48550/arXiv.1908.02265

Lu, X., Wang, B., Zheng, X., & Li, X. (2017). Exploring Models and Data for Remote Sensing Image Caption Generation. *IEEE Transactions on Geoscience and Remote Sensing*, 56(4), 2183-2195. https://doi.org/10.1109/TGRS.2017.2776321.

Luo, J., Yu, P., Yang, J., Shi, Y., Huang, F., & Zhang, Y. (2025). Exploring a vision-language foundation model for novel object captioning. *IEEE Transactions on Circuits and Systems for Video Technology*, 35(1), 91–102. https://doi.org/10.1109/TCSVT.2024.3452437

Moreira, A., Prats-Iraola, P., Younis, M., Krieger, G., Hajnsek, I., & Papathanassiou, K. P. (2013). A tutorial on synthetic aperture radar. *IEEE Geoscience and Remote Sensing Magazine*, 1(1), 6–43. https://doi.org/10.1109/MGRS.2013.2248301

Peng, D., Zhang, C., Guan, H., & Li, J. (2019). End-to-end change detection for high-resolution satellite images using improved UNet++. *Remote Sensing*, 11(11), 1382. https://doi.org/10.3390/rs11111382

Qu, Y., Li, X., Tao, C., & Lu, X. (2016). Deep semantic understanding of high-resolution remote sensing images. *Proceedings of the International Geoscience and Remote Sensing Symposium (IGARSS)*, 1662–1665. https://doi.org/10.1109/IGARSS.2016.7729355

Radford, A., Kim, J. W., Hallacy, C., Ramesh, A., Goh, G., Agarwal, S., Sastry, G., Askell, A., Mishkin, P., Clark, J., Krueger, G., & Sutskever, I. (2021). Learning transferable visual models from natural language supervision. *Proceedings of the International Conference on Machine Learning*. https://doi.org/10.48550/arXiv.2103.00020

Roy, S. K., Deria, A., Hong, D., Rasti, B., Plaza, A., & Chanussot, J. (2022). Multimodal fusion transformer for remote sensing image classification. *arXiv preprint arXiv:2203.16952*. https://doi.org/10.48550/arXiv.2203.16952

Shen, S., Li, C., Hu, X., Xie, Y., Yang, J., Zhang, P., Gan, Z., Wang, L., Yuan, L., Liu, C., Keutzer, K., Darrell, T., Rohrbach, A., & Gao, J. (2022). K-LITE: Learning transferable visual models with external knowledge. *NeurIPS 2022. arXiv*. https://arxiv.org/abs/2204.09222/arXiv

Usha Nandini, D., & Leni, A. E. S. (2015). Shadow identification using ant colony optimization. *Journal of Theoretical and Applied Information Technology*, 78(2), 195–200.

Xia, G., Bai, X., Ding, J., Zhu, Z., Belongie, S., Luo, J., ... & Zhang, L. (2018). DOTA: A large-scale dataset for object detection in aerial images. *Proceedings of the IEEE Conference on Computer Vision and Pattern Recognition (CVPR)*, 3974–3983. https://doi.org/10.1109/CVPR.2018.00418

Xia, G., Hu, J., Hu, F., & Shi, B. (2017). AID: A benchmark dataset for performance evaluation of aerial scene classification. *IEEE Transactions on Geoscience and Remote Sensing*, 55(7), 3965–3981. https://doi.org/10.1109/CVPR.2018.00418

Yang, Y., & Newsam, S. (2010). Bag-of-visual-words and spatial extensions for land-use classification. *Proceedings of the 18th SIGSPATIAL International Conference on Advances in Geographic Information Systems*, 270–279. https://doi.org/10.1145/1869790.1869829

Yu, H., Kong, B., Hou, Y., Xu, X., Chen, T., & Liu, X. (2017). A critical review of applications of hyperspectral remote sensing in crop monitoring. *Experimental Agriculture*.

Yu, J., Xu, H., Gao, J., & Li, B. (2022). CoCa: Contrastive captioners are image-text foundation models. *arXiv preprint arXiv:2205.01917*. https://doi.org/10.48550/arXiv.2205.01917

Zhai, X., Wang, X., Mustafa, B., Beyer, L., Tokmakov, P., Kirillov, A., & Kolesnikov, A. (2023). Sigmoid loss for language–image pre-training. *Proceedings of the IEEE/CVF International Conference on Computer Vision (ICCV)*, 11975–11986. https://doi.org/10.1017/S0014479722000278

Zhan, Y., Xiong, Z., & Yuan, Y. (2023). *RSVG: Exploring data and models for visual grounding on remote sensing data*. IEEE Transactions on Geoscience and Remote Sensing, 61, 1–13. https://doi.org/10.1109/TGRS.2023.3250471

Zhang, W., Cai, M., Zhang, T., Zhuang, Y., & Mao, X. (2024). EarthGPT: A universal multimodal large language model for multisensor image comprehension in the remote sensing domain. *IEEE Transactions on Geoscience and Remote Sensing*, 62, Article 5917820. https://doi.org/10.1109/TGRS.2024.3409624

Zhu, D., Chen, X., Li, B., & Chen, C. (2023). MiniGPT-4: Enhancing vision-language understanding with advanced large language models. *arXiv preprint arXiv:2304.10592*. https://doi.org/10.48550/arXiv.2304.10592

6 Tools and Software Essential Resources for AI Integration

Neha Bhati, Narayan Vyas, and Surendra Yadav

6.1 INTRODUCTION TO AI INTEGRATION

6.1.1 Defining Artificial Intelligence Integration

Artificial intelligence (AI) integration is the process of integrating AI capabilities into existing software products and processes. It involves embedding AI within various working applications to make them intelligent, using AI to make better decisions, and achieving high levels of automation and perception. Integration embeds AI as a functional layer within larger socio-technical ecosystems, uniting data pipelines, algorithms, and application interfaces (Alenezi & Akour, 2025). In modern times, AI integration is about combining software engineering with data science and domain expertise.

It shows up in different forms across industries. When it comes to healthcare, integration involves embedding diagnostic models into an electronic health record (EHR) platform, making these capabilities available to clinicians in real time. For instance, it involves binding prediction algorithms to risk-assessment dashboards. Integration in the manufacturing and innovative agriculture environment involves connecting sensor networks and machine learning controllers for adaptive process optimisation (Ye et al., 2024). Collections, transformations, and analyses of data are performed using standardised frameworks. When the results of an analysis or model are deployed, monitoring is conducted for drift or bias. This approach allows intelligent behaviour to adapt dynamically to the environmental context. AI integration amalgamates learning algorithms, computational infrastructure, and application logic to drive autonomous, context-aware decisions across domains (Lombardo et al., 2023).

6.1.2 Objectives and Structure of the Chapter

This chapter aims to provide a comprehensive understanding of the essential software tools and frameworks required for effective AI integration. It organises the vast technological landscape into coherent categories that mirror the AI lifecycle, from data management and preprocessing to model training, optimisation, deployment, and monitoring. Each subsequent section focuses on a distinct layer of this

 DOI: 10.1201/9781003616207-7

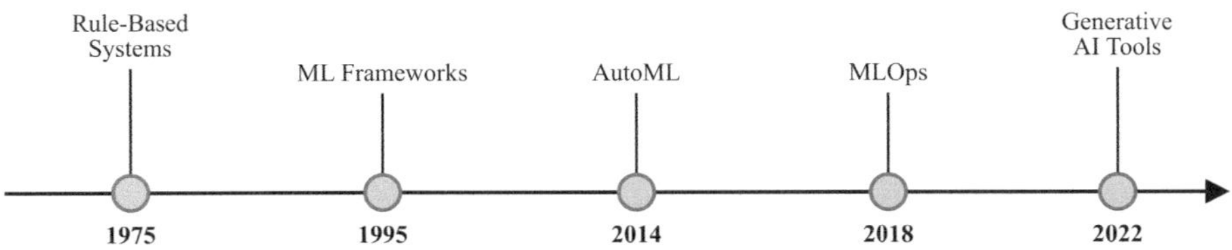

FIGURE 6.1 Evolution of AI tool ecosystem.

ecosystem, including Core AI Development Frameworks that introduce foundational libraries for machine learning and deep learning model creation; Data Management and Processing Tools that review platforms for data collection, transformation, and annotation; Model Training and Deployment Infrastructure that discusses resource optimisation, AutoML systems, and serving frameworks; MLOps and Lifecycle Management that examines continuous integration (CI), monitoring, and governance tools; Emerging Generative and Multimodal Frameworks that explore cutting-edge platforms supporting language–vision and generative applications; and finally, Case Studies and Challenges that illustrate real-world implementations, limitations, and future research avenues. Figure 6.1 shows the evolution of various AI tool ecosystems.

6.2 CORE AI DEVELOPMENT FRAMEWORKS

AI development ultimately depends on various software frameworks and libraries that offer prebuilt functions for data ingestion, model building, and performance tuning. These frameworks abstract the intricacies of mathematical operations, hardware acceleration, and parallel computation to democratise AI implementation (*The AI-Powered Evolution of Big Data*, n.d.). These core AI frameworks can be broadly classified into three categories: machine learning (ML) libraries, deep learning platforms, and reinforcement learning/simulation environments.

6.2.1 Machine Learning Libraries

6.2.1.1 TensorFlow and PyTorch

TensorFlow, created by Google Brain, and PyTorch, developed by Meta AI, are two of the most impactful deep learning frameworks currently offered. Both have tensor support, automatic differentiation, and support for distributed training, but the philosophies and design paradigms are different (Novac et al., 2022).

Execution Model in TensorFlow: TensorFlow employs a static computational graph, allowing compilation and optimisation prior to execution. This method provides a more reliable and scalable solution for massive production environments, especially for models that need to be deployed onto the cloud or edge devices. TensorFlow includes Keras (its high-level API), allowing models to be built and trained with just a few lines of code, while still accessing advanced features, such as TensorFlow Extended (TFX) for production pipelines. Combined with TorchServe for deployment and TorchVision for image processing, PyTorch offers exceptional versatility. PyTorch has gained significant traction in scientific communities, reflected in the rapid growth of its ecosystem. Recent versions introduced TorchScript to optimise models while maintaining flexibility (Leon et al., 2020).

6.2.1.2 Scikit-learn, XGBoost, and LightGBM

TensorFlow and PyTorch are prominent libraries of deep learning, while classical ML still leans on fast and interpretable libraries like Scikit-learn, XGBoost, and LightGBM (Ahn et al., 2023).

Scikit-learn: A Python library to support statistical and traditional ML models such as regression, classification, clustering, and dimensionality reduction. This highlights simplicity with its unified Application Programming Interfaces (APIs) and rich documentation, and has been used extensively for educational and rapid prototyping tasks.

XGBoost (Extreme Gradient Boosting) is an open-source library based on the theory developed by the DMLC group, enabling some of the most efficient gradient-boosted decision trees (GBDT), which are essentially optimal for predictive modelling. It provides parallel tree boosting, support for missing values, and the ability to handle massive datasets at reduced computational cost.

LightGBM, an ML library developed by Microsoft, further extends this paradigm using a leaf-wise growth strategy and histogram-based decision rules to increase both speed and accuracy. It has been specifically optimised for distributed training and low memory consumption. Collectively, these bridge the gap between classical statistical learning and modern deep architectures to establish the foundation for ensemble modelling and hybrid AI systems (Gan et al., 2021) (Table 6.1).

6.2.2 Deep Learning Platforms

6.2.2.1 Keras and MXNet

Keras is a high-level API for building and training neural networks, originally independent but now part of TensorFlow. It concentrates on models rather than low-level code. Even with neural network constructs, it allows for quick prototyping via its sequential and functional APIs and offers extensibility to use your own layers and loss functions. Apache MXNet, which focuses on scalability and performance, is one of the next-generation deep-learning libraries (Kim et al., 2022). It supports both symbolic and imperative programming modes, which is beneficial for research and deployment. MXNet's hybrid feature allows conversion of dynamic graphs into static ones, combining PyTorch-like dynamic capabilities with TensorFlow-like optimisations. AWS put MXNet at the heart of the AWS Deep Learning AMI, showcasing the stability and scalability of MXNet in the cloud.

6.2.2.2 ONNX and Model Interoperability

To overcome this fragmentation, a cross-framework model representation standard called the Open Neural Network Exchange (ONNX) format was developed. ONNX was co-developed by Microsoft and Facebook to avoid reimplementing successful initial ML models across several frameworks ('Faster Scalable ML Model Deployment Using ONNX and Open Source Tools,' 2020). This means executing a model trained in PyTorch on TensorFlow, Caffe2, or MXNet.

TABLE 6.1
Comparative Overview of Major AI Frameworks

Framework	Developer	Primary Language	Core Use Case	Distinctive Features	Licence	Source
TensorFlow	Google	Python, C++	Deep Learning, Production Pipelines	TFX Integration, TPU Support	Apache 2.0	Sami and Ayyan (2025)
PyTorch	Meta	Python	Research, Dynamic Modelling	Dynamic Graphs, TorchServe	BSD	Ansel et al. (2024)
Scikit-learn	Open Source	Python	Classical ML	Simple API, Comprehensive Toolkit	BSD	Pedregosa et al. (2012)
XGBoost	DMLC	Python, R, C++	Gradient Boosting	Parallel Computation, Sparse Handling	Apache 2.0	Nguyen et al. (2019)
LightGBM	Microsoft	Python, R, C++	Fast Tree Models	Histogram-Based Learning, GPU Acceleration	MIT	Ke et al. (2017)

ONNX solves a critical enterprise AI problem: framework lock-in, by enabling this level of interoperability. ONNX provides cross-ecosystem convertibility to give deployability flexibility in cloud and edge devices. ONNX Runtime also provides an optimisation inference engine for hardware acceleration across several backends, including CPU/GPU and dedicated hardware, such as Intel Optane, OpenVINO, and NVIDIA TensorRT. In enterprise contexts, ONNX is a key piece for ensuring that your model operationalisation, AI governance, and migration as frameworks change/age can be sustained/designed to last (Rahrooh et al., 2024).

6.2.3 Reinforcement Learning and Simulation Environments

6.2.3.1 OpenAI Gym

OpenAI Gym: an open-source framework to provide a universal ML harness for developing and comparing reinforcement learning (RL) algorithms. It comprises simple control tasks, challenging games, and robotics simulations, offering a diverse range of environments for algorithms to be fairly compared against each other.

OpenAI Gym frees users from the hassle of working through the environment-agent interaction loop, permitting rapid experimentation with deep learning, policy gradients, and countless actor-critic methods. It is a modular design compatible with TensorFlow and PyTorch. Moreover, the latest versions, Gymnasium and PettingZoo, provide multi-agent environment support. That kind of flexibility is what makes Gym extremely powerful for both academic research and applied RL projects (Laukaitis et al., 2025).

6.2.3.2 Unity ML-Agents

Unity ML-Agents Toolkit creates a scenario involving a 3D simulation and AI for experimentation. It is built on the Unity game engine and enables agents to be trained in rich, simulated environments with reinforcement and imitation learning, using rich visual information. The real power of simulation comes from being able to recreate complex real-world dynamics, such as perception, motion, and physics, cheaply and quickly (Savid et al., 2023).

Another great feature of Unity ML-Agents is domain randomisation, which allows agents to generalise to the real world (sim2real transfer), an essential for robotics and autonomous navigation. It allows for seamless integration into an AI pipeline via its Python API and TensorFlow/PyTorch interoperability. As a result, Unity ML-Agents (Liapis et al., 2025) have become increasingly critical for applications related to spatial reasoning and adaptive control, from robotics to drones and autonomous driving systems.

6.3 DATA MANAGEMENT AND PROCESSING TOOLS

Robust data management forms the backbone of the AI integration journey. As AI models learn their intelligence from data-driven patterns, the quality, scalability, and governance of data are the key factors influencing output reliability. Modern AI workflows rely on a layered data ecosystem that includes data acquisition, data pre-processing, data labelling, storage, and versioning (Saki & Soori, 2025).

6.3.1 Data Acquisition and Cleaning

The initial phase of any AI project includes data collection and data curation from heterogeneous data sources such as databases, sensors, web APIs, or streaming systems. Pandas, NumPy, and Dask are tools that build this analytical foundation for manipulating data in Python (Mekouar et al., 2025).

- NumPy provides high-performance numerical operations on multidimensional arrays and acts as a backend library for many ML libraries.
- In contrast, Dask takes the functionality of Pandas to the distributed computing world and can scale and process terabytes of data on a local cluster or cloud infrastructure.

Apache Spark and Hadoop dominate data engineering pipelines for large-scale enterprise applications. The in-memory distributed computation model of Spark allows for rapid ETL (Extract, Transform, Load) processing and integrates with MLlib for ML at scale. Hadoop still represents the basis of batch-oriented processing and long-term storage (i.e. HDFS and MapReduce) (Šprem et al., 2024). Those frameworks complement each other really well to reframe the core of the modern data engineering architecture behind AI model development.

6.3.2 Data Annotation and Labelling Software

Most supervised learning also relies on labelled datasets to train models. Manual annotation is time-consuming, so efficient software tools have emerged to automate and streamline the process.

LabelImg is a simple graphical interface for annotating object detection datasets, with possible output formats including Pascal VOC and YOLO. CVAT (Computer Vision Annotation Tool) enables collaborative, web-based annotation of images and videos, including polygon segmentation, interpolation, and AI-assisted labelling. Likewise, Label Studio offers more than vision; it provides a multi-domain labelling capability that also works for text, audio, and tabular data. This also allows remote, multi-user workflows by integrating with APIs and cloud storage, both of which can lead to greater efficiency and scalability in the datasets. Quality assurance modules and automated pre-labelling models are increasingly integrated into annotation tools, indicating a shift toward human-in-the-loop AI systems (Karim et al., 2025). Humans supervise annotations produced by models.

6.3.3 Data Versioning and Governance

DVC (Data Version Control) is one of the most commonly used open-source tools that extends Git to handle large datasets and ML projects. It tracks changes to data files, models, and configuration parameters, avoids the duplication of storage, and supports collaborative development of ML. MLflow likewise offers model tracking, experiment management, and parameter logging features that help preserve continuity across iterative experiments. Delta Lake (from Databricks) provides enterprise-level transactional consistency over data lakes by enabling atomic updates, schema enforcement, and integrating audit trails. Embedded metadata and lineage tracking, along with access control policies, add accountability to data governance frameworks (Ruf et al., 2021). Together, they ensure that AI systems are explainable and auditable throughout their development lifecycle, aligning with the ethical and legal requirements of responsible AI (Table 6.2). The stages of the AI data lifecycle are shown in a schematic in Figure 6.2.

All AI innovation is built on data, which must be managed with high precision, consistency, and traceability. The evolution of data-centric tools—from processing engines built on Spark to labelling platforms like CVAT and version control solutions like DVC—indicates a shift away from ad hoc experimentation and toward a strong trend in classical AI engineering. Each layer (acquisition, cleaning, labelling, governance) delivers reliability and repeatability. A good data management practice will guarantee better model performance, but it will also lead to compliance, scalability, and cross-domain usability. Despite the rise in AI adoption, data engineering and governance will remain the quiet backbone of the intelligent enterprise.

6.4 MODEL TRAINING INFRASTRUCTURE

6.4.1 GPU, TPU, and Cloud-Based Training

Contemporary AI models, such as neural networks, require substantial computation, often involving millions or billions of parameters. Thus, practical training relies on heterogeneous hardware acceleration. Deep learning models require massive parallelisation, so Graphics Processing Units (GPUs) are now the standard for training them. The optimised CUDA architecture (NVIDIA's parallel computing platform), combined with the highly efficient routines provided by the cuDNN library for matrix multiplications and convolutions, drastically reduces the training time (Lerat et al., 2022). If workloads demand higher throughput (and are TensorFlow-based), Google offers TPUs, specialised processors for optimising TensorFlow performance at scale in production environments. For example, TPU/GPU instances from cloud providers, including Google Cloud, AWS, and Microsoft Azure, feature auto-scaling, allowing developers to allocate resources dynamically based on training demands. Cloud native training environments also make distributed learning much easier. Tools such as Horovod (from Uber) and Ray provide parallelised model training on multiple nodes, allowing for better efficiency and a level of fault tolerance. Such distributed training strategies play an essential role in large-scale models of computer vision, natural language processing, and scientific computing (Rodríguez Corral et al., 2024).

6.4.2 AutoML Platforms (Google Vertex AI, H2O.ai, DataRobot)

Automated Machine Learning (AutoML) represents a paradigm shift in model training, automating feature selection, hyperparameter tuning, and model selection. Tools such as Google Vertex AI, H2O.ai, and DataRobot give technical and non-technical people alike the ability to implement high-performing models without manual effort (de Oliveira et al., 2024).

TABLE 6.2
Key Data Processing Tools and Their Functional Layers

Category	Tool	Primary Function	Scalability	Integration Level	Source
ETL/Processing	Apache Spark	Distributed computation and in-memory analytics	Very High	Multi-language (Python, Scala, Java)	Manconi et al. (2023)
Batch Processing	Hadoop	Distributed file storage and batch jobs	High	Cluster-based	Ahmed et al. (2020)
Data Cleaning	Great Expectations	Data validation and schema enforcement	Moderate	Python, Cloud APIs	Martins et al. (2025)
Labelling	CVAT	Image/video annotation and automation	Medium	Web-based	Ribeiro et al. (2025)
Versioning	DVC	Data and model version control	High	Git-compatible	Ruf et al. (2021)
Governance	Delta Lake	Transactional data lake management	High	Apache Spark ecosystem	Azzabi et al. (2024)

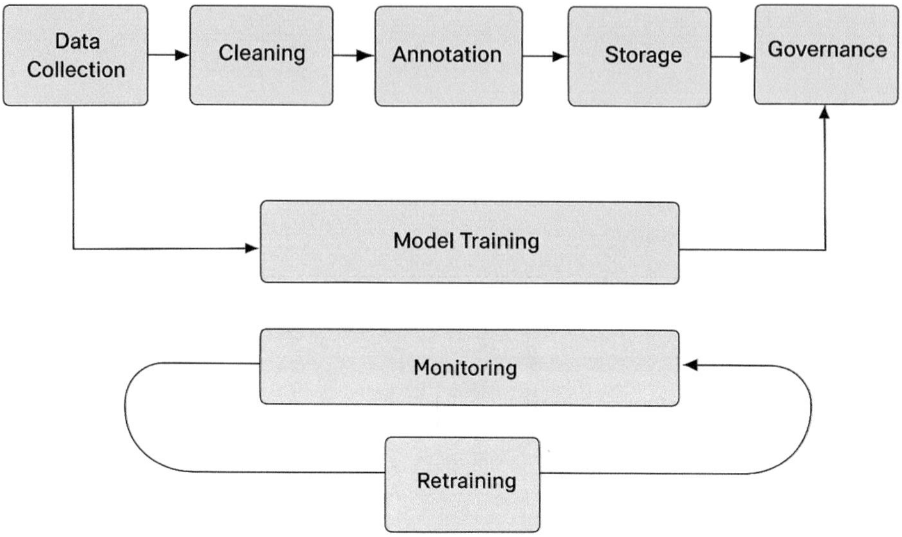

FIGURE 6.2 AI data lifecycle from acquisition to governance.

Google Vertex AI provides data pipelines, training orchestration, and a model registry as part of a unified interface.

- H2O AutoML is an open-source framework for AI that can train different model families and automatically choose the best one according to performance metrics.
- DataRobot, another commercial platform that focuses on enterprise enablement, includes explainability and compliance checks as well as API endpoints for deployment.

The democratisation of AI occurs by lowering entry barriers and adhering to the best practices followed by AutoML platforms. However, they also need to be used with a human in the loop to ensure contextual relevance and interpretability, which are crucial for regulated sectors like healthcare and finance.

6.4.3 Model Deployment Frameworks

6.4.3.1 TensorFlow Serving, TorchServe, ONNX Runtime

Deployment is when the AI models are converted into services that can be used within applications, through APIs, or via embedded systems. TensorFlow Serving is a flexible system for serving ML models, supporting gRPC and REST APIs for updating models without downtime.

Then there is TorchServe, co-developed by AWS and Meta, which serves a similar purpose for PyTorch models by offering seamless integration with AWS services, multi-model management, and logging tools.

ONNX Runtime is a cross-platform ONNX model inference engine. It is backed by hardware, delivering hardware acceleration on GPU, CPU, and other specialised chips (e.g. FPGA) with great interoperability and cross-platform consistency. These tools combined allow for consistent deployment across cloud, edge, and hybrid infrastructures.

6.4.3.2 Edge AI Deployment (NVIDIA Jetson, OpenVINO)

With IoT and embedded intelligence, the need for Edge AI has increased, allowing models to run close to data instead of faraway cloud servers. Doing this at the Edge provides the advantages of reduced latency, lower bandwidth consumption, and increased privacy. GPU-accelerated inference on NVIDIA Jetson platforms (Nano, Xavier, Orin) is ideal for robotics, surveillance, and autonomous systems (García-Pérez et al., 2023). Intel OpenVINO toolkit provides you with some optimised deep learning models for inference on CPUs, VPUs, and FPGAs to be deployed in real-time industry or healthcare solutions. Massive change comes with a trend called Edge AI, which connects cloud analytics to decision-making in the physical world, moving more of the intelligence into local systems. Figure 6.3 shows a layered architecture showing three deployment paradigms: (1) Cloud AI—centralised,

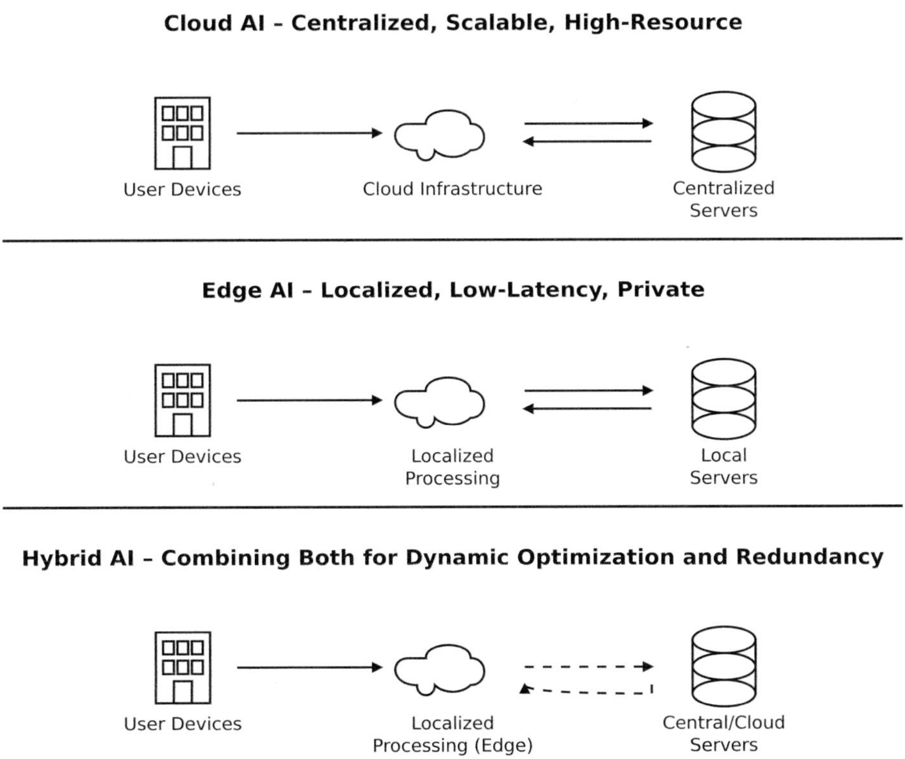

FIGURE 6.3 AI deployment architecture: cloud, edge, and hybrid models.

scalable, high-resource; (2) Edge AI—localised, low-latency, private; (3) Hybrid AI—combining both for dynamic optimisation and redundancy.

6.5 MLOPS AND AI LIFECYCLE MANAGEMENT TOOLS

AI systems advance from experimental prototypes to production-grade applications, and their complexity grows exponentially. These systems have to be automated, traced, scaled, and governed—and all of these are covered within a specific discipline called machine learning operations (MLOps). MLOps unifies data engineering, model development, and deployment into a single operational framework, similar to DevOps in software engineering. It is aimed at making sure that AI solutions are reputable, current, and compliant every step of the way. This is where MLOps comes in; it brings a systematic method of CI/CD/CM for AI models. It provides reproducibility, manages data and model drift, and allows data scientists, engineers, and business stakeholders to collaborate with ease (Pahune & Akhtar, 2025).

6.5.1 Continuous Integration and Delivery (CI/CD) for AI

Data has to be automated throughout the complete workflow from model training to deployment, since AI development is an iterative process. CI automates testing, validation, and version control at the component level, and Continuous Delivery (CD) enables proven models to be deployed reliably into production environments (Farihane et al., 2025).

Automation tools like Jenkins, GitHub Actions, and Azure ML Pipelines are widely used to orchestrate CI/CD workflows in AI projects.

- Jenkins, a long-established automation server, allows pipelines to run training, testing, and deployment scripts in order of execution, and can easily be integrated with containerisation tools like Docker and Kubernetes.
- GitHub Actions offers CI/CD at ease within the Git ecosystem, allowing workflows to be automated on code commits, model updates, or dataset changes.
- With modular input pipeline components, Azure ML Pipelines takes a step further in implementing CI/CD principles at each data preparation step and makes it easy to version the data and retrain and deploy a model into production.

Such tools empower collaborative AI engineering where each code commit, data set change, or hyperparameter tweak is automatically tracked, validated, and incorporated (Manolov et al., 2025).

6.5.2 Model Monitoring and Drift Detection

When deployed, AI models exist in ever-changing environments with data distributions that change over time. This behaviour of changing over time is called data drift (or concept drift), which can cause the model to underperform if not accounted for. Hence, continuous monitoring tools are an important part of making sure our models are reliable. The following are the top open-source solutions for obtaining drift detection and monitoring model performance metrics led by EvidentlyAI and WhyLabs. EvidentlyAI enables users to create dashboards that visualise statistical changes between training and live data. At the same time, WhyLabs uses AI to raise alerts if anything in the live data deviates from the norm (Kustitskaya et al., 2025).

Similar to DevOps scenarios, Prometheus and Grafana have become popular monitoring solutions, and they have also been adapted for AI monitoring, capturing performance metrics such as latency and inference throughput in real time. These tools, when used in conjunction with alerting systems, can trigger model retraining or rollback, so the models do not drift from their deployment data constellation. It extends the definition of performance monitoring beyond the usual accuracy metrics to encompass fairness, explainability, and compliance dimensions, turning performance management into a rich AI assurance process.

6.5.3 AI Governance, Explainability, and Fairness Tools

With the increasing ubiquity of AI, governance and ethics are now critical aspects of lifecycle management. Tools for explainability, like LIME (Local Interpretable Model-Agnostic Explanations), SHAP (Shapley Additive Explanations), and ELI5, help visualise how input features impact model outputs, enhancing trust and accountability (Ferrara, 2024). On the other hand, fairness and bias detection frameworks, such as IBM AI Fairness 360 (AIF360) and Fairlearn, help measure differences between demographic groups and address model issues. Integrating such tools in the MLOps pipeline ensures that explainability and fairness are not left as afterthoughts but are core elements of the AI lifecycle. Governance platforms such as MLflow and Neptune.ai provide a stronger focus on compliance through experiment tracking, audit logs, and lineage management. Such capabilities are essential in regulated industries such as finance, defence, and health care, where explainability or traceability is a legal necessity, among other drivers (Angerschmid et al., 2022).

6.5.4 Cloud-Native AI Services

A large part of the major cloud providers now have had the time to build more or less comprehensive MLOps ecosystems with most of the tools we need for training, deploying, monitoring, and governing our models in one place (Kratzke, 2022).

- AWS SageMaker also offers lifecycle end-to-end automation features for in-place tools such as Model Register, Model Monitor, and Pipelines.
- Azure Machine Learning provides managed MLOps environments that support automated retraining and drift detection.

CI/CD pipelines, model governance, and AI explainability dashboards are integrated in Google Cloud AI Platform (Vertex AI).

They provide APIs and SDKs for custom extensions, emphasising platform abstractions that focus on scalability, security, and compliance. They are gradually

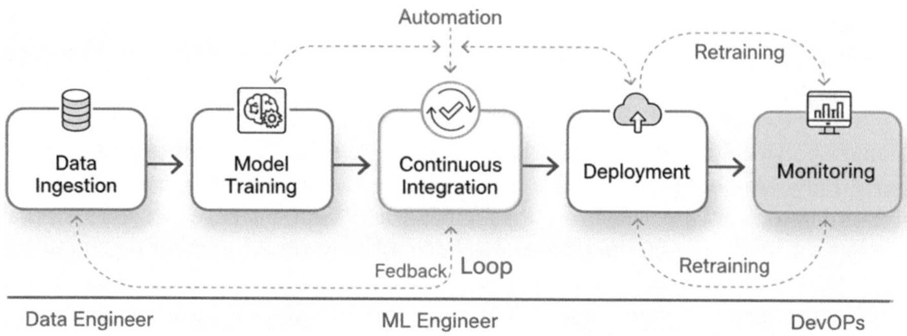

FIGURE 6.4 Complete MLOps pipeline.

transforming into the backbone of enterprise-grade AI-driven operations. The process flow diagram for MLOps lifecycle is shown in Figure 6.4. The diagram shows automation at each stage and integration between data engineers, ML engineers, and the DevOps team (Table 6.3).

6.6 EMERGING TOOLS FOR GENERATIVE AI AND MULTIMODAL SYSTEMS

Over the past few years, the development of AI has been revolutionised by the new trend set by generative AI (GenAI) and multimodal intelligence—technology capable of synthesising text, images, audio, video, and more under a common framework. Generative systems do not just analyse patterns; they generate new content, designs, and solutions, which represents a fundamentally different paradigm from traditional predictive AI. We pave this paradigm shift using dedicated frameworks, APIs, and data infrastructures to train, infer, and retrieve large-scale models (Fitriawijaya & Jeng, 2024). This integration of technologies represents a new frontier for AI, where creativity integrates with an understanding of context and cross-domain reasoning capabilities.

6.6.1 Large Language Models (LLMs) and APIs

Large Language Models (LLMs) are primary enablers of generative intelligence, with few others surpassing them: OpenAI's GPT series, Anthropic's Claude, and Google's Gemini. State-of-the-art natural language understanding and generation performance is achieved by these models, which are trained on trillions of tokens (Brohi et al., 2025).

APIs and Open Frameworks have made LLMs readily available for developers and researchers alike.

The OpenAI API provides REST endpoints for generating text, images, and embeddings, useful in chatbots, summarisation applications, and automatic agents.

- Hugging Face Transformers, an extensive collection of state-of-the-art pre-trained models for natural language processing, which can be used in translation, summarisation, text generation, word embeddings, and more!

TABLE 6.3
MLOps Tools for End-to-End Lifecycle Automation

Lifecycle Stage	Tool	Primary Function	Cloud Integration	Source
CI/CD	Jenkins	Workflow automation, testing, and deployment orchestration	AWS, Azure, GCP	Hyun et al. (2024), Prabu (2025)
Explainability	SHAP	Model interpretability and feature attribution	All major frameworks	Nazir et al. (2025)
Governance/Tracking	MLflow	Experiment tracking, model registry, artefact management	Multi-cloud	Pahune and Akhtar (2025)
Fairness Assessment	IBM AIF360	Bias detection and mitigation toolkit	API/Notebook integration	Chen et al. (2023)
Cloud Platform	Azure ML/SageMaker	Full-stack MLOps integration	Native cloud services	Choi et al. (2023)

- LangChain is a modular framework to develop LLM applications, wrapping prompts, memory, and external tools interactions in structured pipelines.

These platforms enable experimentation and industrial deployment, allowing organisations to embed natural language functionality into workflows with minimal infrastructure overhead. Tools for orchestrating LLMs, such as LlamaIndex (recently rebranded from GPT Index) or AutoGen, enhance the system's intelligence by linking multiple model calls, enabling them to perform analysis, decision-making, and reference private data.

6.6.2 Visual and Multimodal AI Frameworks

The current trend in AI is to have systems interpret and generate across many modalities, such as images, text, speech, and video. Recent advancements have led to so-called multimodal AI systems, which combine perception and cognition into a single architecture (Skaug & Nojoumian, 2025).

- CLIP—Contrastive language–image pre-training: A model bridging vision and language by learning joint representations of text and images (OpenAI). This allows for zero-shot classification and text-guided visual search with no task-specific fine-tuning (Dobrzycki et al., 2024).
- BLIP-2 (Bootstrapped Language–Image Pretraining) expands this idea by connecting vision encoders with frozen LLMs to enable efficient multimodal reasoning (Wang et al., 2025).
- Stable Diffusion, Midjourney, and DALL·E using latent diffusion models have transformed image generation by enabling artistic input from text.

These tools are not only for artistic purposes but have also gradually found applications in healthcare imaging, remote sensing, industrial inspection, and robotics for multimodal analysis and decision support. Such paired representations of sight and language emphasise the concept on which the potential new generation of AI assistants, designed to understand context in deep perceptual spaces, rely.

6.6.3 Vector Databases and Retrieval-Augmented Generation (RAG)

The contextual strength of generative AI relies significantly on retrieval for information redundancy. Semantic similarity search against arbitrarily large unstructured datasets is something that traditional databases cannot handle. Thus, vector databases have become a core component of AI integration pipelines. They support RAG, which is where external knowledge is retrieved in real time to anchor model outputs to reality (Pawlik, 2025).

So, for example, in RAG pipelines, they utilise the embedding to match a user query with relevant document vectors, which are then used as provided context to a generative model. It helps improve factual correctness, reduces hallucination, and enables organisations to use their private knowledge bases safely. Now, with the addition of vector stores and orchestration frameworks such as Lang Chain and

LlamaIndex, it has led to reasonably powerful enterprise AI with memory, context coverage, and retrieval capabilities.

6.7 CASE STUDIES AND IMPLEMENTATION SCENARIOS

To appreciate the practical significance of the tools and frameworks outlined in the earlier sections, it must be contextualised in actual deployment. From healthcare to agriculture to manufacturing, the integration of AI has reshaped efficiency, prediction, and decision support. These case studies highlight solutions to domain-specific problems by leveraging the convergence of data processing, model training with MLOps, and deployment frameworks, while adhering to scalability and compliance.

6.7.1 AI Integration in Agriculture

AI is having an equally transformative impact on another area called precision agriculture. Accurate estimation of yield, soil scanning, and crop disease using satellite images, sensor networks, and ML.

One implementation of such a model involved an agricultural analytics startup training convolutional neural networks with Sentinel-2 satellite data to determine whether their crops were experiencing water stress using PyTorch. The data pipeline was built using Apache Spark for scalable preprocessing and tracked our model metadata and hyperparameters using MLflow (Giansanti & Pirrera, 2025). The optimised model was then pruned and quantised using ONNX Runtime to deploy on edge devices that were spread out across the agricultural fields. Google Vertex AI was also used to automate retraining cycles with sensor data collected to keep models up to date. This led to a hybrid AI system that is native to both the cloud and edge levels, working in near real time for field analysis.

6.7.2 AI Integration in Manufacturing and Automation

Industrial automation coupled with AI provides for predictive maintenance, process optimisation, and the assurance of efficiency. Models were developed using historical data to assess whether the machine would fail. Upon detecting drift, it initiated an automated retraining via Azure Pipelines. SHAP explainability modules were integrated into the MLOps workflow to explain predictions and ensure regulatory and operational transparency. These trained models were then deployed on the Edge AI devices with OpenVINO, resulting in near-zero downtime, lower maintenance costs, and reduced energy costs to the factory. The AI-IIoT hybrid case covered in this paper is an industry use case where AI acts as a smart link between data collection and real physical process manipulation (Rahman et al., 2025).

6.8 CHALLENGES, LIMITATIONS, AND FUTURE DIRECTIONS

While point solutions for AI integration have undoubtedly accelerated innovation in industries, the journey towards democratising AI integration in a continuous, scalable, and responsible manner is long. However, despite these processes operating around the clock, we still face technical and ethical challenges that hinder our ability to go beyond the last 5% to reach the potential of our AI.

6.8.1 Interoperability and Standardisation Issues

One of the significant challenges for AI integration is interoperability across different systems, tools, frameworks, and platforms. Following the rapid proliferation of ML and deep learning libraries, a more fragmented ecosystem was created where the data formats, model representations, and pipeline workflows have grown extremely diverse.

Although efforts have been made to standardise model interchange formats like ONNX, the differences in what each framework actually supports have made it challenging to move between TensorFlow, PyTorch, and so on. Similarly, data versioning tools and storage systems lack a standard metadata schema, leading to concerns about efficient integration and reproducibility. Multi-cloud and hybrid deployments are where organisations strive for an optimal trifecta of performance, cost, and compliance, but the absence of uniform standards complicates this effort. A flourishing AI ecosystem will require future research and industry work on interoperable data schemas, shared APIs, and cross-framework governance protocols to enable plug-and-play AI.

6.8.2 Ethical, Regulatory, and Governance Considerations

With AI systems starting to make important decisions in health care, finance, and even governance, ethical, bias, and transparency issues have crept to the forefront. When models are trained on biased/incomplete data, they can inadvertently boost discrimination, resulting in biased solutions. Though there are already fair auditing tools available (e.g. IBM AIF360, Fairlearn), their success is dependent partly on the diversity and reflectiveness of the input datasets. We have regulatory frameworks to work with, such as the EU AI Act, GDPR, and sectoral standards (HIPAA) that provide explainability, accountability, and data protection. Nevertheless, compliance across jurisdictions represents a top problem for global organisations. Additionally, using explainability tools like LIME and SHAP in production pipelines typically incurs greater computational costs and requires domain-specific expertise to understand.

6.9 CONCLUSION

In fact, there are few trends in the 21st century as surely as the embedding of AI across operational ecosystems. However, the sophistication and reliability of this integration primarily depend on the tools, software frameworks, and lifecycle management systems that interface in the development and deployment of AI. We discussed these core building blocks, data processing libraries, deep learning frameworks, MLOps pipelines, and generative AI platforms in this chapter. We explained how they interconnect to unlock the full potential to develop scalable, explainable, and sustainable AI solutions. In later sections, we also explored how libraries such as TensorFlow,

PyTorch, and Scikit-learn boost native model building, and how Spark, DVC, and MLflow provide data reliability and lifecycle traceability, respectively. While technological advances are important in preparing for AI in the future, a greater challenge awaits humanity: linking the creation of intelligence with transparency, inclusiveness, and sustainability. In other words, the integration of AI is not merely a disruption; it is a paradigm shift in human progress. Thus, adaptive ecosystems, responsible intelligence, and various forms of human-machine collaboration will characterise the next era of AI, facilitating the transition to a knowledge society.

REFERENCES

Ahmed, N., Barczak, A. L. C., Susnjak, T., & Rashid, M. A. (2020). A comprehensive performance analysis of Apache Hadoop and Apache Spark for large scale data sets using HiBench. *Journal of Big Data*, 7(1). https://doi.org/10.1186/s40537-020-00388-5

Ahn, J. M., Kim, J., & Kim, K. (2023). Ensemble Machine Learning of Gradient Boosting (XGBoost, LightGBM, CatBoost) and Attention-Based CNN-LSTM for Harmful Algal Blooms Forecasting. *Toxins*, *15*(10), 608. https://doi.org/10.3390/toxins15100608

Alenezi, M., & Akour, M. (2025). AI-Driven Innovations in Software Engineering: A Review of Current Practices and Future Directions. *Applied Sciences*, *15*(3), 1344. https://doi.org/10.3390/app15031344

Angerschmid, A., Zhou, J., Theuermann, K., Chen, F., & Holzinger, A. (2022). Fairness and Explanation in AI-Informed Decision Making. *Machine Learning and Knowledge Extraction*, *4*(2), 556–579. https://doi.org/10.3390/make4020026

Ansel, J., Yang, E., He, H., Gimelshein, N., Jain, A., Voznesensky, M., Bao, B., Bell, P., Berard, D., Burovski, E., Chauhan, G., Chourdia, A., Constable, W., Desmaison, A., DeVito, Z., Ellison, E., Feng, W., Gong, J., Gschwind, M., … Chintala, S. (2024). PyTorch 2: Faster Machine Learning Through Dynamic Python Bytecode Transformation and Graph Compilation. *Proceedings of the 29th ACM International Conference on Architectural Support for Programming Languages and Operating Systems*, *2* (2), 929–947.

Azzabi, S., Alfughi, Z., & Ouda, A. (2024). Data Lakes: A Survey of Concepts and Architectures. *Computers*, 13(7), 183. https://doi.org/10.3390/computers13070183.

Brohi, S., Mastoi, Q., Jhanjhi, N. Z., & Pillai, T. R. (2025). A Research Landscape of Agentic AI and Large Language Models: Applications, Challenges, and Future Directions. *Algorithms*, *18*(8), 499. https://doi.org/10.3390/a18080499

Chen, P., Wu, L., & Wang, L. (2023). AI Fairness in Data Management and Analytics: A Review on Challenges, Methodologies, and Applications. *Applied Sciences*, *13*(18), 10258. https://doi.org/10.3390/app131810258

Choi, W., Choi, T., & Heo, S. (2023). A Comparative Study of Automated Machine Learning Platforms for Exercise Anthropometry-Based Typology Analysis: Performance Evaluation of AWS SageMaker, GCP VertexAI, and MS Azure. *Bioengineering*, *10*(8), 891. https://doi.org/10.3390/bioengineering10080891

de Oliveira, S., Topsakal, O., & Toker, O. (2024). Benchmarking Automated Machine Learning (AutoML) Frameworks for Object Detection. *Information*, *15*(1), 63. https://doi.org/10.3390/info15010063

Dobrzycki, A. D., Bernardos, A. M., Bergesio, L., Pomirski, A., & Sáez-Trigueros, D. (2024). Exploring the Use of Contrastive Language-Image Pre-Training for Human Posture Classification: Insights from Yoga Pose Analysis. *Mathematics*, *12*(1), 76. https://doi.org/10.3390/math12010076

Farihane, R., Chlioui, I., & Radgui, M. (2025). CI/CD Pipeline Optimization Using AI: A Systematic Mapping Study. *Engineering Proceedings*, *112*(1), 32. https://doi.org/10.3390/engproc2025112032

Faster Scalable ML Model Deployment Using ONNX and Open Source Tools. (2020). *2020 IEEE Infrastructure Conference*, i–i. https://doi.org/10.1109/IEEECONF47748.2020.9377615

Ferrara, E. (2024). Fairness and Bias in Artificial Intelligence: A Brief Survey of Sources, Impacts, and Mitigation Strategies. *Sci*, *6*(1), 3. https://doi.org/10.3390/sci6010003

Fitriawijaya, A., & Jeng, T. (2024). Integrating Multimodal Generative AI and Blockchain for Enhancing Generative Design in the Early Phase of the Architectural Design Process. *Buildings*, *14*(8), 2533. https://doi.org/10.3390/buildings14082533

Gan, M., Pan, S., Chen, Y., Cheng, C., Pan, H., & Zhu, X. (2021). Application of the Machine Learning LightGBM Model to the Prediction of the Water Levels of the Lower Columbia River. *Journal of Marine Science and Engineering*, *9*(5), 496. https://doi.org/10.3390/jmse9050496

García-Pérez, A., Miñón, R., Torre-Bastida, A. I., & Zulueta-Guerrero, E. (2023). Analysing Edge Computing Devices for the Deployment of Embedded AI. *Sensors*, *23*(23), 9495. https://doi.org/10.3390/s23239495

Giansanti, D., & Pirrera, A. (2025). Integrating AI and Assistive Technologies in Healthcare: Insights from a Narrative Review of Reviews. *Healthcare*, *13*(5), 556. https://doi.org/10.3390/healthcare13050556

Hyun, G., Oak, J., Kim, D., & Kim, K. (2024). The Impact of an Automation System Built with Jenkins on the Efficiency of Container-Based System Deployment. *Sensors (Basel, Switzerland)*, *24*(18), 6002. https://doi.org/10.3390/s24186002

Karim, M. M., Khan, S., Van, D. H., Liu, X., Wang, C., & Qu, Q. (2025). Transforming Data Annotation with AI Agents: A Review of Architectures, Reasoning, Applications, and Impact. *Future Internet*, *17*(8), 353. https://doi.org/10.3390/fi17080353

Ke, G., Meng, Q., Finley, T., Wang, T., Chen, W., Ma, W., Ye, Q., & Liu, T.-Y. (2017). *LightGBM: A Highly Efficient Gradient Boosting Decision Tree*. In *Proceedings of the 31st International Conference on Neural Information Processing Systems (NeurIPS 2017)*, 3149–3157). Curran Associates Inc.

Kim, S., Wimmer, H., & Kim, J. (2022). Analysis of Deep Learning Libraries: Keras, PyTorch, and MXnet. *2022 IEEE/ACIS 20th International Conference on Software Engineering Research, Management and Applications (SERA)*, 54–62. https://doi.org/10.1109/SERA54885.2022.9806734

Kratzke, N. (2022). Cloud-Native Applications and Services. *Future Internet*, *14*(12), 346. https://doi.org/10.3390/fi14120346

Kustitskaya, T. A., Esin, R. V., & Noskov, M. V. (2025). Model Drift in Deployed Machine Learning Models for Predicting Learning Success. *Computers*, *14*(9), 351. https://doi.org/10.3390/computers14090351

Laukaitis, A., Šareiko, A., & Mažeika, D. (2025). Facilitating Robot Learning in Virtual Environments: A Deep Reinforcement Learning Framework. *Applied Sciences*, *15*(9), 5016. https://doi.org/10.3390/app15095016

Leon, V., Mouselinos, S., Koliogeorgi, K., Xydis, S., Soudris, D., & Pekmestzi, K. (2020). A TensorFlow Extension Framework for Optimized Generation of Hardware CNN Inference Engines. *Technologies*, *8*(1), 6. https://doi.org/10.3390/technologies8010006

Lerat, J.-S., Mahmoudi, S. A., & Mahmoudi, S. (2022). Distributed Deep Learning: From Single-Node to Multi-Node Architecture. *Electronics*, *11*(10), 1525. https://doi.org/10.3390/electronics11101525

Liapis, G., Vordou, A., Nikolaidis, S., & Vlahavas, I. (2025). Reinforcement Learning Methods for Emulating Personality in a Game Environment. *Applied Sciences*, *15*(14), 7894. https://doi.org/10.3390/app15147894

Lombardo, G., Couvert, C., Kose, M., Begum, A., Spiertz, C., Worrell, C., Hasselbaink, D., Didden, E.-M., Sforzini, L., Todorovic, M., Lewi, M., Brown, M., Vaterkowski, M., Gullet, N., Amasi-Hartoonian, N., Griffon, N., Pais, R., Rodriguez Navarro, S., Kremer, A., … Kalko, S. (2023). Electronic Health Records (EHRs) in Clinical Research and

Platform Trials: Application of the Innovative EHR-Based Methods Developed by EU-PEARL. *Journal of Biomedical Informatics*, *148*, 104553. https://doi.org/10.1016/j.jbi.2023.104553

Manconi, A., Gnocchi, M., Milanesi, L., Marullo, O., & Armano, G. (2023). Framing Apache Spark in Life Sciences. *Heliyon*, *9*(2), e13368. https://doi.org/10.1016/j.heliyon.2023.e13368

Manolov, V., Gotseva, D., & Hinov, N. (2025). Practical Comparison Between the CI/CD Platforms Azure DevOps and GitHub. *Future Internet*, *17*(4), 153. https://doi.org/10.3390/fi17040153

Martins, P., Cardoso, F., Váz, P., Silva, J., & Abbasi, M. (2025). Performance and Scalability of Data Cleaning and Preprocessing Tools: A Benchmark on Large Real-World Datasets. *Data*, *10*(5), 68. https://doi.org/10.3390/data10050068

Mekouar, Y., Lahmer, M., & Karim, M. (2025). Optimizing Data Pipelines for Green AI: A Comparative Analysis of Pandas, Polars, and PySpark for CO2 Emission Prediction. *Computers*, *14*(8), 319. https://doi.org/10.3390/computers14080319

Nazir, M. A., Evangelista, E., Bukhari, S. M. S., & Sharma, R. (2025). A Survey of Feature Attribution Techniques in Explainable AI: Taxonomy, Analysis and Comparison. *Annals of Mathematics and Computer Science*, *28*, 115–126. https://doi.org/10.56947/amcs.v28.547

Nguyen, G., Dlugolinsky, S., Bobák, M., Tran, V., López García, Á., Heredia, I., Malík, P., & Hluchý, L. (2019). Machine Learning and Deep Learning Frameworks and Libraries for Large-Scale Data Mining: A Survey. *Artificial Intelligence Review*, *52*(1), 77–124. https://doi.org/10.1007/s10462-018-09679-z

Novac, O.-C., Chirodea, M. C., Novac, C. M., Bizon, N., Oproescu, M., Stan, O. P., & Gordan, C. E. (2022). Analysis of the Application Efficiency of TensorFlow and PyTorch in Convolutional Neural Network. *Sensors*, *22*(22), 8872. https://doi.org/10.3390/s22228872

Pahune, S., & Akhtar, Z. (2025). Transitioning from MLOps to LLMOps: Navigating the Unique Challenges of Large Language Models. *Information*, *16*(2), 87. https://doi.org/10.3390/info16020087

Pawlik, L. (2025). LLM Selection and Vector Database Tuning: A Methodology for Enhancing RAG Systems. *Applied Sciences*, *15*(20), 10886. https://doi.org/10.3390/app152010886

Pedregosa, F., Varoquaux, G., Gramfort, A., Michel, V., Thirion, B., Grisel, O., Blondel, M., Prettenhofer, P., Weiss, R., Dubourg, V., Vanderplas, J., Passos, A., Cournapeau, D., Brucher, M., Perrot, M., & Duchesnay, E. (2012). Scikit-learn: Machine Learning in Python. arXiv. https://doi.org/10.48550/ARXIV.1201.0490

Prabu, V. P. (2025). CI/CD in a Multi-Cloud World: Challenges and Solutions. Zenodo. https://doi.org/10.5281/ZENODO.15154848

Rahman, M. A., Shahrior, M. F., Iqbal, K., & Abushaiba, A. A. (2025). Enabling Intelligent Industrial Automation: A Review of Machine Learning Applications with Digital Twin and Edge AI Integration. *Automation*, *6*(3), 37. https://doi.org/10.3390/automation6030037

Rahrooh, A., Garlid, A. O., Bartlett, K., Coons, W., Petousis, P., Hsu, W., & Bui, A. A. T. (2024). Towards a Framework for Interoperability and Reproducibility of Predictive Models. *Journal of Biomedical Informatics*, *149*, 104551. https://doi.org/10.1016/j.jbi.2023.104551

Ribeiro, A. G., Vilaça, L., Costa, C., Soares da Costa, T., & Carvalho, P. M. (2025). Automatic Visual Inspection for Industrial Application. *Journal of Imaging*, *11*(10), 350. https://doi.org/10.3390/jimaging11100350

Rodríguez Corral, J. M., Civit-Masot, J., Luna-Perejón, F., Díaz-Cano, I., Morgado-Estévez, A., & Domínguez-Morales, M. (2024). Energy Efficiency in Edge TPU vs. Embedded GPU for Computer-Aided Medical Imaging Segmentation and Classification. *Engineering Applications of Artificial Intelligence*, *127*, 107298. https://doi.org/10.1016/j.engappai.2023.107298

Ruf, P., Madan, M., Reich, C., & Ould-Abdeslam, D. (2021). Demystifying MLOps and Presenting a Recipe for the Selection of Open-Source Tools. *Applied Sciences*, *11*(19), 8861. https://doi.org/10.3390/app11198861

Saki, S., & Soori, M. (2025). Artificial Intelligence, Machine Learning and Deep Learning in Advanced Transportation Systems, A Review. *Multimodal Transportation*, 100242. https://doi.org/10.1016/j.multra.2025.100242

Sami, M., & Ayyan, H. (2025). Deep Learning Platforms: A Comprehensive Research Overview. *Journal of Computer Science*, *1*, 15.

Savid, Y., Mahmoudi, R., Maskeliūnas, R., & Damaševičius, R. (2023). Simulated Autonomous Driving Using Reinforcement Learning: A Comparative Study on Unity's ML-Agents Framework. *Information*, *14*(5), 290. https://doi.org/10.3390/info14050290

Skaug, L., & Nojoumian, M. (2025). A Multimodal Artificial Intelligence Framework for Intelligent Geospatial Data Validation and Correction. *Inventions*, *10*(4), 59. https://doi.org/10.3390/inventions10040059

Šprem, Š., Tomažin, N., Matečić, J., & Horvat, M. (2024). Building Advanced Web Applications Using Data Ingestion and Data Processing Tools. *Electronics*, *13*(4), 709. https://doi.org/10.3390/electronics13040709

The AI-Powered Evolution of Big Data. (n.d.). Retrieved October 17, 2025, from https://www.mdpi.com/2076-3417/14/22/10176

Wang, T., Zhang, B., Jiang, D., & Li, D. (2025). A Multimodal Large Language Model Framework for Intelligent Perception and Decision-Making in Smart Manufacturing. *Sensors*, *25*(10), 3072. https://doi.org/10.3390/s25103072

Ye, J., Woods, D., Jordan, N., & Starren, J. (2024). The Role of Artificial Intelligence for the Application of Integrating Electronic Health Records and Patient-Generated Data in Clinical Decision Support. *AMIA Summits on Translational Science Proceedings*, *2024*, 459–467.

Part II

Applications and Case Studies of Gen AI in Remote Sensing

7 Remote Sensing Satellite Datasets, Preprocessing Techniques, and Tools for Agricultural Land Cover Classification

Nisha Sharma, Kawaljit Kaur, and Sartajvir Singh

7.1 INTRODUCTION

Modern agriculture involves agricultural land cover classification as an essential element. It entails a process of detection and layout of various land covers in farm plots or wide-ranging farmland. Trustworthy data about land cover is also valuable to many agricultural applications, including health monitoring of crops, forecasting of crop productivity, and fine-tuning of farm management. Farmers and policymakers make use of these categories to apply interventions like irrigation, fertilization, pest management, and optimal land use planning, all of which help enhance overall crop productivity and sustainability (Hudait & Patel, 2022).

Several studies have emphasized the potential of satellite datasets, particularly Landsat and Sentinel-2, for the mapping of agricultural land cover due to their distinct attributes, including high spectral information and frequent revisit periods (Schmitt et al., 2019). Earlier studies have shown that these datasets, when processed appropriately, can provide accurate maps of vegetation health and agricultural land use trends. However, despite the progress, existing literature often treats these datasets separately, and there is limited comparative research that critically evaluates their relative strengths, limitations, and combined utility for land cover classification across diverse agricultural contexts.

This indicates an area of research. While satellites offer significant benefits, there is a lack of guidance on applying multi-sensor data and preprocessing workflows to enhance classification accuracy across various scales. In addition, software such as Environment for Visualizing Images (ENVI) and Google Earth Engine (Van Leeuwen et al., 2020) facilitates practical analysis, capabilities, and requirements for preprocessing remain unexplored in this particular domain of application. For this research, we aim to address these gaps by systematically considering the most widely used agricultural satellite data sources (Landsat and Sentinel-2) (Sharma & Kaur, 2024).

DOI: 10.1201/9781003616207-9

Their preprocessing methods include radiometric and atmospheric correction, and they utilize popular processing software tools. We seek to (i) evaluate the applicability of these data sources for agricultural land cover mapping (Schmitt et al., 2019), (ii) identify critical preprocessing steps enhancing data quality, and (iii) assess practical tools for their application. The results will enable researchers and practitioners to make informed choices regarding the use of proper data, workflows, and tools to enable correct agricultural mapping and improve sustainable agriculture worldwide.

7.2 SATELLITE DATASETS

Satellite datasets have been used for agriculture to monitor crop health, yield prediction, housing projects, road projects, and seawater studies. Several satellite datasets are available, such as Landsat series, Sentinel series, MODIS, and SMAP (Sharma & Kaur, 2024), among others. The Landsat satellite series is a program of Earth observation satellites launched by the United States, through NASA and the United States Geological Survey (USGS). This program provides important data that has been used since 1972 and provides important data that is used for research purposes. Table 7.1 shows detailed information on the Landsat Satellite (Theres & Selvakumar, 2022). It includes sensor types that are working in the satellite, working duration, launch date, and application areas of different satellites. The Sentinel dataset comprises Earth observation data gathered by the Sentinel satellites (Schmitt et al., 2019). It is part of the Copernicus program, which the European Union manages.

TABLE 7.1
Detailed Information on the Landsat Series

Satellite Dataset	Launch Date	Sensor	Period	Applications
Landsat 1	Jul 23, 1972	MSS	1972–1983	Agricultural monitoring, landcover mapping, and resource management
Landsat 2	Jan 2, 1975			
Landsat 3	Mar 5, 1978			
Landsat 4	Jul 16, 1982	MSS, TM	1982–2013	LULC change detection, water resource management, forest resource management, disaster management
Landsat 5	Mar 1, 1984	MSS		
Landsat 7	Apr 15, 1999	ETM+	1999–present	Same as Landsat 4 and 5, with improved data quality and reliability
Landsat 8	Feb 11, 2013	OLI, TIRS	2013–present	Environment monitoring, climate change studies, urban planning, natural disaster response
Landsat 9	Sep 27, 2021	OLI-2, TIRS-2	2021–present	Improved image quality and consistency supporting agriculture, water resource management, and disaster management

Sentinel satellites provide huge datasets that are used for various applications, including radar imaging, optical imaging, ocean and land monitoring, atmospheric monitoring, and sea surface height monitoring. Table 7.2 provides detailed information on the Sentinel program series, including launch dates, specifications, and applications.

7.2.1 MODIS Satellite

MODIS stands for Moderate Resolution Imaging Spectroradiometer. It is NASA's satellite that offers valuable data for a wide range of Earth observation applications. It is a multispectral imaging radiometer and uses 36 spectral bands ranging from visible to infrared (0.4–14.4μm). Spatial resolution differs by band: 250 meters (bands 1–2), 500 meters (bands 3–7), and 1 kilometer (bands 8–36). Temporal resolution is characterized by a very high revisit frequency, providing one to two days of global coverage. It allows near-global daily coverage of up to 2330 kilometers. It is used for various applications like Land Cover Land Use (LCLU), agriculture, climate and weather, disaster management, and environment monitoring (Zhao et al., 2005).

7.2.2 SMAP Satellite

SMAP stands for Soil Moisture Active Passive. It is a NASA mission aimed at measuring and mapping Earth's soil moisture and freeze-thaw state (Yang et al., 2022). It utilizes two types of sensors: an L-band radar for active sensing and an L-band radiometer for passive sensing. Its primary function is to measure soil moisture content and detect the freeze-thaw state of the ground. Spatial resolution for the radiometer is 40 kilometers, for the radar is 3 kilometers, and combined is 9 kilometers. Temporal resolution for global coverage is two to three days revisit time. It allows 1000 kilometers of frequent global coverage. It is used for agriculture, hydrology, climate science, and environmental monitoring.

7.2.3 SPOT 5

The SPOT 5 dataset comprises high-resolution satellite images captured by the SPOT 5 (Satellite Pour l'Observation de la Terre) satellite. Launched by the French Space Agency (CNES) on May 4, 2002, SPOT 5 was designed to deliver detailed imagery for applications including cartography, agriculture, forestry, environmental monitoring, and disaster management. It works on five spectral bands: one panchromatic (black and white) and four multispectral bands. In the panchromatic band, the image is from 2.5 meters, and for the multispectral band, the spatial resolution is 10 meters (Badiola-Bengoa & Méndez-Zorrilla, 2021).

In the present study, the author discusses various satellite datasets available for agricultural applications and provides a comparative analysis of Table 7.3 of these datasets. Once a suitable satellite dataset is selected, the next step is to capture the dataset. After capturing the dataset, it is necessary to preprocess the data to make it suitable for research purposes. To effectively preprocess the dataset, one must be familiar with preprocessing techniques and the tools used for preprocessing satellite data.

TABLE 7.2
Detailed Information on the Sentinel Series

S. No	Satellite	Launch Date	Specification	Sensor	Applications
1	Sentinel-1A	Apr 3, 2014	Radar Imaging	C-band Synthetic Aperture Radar (SAR)	All-weather, day and night radar imaging, ocean monitoring, sea-ice observations, forest cover monitoring
2	Sentinel-1B	Apr 25, 2016			
3	Sentinel-2A	Jun 23, 2015	Optical imaging	MSI with 13 spectral bands from visible to shortwave infrared	Land cover classification, agriculture monitoring, forest management, and disaster monitoring
4	Sentinel-2B	Mar 7, 2017			
5	Sentinel-3A	Feb 16, 2016	Ocean and Land Monitoring	OLCI, SLSTR, SRAL, MWR	Monitoring sea surface, land, and sea surface temperature, and atmospheric composition
6	Sentinel-3B	Apr 25, 2018			
7	Sentinel-4	Planned for launch in 2023	Atmospheric Composition Monitoring	Ultraviolet, Visible, and Near-infrared (UVN) spectrometer	Atmospheric composition monitoring
8	Sentinel-5	Oct 13, 2017	Atmospheric Composition Monitoring	TROPOMI	High-resolution monitoring of atmospheric trace gases and clouds to support air quality forecasting and climate monitoring
9	Sentinel-6	Nov 21, 2020	Sea Surface Height Monitoring	Poseidon-4, AMR-C	Continuation of high-precision sea surface height measurements for climate monitoring

TABLE 7.3
Detailed Information on Satellite Datasets Used in Agriculture

S. No	Satellite Dataset	Sensor Type	Spectral Bands	Spatial Resolution	Temporal Resolution	Swath Width
1	Sentinel-1	Synthetic aperture radar (SAR)	C-band (5.405 GHz)	5–40 meters	6 days	250 km
2	Sentinel-2	Multispectral imager (MSI)	13 spectral bands visible to infrared	10 meters (VIS/NIR bands), 20 meters (red-edge and SWIR bands), and 60 meters (atmospheric correction bands)	5 days	290 km
3	MODIS	Multispectral imaging radiometer	36 spectral bands (0.4µm–14.4 µm) visible to infrared	250 meters (1–2 bands), 500 meters (3–7 bands), 1 kilometer (8–36 bands)	1–2 days (high revisit frequency)	2330 km
4	SMAP	L-band radar and L-band radiometer	2 bands (active and passive), a combined band	Radiometer: 40 kilometers Radar 3 kilometers	2–3 days	1000 km
5	SPOT 5	2 high-resolution geometric (HRG)	5 spectral bands Panchromatic, green (0.50–0.59 µm) Red (0.61–0.68 µm) Near-infrared (0.78–0.89 µm) Short-wave infrared (1.58–1.75 µm)	Panchromatic with 2.5-meter resolution, Multispectral with 10-meter resolution	2–3 days	60 km
6	Landsat 7	OLI, TIRS	8 spectral bands (blue, green, red, Near-infrared, thermal infrared, shortwave infrared, panchromatic) (0.45–0.90 µm)	Band 1–5 and 7–30 meters Band 6–60 meter Band 8–15 meters	16 days	185 km
7	Landsat 8	OLI-2, TIRS-1	It works on 11 spectral bands. OLI-2 works on spectral bands 1–9 (0.43–1.38 µm), TIRS works on band 10 (10.6+0-11-19 µm), and band 11 (11.5–12.51 µm).	Band 1–7 and 9: 30 meters, band 8:15 meters, band 10–11: 100 meters	16 days	185 km
8	Landsat 9	OLI-2, TIRS-2	It works on 11 multispectral bands. OLI-2 works on spectral bands 1–9 (0.43–1.38 µm), TIRS works on band 10 (10.6+0-11-19 µm), and band 11 (11.5–12.51 µm).	Band 1–7 and 9: 30 meters, band 8:15 meters, band 10–11: 100 meters	16 days	185 km

7.3 DATA ACQUISITION

Satellite imagery data relevant to agricultural applications is available through two primary approaches: free and open-access data, and commercially available options. Table 7.4 presents a comparative analysis of free and open access data versus commercial satellite data providers across several factors. Freely available data, accessible through platforms like USGS EarthExplorer or ESA SciHub, offers a cost-effective entry point, particularly for those new to the field. Some providers even boast user-friendly interfaces to simplify data acquisition. However, challenges associated with free data include the sheer volume of information to download, which can be time-consuming, and the potential need for preprocessing the data before analysis (Kamusoko, 2019). Commercial satellite data providers present an alternative approach. These providers offer high-resolution imagery with revisit times tailored to your specific requirements. They may even allow custom satellite data capture over designated areas of interest. Additionally, some providers streamline the

TABLE 7.4
Comparative Analysis of Free and Open Access Data Versus Commercial Satellite Data

Factor	Free and Open Access Data	Commercial Satellite Data Providers
Data Source	**USGS EarthExplorer**: (https://earthexplorer.usgs.gov/) Offers Landsat data and imagery from other missions **ESA SciHub**: (https://scihub.copernicus.eu/) Provides access to Sentinel-1 and Sentinel-2 data **NASA EarthData Download**: (https://search.earthdata.nasa.gov/) Offers a variety of satellite imagery data	Planet Labs (https://www.planet.com/) offers frequent revisit times with high-resolution imagery. DigitalGlobe (https://www.maxar.com/about) provides a variety of multispectral and high-resolution data options. Maxar Technologies (https://www.maxar.com/products/satellite-imagery) offers tailored solutions and on-demand tasking for specific areas.
Cost	Free	Varies depending on resolution, coverage, and revisit frequency
Data Volume	Large datasets	Can be tailored
Data Processing	May require preprocessing	May offer preprocessed options
Resolution	Moderate-resolution (Landsat, Sentinel)	High-resolution options available
Revisit Time	Regular revisit times (Landsat, Sentinel)	Can be tailored for on-demand tasking
Spectral Bands	Suitable for general agricultural applications	Options for specific needs (e.g., vegetation health)
Ease of Use	User-friendly platforms available	Requires integration with separate software (may vary)
Suitability for Beginners	Good starting point	May require more technical expertise

workflow by offering preprocessed data and user-friendly analysis tools. Naturally, this level of convenience comes at a cost, with pricing structures that vary based on the specific data needs.

Choosing the optimal data source requires a clear understanding of research needs. Key factors to consider include the required resolution, relevant spectral bands for the agricultural application (such as vegetation health monitoring or soil moisture estimation), necessary data update frequency, and budget constraints (Sharma & Kaur, n.d.). By thoroughly evaluating these aspects and exploring both free and commercial options, one can make an informed decision about the satellite imagery data that will best support your agricultural goals.

7.4 PREPROCESSING TECHNIQUES AND DIFFUSION MODELS

Satellite data is crucial to prepare raw satellite imagery for analysis. This process involves multiple steps to correct, enhance, and format the data, ensuring it is suitable for various applications. Figure 7.1 outlines the preprocessing techniques applied to the satellite dataset. These techniques are summarized next.

1. **Radiometric Corrections:** The corrections to the responses of the sensor and effects of the atmosphere are performed through the use of radiometric correction techniques to ensure that the pixel value reflects the surface reflectance. In radiometric correction, it calibrates raw data in correspondence with the specificities of sensors (Kamusoko, 2019).

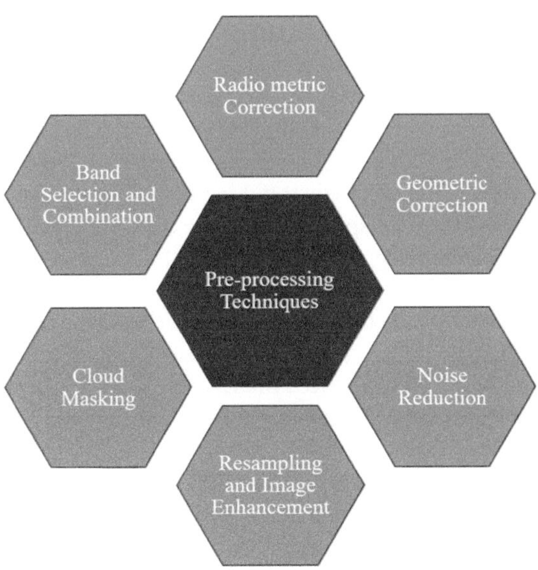

FIGURE 7.1 The preprocessing techniques applied to the satellite dataset.

2. **Geometric Correction:** Geometric correction makes the spatial location of image pixels correct and in line with those of the real world and consists of two major procedures that are shared between satellite image processing and GIS analysis (Zaki et al., 2022).
   a. **Correcting Terrain-Induced Distortions:** The operations and procedures consist of whether the images of the satellite in digital form are placed upon the true shape of the terrain, and this is done by means of the DEM. DEM is a topographical map of terrain height data of the Earth as stored in three-dimensional form. The consequent adjustment provides more precise and realistic images that precisely depict the actual positions of features in the ground.
   b. **Georeferencing:** Georeferencing consists of warping a satellite image to a known coordinate system or map projection so that the image can be used with other geospatial data. It entails determining specific familiar reference points in a satellite image and DEM, correlating these points to their location in a coordinate system. This is executed by means of control points, or points that can be defined accurately in the image and coordinate system. In this process, various computer programs have been employed, including GIS software.
3. **Noise Reduction:** Noise suppression methods are applied to remove undesired signals or artifacts on imagery. In noise reduction, restriping may be applied, and speckle filtering can be done (Khan et al., 2020).
   a. **De-striping:** It is the action of getting rid of deliberate stripe artifacts through imagery to increase its visual-perceptual quality as well as precision.
   b. **Speckle Filtering:** This is a procedure to be employed in radar images to lessen the noise via a filter (e.g., median filter or adaptive filter).
4. **Resampling and Image Enhancement:** It adjusts the spatial resolution of the image used as data in research projects. A resampling process is necessary for geometric corrections. This technique uses nearest neighbor, bilinear interpolation, and cubic convolution methods (Gudavalli et al., 2023). It enhances the visual appearance of the imagery for better interpretation, utilizing histogram and contrast stretching techniques. Histograms improve the quality of an image by spreading out the most frequent intensity values, and the contrast stretching method stretches the range of intensity values to improve the image contrast (Arici et al., 2009).
5. **Cloud Masking:** This technique identifies cloud-covered areas from optical imagery data and excludes them so that incorrect interpretation is not done (N. Sharma & Chawla, 2023; Sharma et al., 2025). It will use machine learning algorithms and thresholding for more accurate cloud detection (Banks & Mélin, 2015).
6. **Band Selection and Combination:** Selecting and combining spectral bands can enhance certain features for analysis. It uses NDVI (Normalized Difference Vegetation Index) and PCA (Principal Component Analysis) (Jeon et al., 2020).

7.5 DIFFUSION MODELS AND TRANSFORMERS

Diffusion models and transformers play an essential role in remote sensing. Diffusion models, as generative models, enhance the quality of satellite images by improving their resolution, generating augmented data for deep learning training, removing cloud cover, and detecting changes over time. DDPM is a standard diffusion model that removes noise from the satellite image, whereas a stable diffusion model enhances the image's quality, enabling better results from the implementation. The SR3 model also improves the quality of the satellite image through repeated refinements. These models compare generated clean images with noisy real ones to detect changes in landscapes.

Transformers are very quick in extended connections for satellite datasets. There are various transformers used for satellite datasets. The ViT transformer is used for landcover classifications for multispectral datasets. SATViT is another example of a vision transformer for satellite imagery. A DETR transformer is a detection transformer used to detect buildings, roads, and other objects in remote sensing. Siamese is another example of a transformer that is used for change detection tasks. Time series analysis can also be done with it by handling multi-temporal datasets for applications like crop monitoring.

7.6 SATELLITE DATASET TOOLS

Various tools and software platforms are available for preprocessing and analyzing satellite datasets. These tools are essential for image processing, analysis, and visualization tasks. Here are some commonly used options, such as ENVI, ArcGIS, ERDAS Imagine, SNAP (Sentinel Application Platform), Google Earth Engine, and QGIS (Geospatial Data Abstraction Library), among others. Table 7.5 represents all the features of the satellite dataset tools (Pawar et al., 2023).

These tools and platforms address diverse aspects of satellite data analysis, covering basic visualization to advanced processing and modeling. The selection of a tool usually depends on the specific analysis needs, the type of satellite data utilized, and the user's familiarity with the tool's interface and capabilities.

7.7 ANALYSIS AND DISCUSSION

The application of remote sensing technologies in agriculture has numerous benefits, such as enhanced crop management, yield forecasting, and disaster management. The range of coverage of different sensors adds prominent options uniquely to a holistic view of agricultural monitoring, with each sensor adding to a holistic view of the agricultural region. The diffusion model is used to preprocess the satellite data, ensuring the optimal output is obtained.

Effective exploitation of high-volume datasets relies on the presence of strong preprocessing methods and good software packages. The continued advancements of these technologies, as well as access to academics and practitioners, are instrumental in ensuring the full potential of remote sensing is achieved in agriculture. Google

TABLE 7.5
Satellite Dataset Tools with Features

Satellite Dataset Tools	Description	Uses/Advantages	References
ENVI	A commercial software used for processing and analyzing geospatial imagery, including satellite data.	• Advanced image processing algorithms • Spectral analysis tools • Integration with GIS workflows	Singh et al. (2022)
ArcGIS	A widely used commercial GIS software suite developed by Esri, offering tools for spatial analysis, mapping, and data management.	• Supports analysis and visualization of satellite imagery • Integration with various data sources • Geoprocessing capabilities.	Basheer et al. (2022)
ERDAS Imagine	ERDAS Imagine is widely recognized for its user-friendly interface and capabilities across various industries, including agriculture, forestry, environmental monitoring, urban planning, and defense, thanks to its strong capabilities in remote sensing and geospatial analysis.	• LULC classification • Change detection • Data import and export • Spatial Analysis • 3D visualization • Geospatial data fusion	Singh and Sethi (2019)
SNAP	A toolbox was developed for the analysis of Sentinel satellite data, including data from Sentinel-1 (radar) and Sentinel-2 (optical).	• Offers tools for data processing • Visualization • Analysis, including algorithms for image classification • Change detection and time-series analysis	De Luca et al. (2022)
GEE	A cloud-based platform for planetary-scale environmental data analysis. It provides a vast collection of satellite imagery and geospatial datasets.	• Allows for analysis of multi-temporal imagery • Machine learning models for classification • Extensive geospatial processing capabilities	Basheer et al. (2022)
QGIS	An open-source GIS platform enabling users to generate, modify, visualize, analyze, and share geospatial data.	• Supports various satellite data formats • Plugins for advanced analysis • Integration with other geospatial tools	García-Álvarez et al. (2022)
GDAL	GDAL is an open-source library developed for processing and converting raster and vector geospatial data formats. It supports numerous file formats and includes utilities for data transformation and processing.	• Format conversion • Georeferencing and projection transformations • Image sub-setting and mosaicking • Command-line interface	Appel et al. (2018)
Sentinel Hub	Sentinel Hub is a cloud-based platform that enables access to and processing of satellite data, particularly from the Sentinel missions. It provides APIs and web interfaces for straightforward data retrieval and analysis.	• On-the-fly data processing • Access to Sentinel, Landsat, and other datasets • Python API for custom workflows • Integration with GIS tools	Dyke et al. (2021)

Earth Engine or QGIS with the SCP plugin would suit as a choice for the inexperienced user, and both are free. Advanced research users may use SNAP, GDAL, or Python to script with the respective libraries. There are ArcGIS and ENVI, which are commercial packages enhanced in functionality. Based on the exact needs, it is possible to select the right tool and apply the necessary preprocessing procedures to the dataset.

Moreover, the discrepancy between one instrument and another, as well as the inability to use the right information at the right time, can be overcome by the interoperability of various instruments and information, making agricultural monitoring more proficient and precise. The comprehensive spectral coverage of remote sensing instruments, excellent quality of pre- and post-processing, and effective software tools provided by them constitute a very strong basis for mapping agricultural land use. Through these tools and instruments, more precise and timely data on agricultural processes is available, which is a significant factor in the sustainability of agriculture and food safety.

7.8 CONCLUSION

This chapter has examined the prospects of remote sensing because it provides possibilities for agricultural land cover classification using satellite data. It highlighted the importance of selecting a suitable dataset based on sensor type, spectral bands, spatial resolution, and temporal resolution. The chapter compared some datasets, such as Landsat and Sentinel, to sensitize the researchers on the most appropriate data to use. Raw satellite imagery preprocessing involves correcting radiometric, atmospheric, and geometric issues, as well as reducing noise, to ensure data accuracy. Capabilities of commonly used processing tools, ENVI and Google Earth Engine, have also been overviewed, enabling the researchers to make informed decisions when classifying agricultural land cover. The satellite data processing can be completed quite successfully with the help of diffusion models and transformers.

Satellite remote sensing has transformative prospects in the agricultural sector, as it provides data-driven information to support crop management, forecast yield, and direct or allocate resources. Remote sensing in global food security will become vital as technology advances in areas such as preprocessing applications and user-friendly interfaces. Future studies should integrate satellite information with additional data, such as weather and soil sensors, to provide a more robust understanding and ensure a sustainable future of food production worldwide.

REFERENCES

Appel, M., Lahn, F., Buytaert, W., & Pebesma, E. (2018). Open and scalable analytics of large Earth observation datasets: From scenes to multidimensional arrays using SciDB and GDAL. *ISPRS Journal of Photogrammetry and Remote Sensing*, *138*, 47–56.

Arici, T., Dikbas, S., & Altunbasak, Y. (2009). A histogram modification framework and its application for image contrast enhancement. *IEEE Transactions on Image Processing*, *18*(9), 1921–1935.

Badiola-Bengoa, A., & Méndez-Zorrilla, A. (2021). A systematic review of the application of camera-based human pose estimation in the field of sport and physical exercise. *Sensors*, *21*(18), 5996.

Banks, A. C., & Mélin, F. (2015). An assessment of cloud masking schemes for satellite ocean colour data of marine optical extremes. *International Journal of Remote Sensing*, *36*(3), 797–821.

Basheer, S., Wang, X., Farooque, A. A., Nawaz, R. A., Liu, K., Adekanmbi, T., & Liu, S. (2022). Comparison of land use land cover classifiers using different satellite imagery and machine learning techniques. *Remote Sensing*, *14*(19), 1–18. https://doi.org/10.3390/rs14194978

De Luca, G., Silva, J., Di Fazio, S., & Modica, G. (2022). Integrated use of Sentinel-1 and Sentinel-2 data and open-source machine learning algorithms for land cover mapping in a Mediterranean region. *European Journal of Remote Sensing*, *55*(1), 52–70. https://doi.org/10.1080/22797254.2021.2018667

Dyke, G., Rosenqvist, A., Killough, B., & Yuan, F. (2021). Intercomparison of Sentinel-1 datasets from Google Earth Engine and the Sinergise Sentinel Hub Card4L Tool. *2021 IEEE International Geoscience and Remote Sensing Symposium IGARSS*, 1796–1799.

García-Álvarez, D., Camacho Olmedo, M. T., Paegelow, M., & Mas, J. F. (2022). *Land Use Cover Datasets and Validation Tools: Validation Practices with QGIS*. Springer International Publishing. https://doi.org/10.1007/978-3-030-90998-7

Gudavalli, C., Goebel, M., Nanjundaswamy, T., Nataraj, L., Chandrasekaran, S., & Manjunath, B. S. (2023). Resampling estimation based rpc metadata verification in satellite imagery. *IEEE Transactions on Information Forensics and Security*, *18*, 3212–3221.

Hudait, M., & Patel, P. P. (2022). Crop-type mapping and acreage estimation in smallholding plots using Sentinel-2 images and machine learning algorithms: Some comparisons. *Egyptian Journal of Remote Sensing and Space Science*, *25*(1), 147–156. https://doi.org/10.1016/j.ejrs.2022.01.004

Jeon, H., Kim, D., Kim, J., Vadivel, S. K. P., Kim, J., Kim, T., & Jeong, S. (2020). Selection of optimal band combination for machine learning-based water body extraction using SAR satellite images. *Journal of the Korean Association of Geographic Information Studies*, *23*(3), 120–131.

Kamusoko, C. (2019). Pre-processing. In *Springer Geography*, 25–66. Springer, Singapore. https://doi.org/10.1007/978-981-13-8012-9_2

Khan, S., Alam, M. A., Ram, N. R. S., Mirza, K., & Chowdary, V. (2020). Noise reduction of time-series satellite data using various de-noising algorithms. *International Journal of Technical Research & Science*, *3*, 55–69.

Pawar, B., Prakash, V., Garg, L., Galdies, C., Buttigieg, S., & Calleja, N. (2023). A review of satellite image analysis tools. *International Conference on Recent Trends in Machine Learning, IOT, Smart Cities & Applications*, 759–769.

Schmitt, M., Hughes, L. H., Qiu, C., & Zhu, X. X. (2019). SEN12MS--A curated dataset of georeferenced multi-spectral sentinel-1/2 imagery for deep learning and data fusion. *ArXiv Preprint ArXiv:1906.07789.*

Sharma, I., & Kaur, K. (2024). Datasets for satellite remote sensing in agriculture: An overview and analysis. *Proceedings of the 2024 3rd Edition of IEEE Delhi Section Flagship Conference, DELCON 2024*, 1–7. IEEE. https://doi.org/10.1109/delcon64804.2024.10866421

Sharma, N., & Chawla, S. (2023). Digital change detection analysis criteria and techniques used for land use and land cover classification in agriculture. *2023 3rd International Conference on Advance Computing and Innovative Technologies in Engineering (ICACITE)*, 331–335. IEEE. https://doi.org/10.1109/icacite57410.2023.10182604

Sharma, N., & Kaur, K. (n.d.). Satellite data and machine learning in agricultural land use: A comprehensive review of LULC classification techniques. *Progressive Computational Intelligence, Information Technology and Networking*, 357–363.

Sharma, N., Singh, S., & Kaur, K. (2025). Optimised DNN-based agricultural land cover mapping using Sentinel-2 and Landsat-8 with Google Earth Engine. *Land*, *14*(8), 1578. https://doi.org/10.3390/land14081578

Singh, G., & Sethi, G. K. (2019). Automatic land cover classification using learning techniques with dynamic features. *International Journal of Innovative Technology and Exploring Engineering*, *8*(8 Special Issue 3), 499–503.

Singh, G., Singh, S., Sethi, G., & Sood, V. (2022). Deep learning in the mapping of agricultural land use using Sentinel-2 Satellite data. *Geographies*, *2*(4), 691–700. https://doi.org/10.3390/geographies2040042

Theres, B. L., & Selvakumar, R. (2022). Comparison of landuse/landcover classifiers for monitoring urban dynamics using a spatially enhanced Landsat dataset. *Environmental Earth Sciences*, *81*(5), 142.

Van Leeuwen, B., Tobak, Z., & Kovács, F. (2020). Machine learning techniques for land use/land cover classification of medium resolution optical satellite imagery, focusing on temporary inundated areas. *Journal of Environmental Geography*, *13*(1–2), 43–52.

Yang, H., Wang, Q., Zhao, W., Tong, X., & Atkinson, P. M. (2022). Reconstruction of a global 9 km, 8-day SMAP surface soil moisture dataset during 2015--2020 by spatiotemporal fusion. *Journal of Remote Sensing*. https://doi.org/10.34133/2022/9871246

Zaki, N. H. M., Chong, W. S., Muslim, A. M., Reba, M. N. M., & Hossain, M. S. (2022). Assessing optimal UAV-data pre-processing workflows for quality ortho-image generation to support coral reef mapping. *Geocarto International*, *37*(25), 10556–10580.

Zhao, M., Heinsch, F. A., Nemani, R. R., & Running, S. W. (2005). Improvements of the MODIS terrestrial gross and net primary production global data set. *Remote Sensing of Environment*, *95*(2), 164–176.

8 Application of Remote Sensing and Artificial Intelligence for Environmental Monitoring

Santanu Mallik, Ruma Debnath, and Sharbari Deb

8.1 INTRODUCTION

Environmental monitoring involves systematically collecting, processing, and interpreting ecological data to assess spatiotemporal changes within geospatial datasets. Traditionally, ecological information has been obtained through ground-based surveys combined with manual data collection. However, such methods are labour-intensive and time-consuming and provide limited spatial coverage. RS technologies (satellite imagery, aerial vehicles, UAVs) facilitate persistent and large-scale environmental monitoring. Datasets from numerous satellite missions are readily available and used in various applications such as vegetative changes, hydrological changes, land surface temperature, and urban expansion. However, due to the magnitude, RS data are complex and require robust processing architectures. In this case, the extraction of information from multi-source data is preferred using Artificial Intelligence (AI), Machine Learning (ML), and Deep Learning (DL) to increase the aspects of automation, sensitivity, and coverage of environmental monitoring. ML techniques commonly used in ecological applications include Support Vector Machines (SVMs), Random Forests (RF), Gradient Boosting Models/Extreme Gradient Boosting (GBM/XGBoost), and Convolutional Neural Networks (CNNs): namely, as classifiers, regressors, feature extractors, and predictive models. Talukdar et al. (2023) report that RF and GBM models and ALE and Shapley Additive exPlanations (SHAP) interpretation approaches were recorded to produce the best Water Quality Index (WQI) modelling outcomes in a Lake Loktak, India case study. This is because the XGBoost RF GPR ensemble ML models demonstrated their efficiency in minimising prediction error and uncertainty when evaluating the water quality of Cork Harbour in Ireland, as Uddin et al. (2022) argued. The compatibility of SVM and optimised RF paradigms with the validity of Receiver Operating Characteristic (ROC) curves assisted Naghibi et al. (2017) and Sajedi-Hosseini et al. (2018) in

 DOI: 10.1201/9781003616207-10

accurately demarcating the potential groundwater areas. The CNN models reported high performance in agricultural surveillance, where spectral data were utilised to translate soil properties directly (Ng et al., 2019). They retrieved Landsat and SMAP high-resolution soil moisture using XGBoost and RF models (Zhang et al., 2022).

Additionally, some models implemented on multi-source inputs within Google Earth Engine (GEE) made global 50 m-resolution soil moisture predictions (Greifeneder et al., 2021), showcasing the use of AI at large scales for environmental data assimilation. Karandish and Šimůnek (2016) combined HYDRUS-2D using SVM and ANFIS to predict soil water content under scarce input conditions for irrigation efficiency integrated modelling. AI approaches have also been crucial to air pollution monitoring, particularly in forecasting and estimating PM2.5 concentrations. Joharestani et al. (2019) employed RF, XGBoost, and DL models from satellite-based AOD and meteorological inputs to effectively forecast PM2.5 in Tehran, achieving R^2 values up to 0.81. Liu and Dong (2020) proposed a hybrid ensemble DL structure that achieved an extremely high correlation ($r = 0.9917$) when forecasting PM2.5 for Shanghai. Luo et al. (2020) demonstrated that ML models such as Gaussian Process Regression (GPR), Artificial Neural Network (ANN), and SVM could effectively forecast PM2.5 and PM10 across cities of Bangladesh from meteorological and pollutant concentration inputs. A study by Deters et al. (2017) demonstrated that boosted trees and neural networks produced robust prediction results regarding the level of PM2.5 in the air when urban conditions were considered, even using raw meteorological data. Other important uses are land surface dynamics and urban growth modelling, where AI and remote sensing (RS) are increasingly coming together. Avdan and Jovanovska (2016) conducted research. They developed an automated algorithm that utilises Landsat 8 bands to map the Land Surface Temperature (LST) with a high degree of accuracy, as compared to that provided by the meteorological stations. Wang et al. (2015) used Normalised Difference Impervious Index (NDII) to achieve an incredible accuracy of 91.4% and Kappa figures.

In the same way, Yang and Liu (2005) developed a landscape imperviousness index to measure urban extent from Landsat and Digital Orthophoto Quarter Quadrangle (DOQQ) datasets and conducted a thorough analysis of urban growth. Anand et al. (2018) used Land Change Modeller and Markov and MLP for long-term land use forecasting in the Ganga basin, predicting spatial change up to 2090 under different scenarios. Moreover, Balha et al. (2021) compared ML algorithms, including RF, SVM, ANN, and pixel/object-based classification for LULC mapping of Delhi and established that RF and object-based analysis yielded the best classification accuracy (~90%).

These advancements are aligned with the surge in global academic interest in RS and Geographical Information System (GIS) applications. As evidenced in the bibliometric analysis, publications on environmental monitoring using RS and GIS have consistently risen from 2004 to 2024, with a sharp spike post-2020 (Figure 8.1a). The co-word network analysis underscores the growing research synergy among critical topics like water quality, climate change, soil erosion, urbanisation, and ML, highlighting the interdisciplinary growth and prominence of AI-driven environmental monitoring research (Figure 8.1b). Furthermore, key bibliometric metrics indicate significant scholarly output and international collaboration, with an annual growth

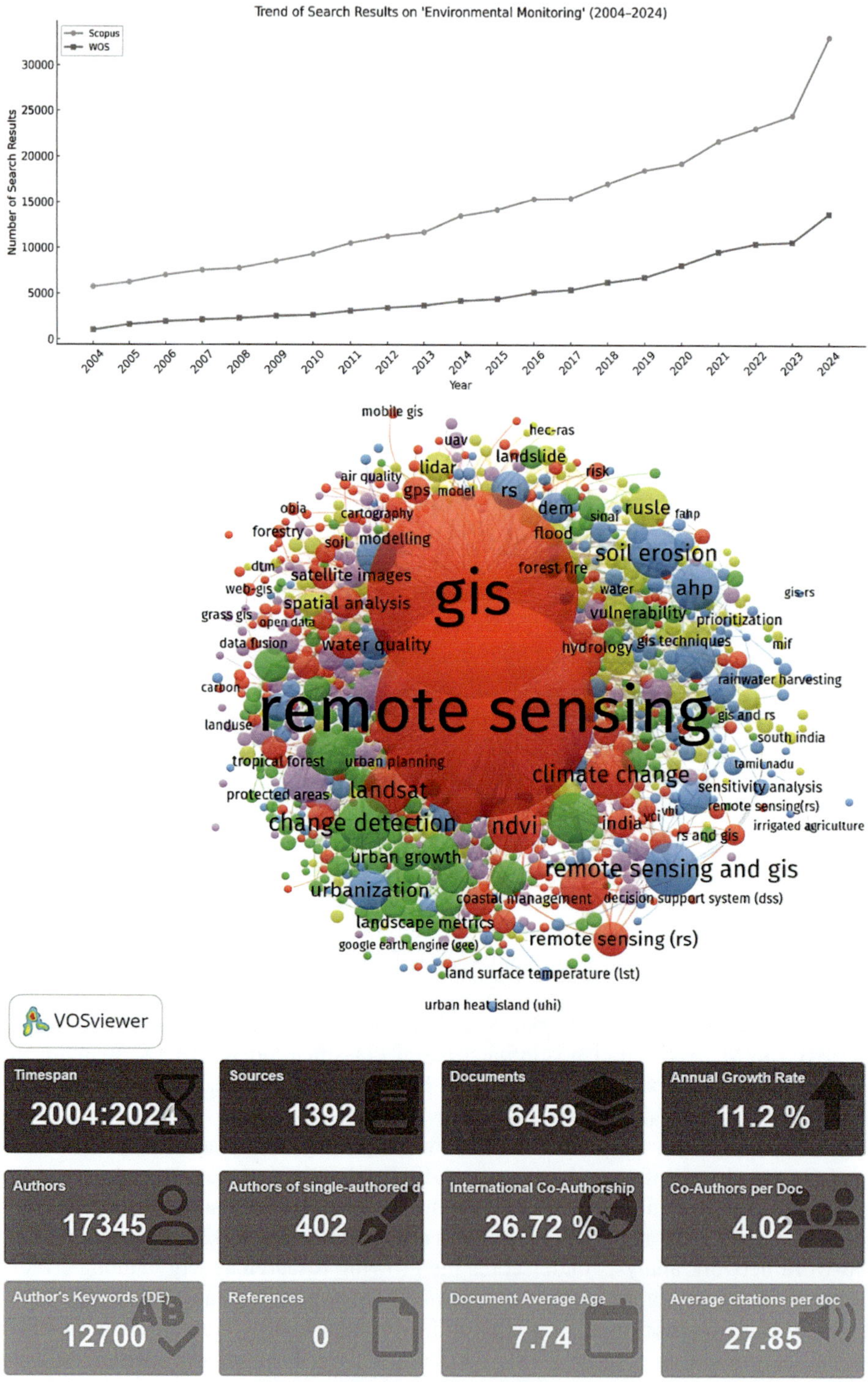

FIGURE 8.1 (a) Trend of publications, (b) co-word network visualisation, (c) summary of bibliometric analysis.

rate of 11.2% in relevant publications, reflecting growing global reliance on AI-integrated RS approaches (Figure 8.1c).

This chapter, therefore, provides a structured exploration of the role of AI in RS-based environmental monitoring. It includes theoretical bases, algorithmic operations flow, and practical uses in five thematic domains: Air Pollution Assessment, Water Quality Evaluation and Treatment, Land Surface Dynamics and Urban Growth Analysis, Soil and Agricultural Monitoring, and Disaster Management. For each domain, we assess key input parameters, AI models applied to predictive outcomes, and validation metrics, all supported by empirical case studies. In doing so, the chapter showcases successful implementations. It reflects on critical limitations such as data quality, transferability, and ethical AI concerns, thus offering a holistic perspective for researchers, policymakers, and practitioners engaged in a sustainable environment.

8.2 FUNDAMENTALS OF REMOTE SENSING

RS is the science of remotely collecting information about the Earth's surface, typically by detecting reflected or emitted electromagnetic radiation. As electromagnetic waves pass through the atmosphere, they are scattered, absorbed, and transmitted before interacting with surface materials like water, soil, or vegetation. The energy then reflected or emitted from these surfaces is captured by sensors, either passive (e.g., optical, thermal) or active (e.g., Synthetic Aperture Radar), on satellites or aircraft. The process also includes the transmission, reception, preprocessing, and interpretation of data to extract meaningful insights about Earth's systems (Jasmin & Mallikarjuna, 2011). The central concept in RS is the electromagnetic spectrum, where sensors operate across visible, infrared, and microwave bands. The capability to distinguish Earth features depends on the sensor's spatial, spectral, radiometric, and temporal resolutions. Figure 8.2 visually compares single-band, multispectral, and hyperspectral imaging, highlighting the increasing detail and classification potential with higher spectral resolution.

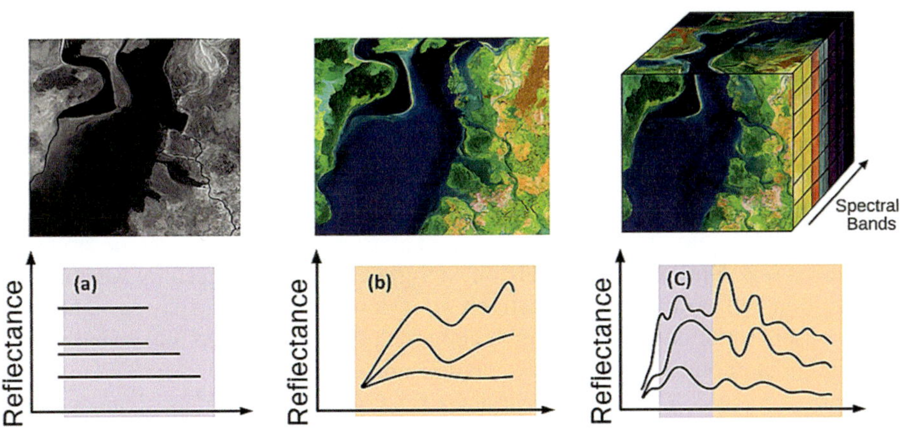

FIGURE 8.2 Types of spectral imaging.

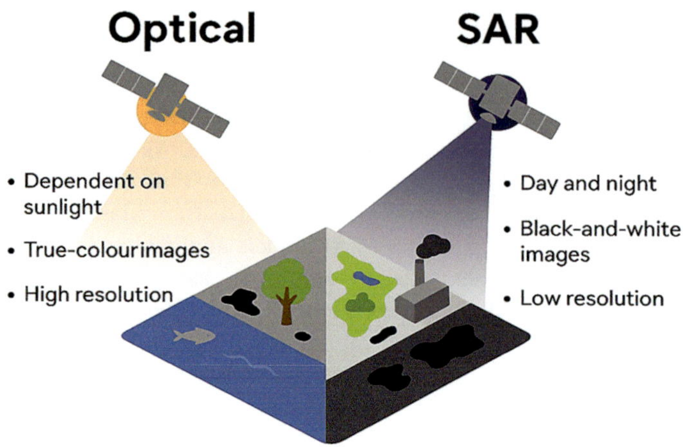

FIGURE 8.3 Remote sensing platforms.

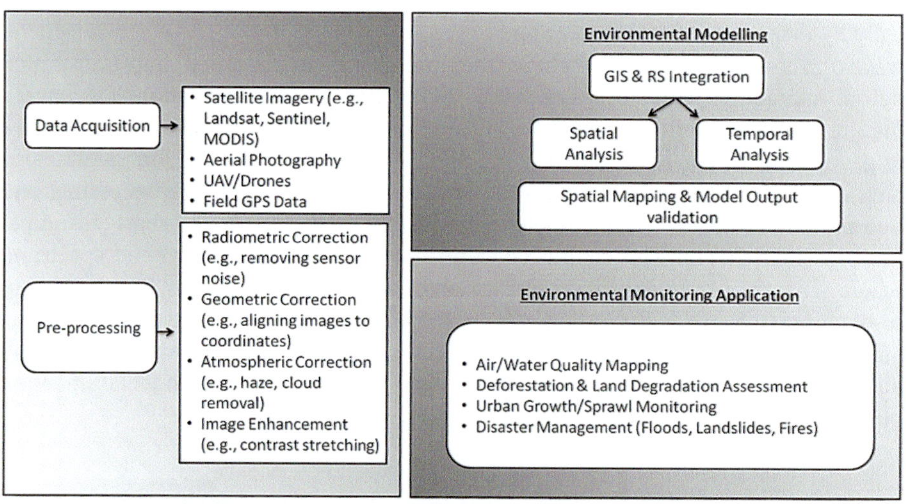

FIGURE 8.4 Conceptual workflow for environmental modelling using GIS & RS.

Hyperspectral sensors, which can capture hundreds of narrow bands, are especially valuable for precise material differentiation, including subtle variations in soil, crops, or urban surfaces (Weng, 2011). Moreover, Figure 8.3 shows the various broad categories of RS into optical and radar-based platforms, each with specific strengths and limitations. Optical systems rely on sunlight and are influenced by weather and illumination, while radar systems actively emit signals and operate day or night, even in the presence of clouds. RS in combination with GIS has evolved into an indispensable tool in environmental assessment and resource management.

Figure 8.4 shows a structured overview of the key processes of amalgamating RS and GIS for environmental modelling and monitoring applications. GIS complements RS by providing powerful analytical tools and capabilities for handling large spatial datasets. GIS performs spatial-temporal analysis, enabling users to visualise, analyse, and interpret environmental changes using RS datasets. This synergy helps deliver near-real-time, spatially continuous information that supports policy-making in disaster preparedness, climate adaptation, and sustainable development planning.

8.2.1 Machine Learning versus Deep Learning

ML and DL are two subsets of AI that are frequently used for environmental monitoring using RS data. ML comprises algorithms that identify patterns and relationships within datasets to make predictions or classifications without explicit programming. Some of the popular ML models, such as RF, SVM, and XGBoost, are easy to interpret and robust. These models are typically grounded on manual feature extraction, whereby domain expertise is leveraged to preprocess and select features. Haggerty et al. (2023) observed that ML models performed satisfactorily in groundwater quality modelling by capturing complex, non-linear relationships between parameters such as pH, nitrate, and electrical conductivity. Chen et al. (2023) further observed that ML approaches are beneficial when the dataset is limited or when interpretability matters, mainly when predicting pollution.

DL, however, automatically extracts features through multilayer neural networks such as CNN, Autoencoders (AE), and Deep Belief Networks (DBN). Unstructured, high-dimensional data such as satellite imagery, hyperspectral imagery, and sensor streams are processed by DL models. Zhu et al. (2017) discussed why CNNs perform better than traditional ML models in challenging RS scenarios; that is to say, CNNs enable machines to learn mid- and high-level features from imagery automatically. Bhagat et al. (2022) also noted that DL models minimise preprocessing requirements and achieve high accuracy over highly non-linear data. The problem is that DL models are computationally expensive, requiring massive, labelled datasets and GPUs. So, DL is always preferred in modules where the data is big and the task is complex, spatial, or vision pattern-based, whereas ML is cheaper to use, data is less, and transparency and very low computation are required. Figure 8.5 displays the workflow comparison for the ML and DL models' approaches.

Understanding the differences between the structural and functional aspects of ML and DL is crucial for selecting the most suitable model for evaluating environmental monitors. Figure 8.6 depicts the skeleton design for ML-based and DL-based models, which are included in this article. Further details about all the models will be discussed in the subsequent part of the article.

8.2.1.1 Random Forests

RF is an ensemble learning algorithm that provides more accurate predictions. It uses multiple decision trees, takes the output of the trees, and averages it. Initially introduced by Breiman (2001), RF achieves this by randomly sampling subsets (Bootstrap) and randomly selecting a subset of features per split, ensuring diversity amongst trees and preventing overfitting. This approach offers resilience in dealing with high-dimensional data and non-linearity and is appropriate in land covering separation, groundwater potential assessment, and disaster prediction (Breiman, 2001; Hoyos et al., 2015). Compared to a single decision tree model, RF effectively

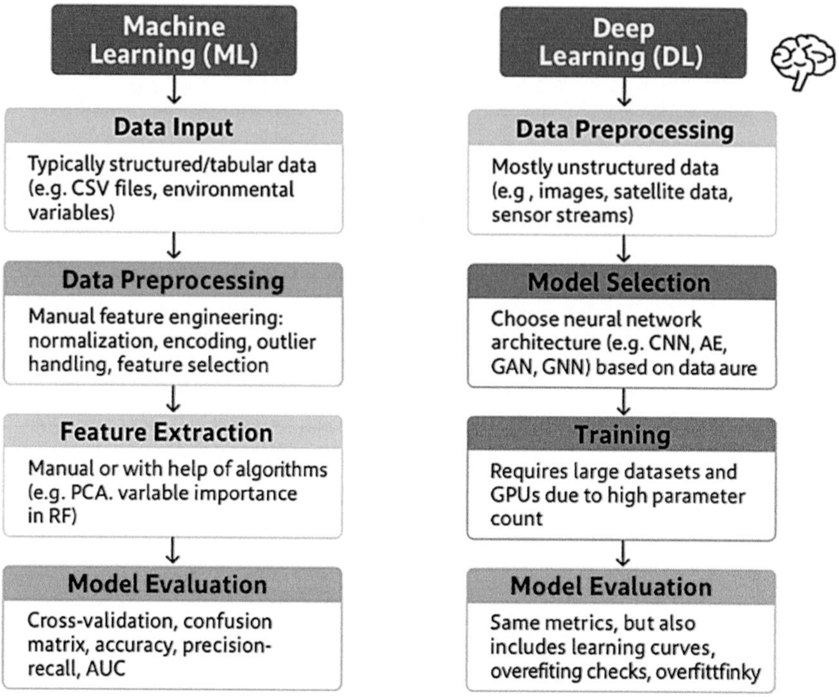

FIGURE 8.5 Workflow comparison between machine learning and deep learning approaches for environmental monitoring.

improves the ability to generalise by averaging predictions from ensembles of trees, decreasing variance, and enhancing stability. It also provides a built-in method for computing variable importance; thus, interpretability and variable selection have been simplified when auditing an environment (Golkarian et al., 2018; Naghibi et al., 2017).

RF requires hyperparameter tuning so that its performance can be maximised. Parameters such as the number of trees (n_estimators), the number of features to consider at each node (max_features), and the maximum depth of the tree (max_depth) are among the most important. Whether particular values of these parameters are correctly adjusted, it will be the optimal compromise between model complexity and computing expense, resulting in greater accuracy and reduced overfitting. Missing values and noisy features have no bearing on RF operations, and this aspect renders it preferable to conventional ML techniques. To conclude, RF flexibility, robustness, and explainability bring to the fore the significance of RF regarding large-scale environmental surveillance and decision-making.

8.2.1.2 Support Vector Machines

SVMs are strong supervised learning techniques, used for classification and regression, especially when the data are in high dimensions. SVMs are implemented to determine the best-separating hyperplane that will be used to segregate various

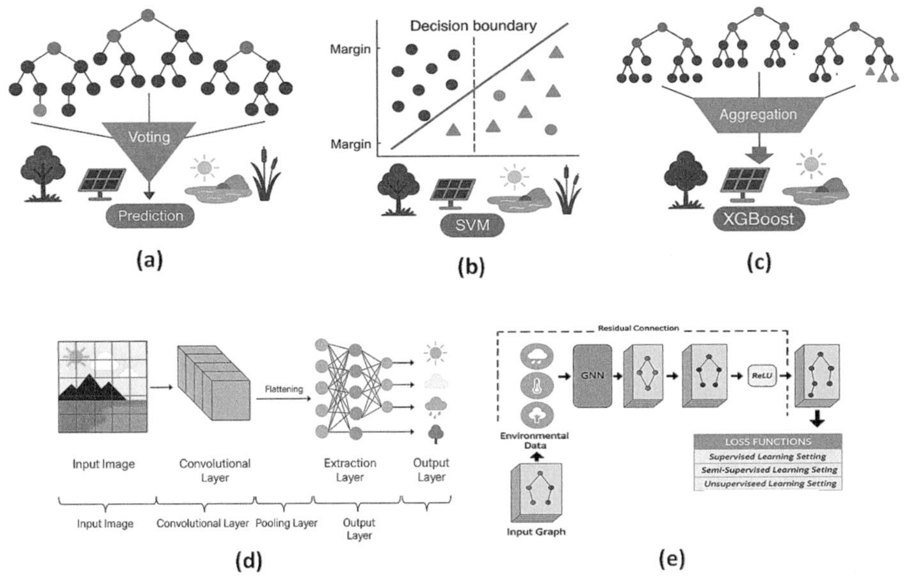

FIGURE 8.6 Architectures of advanced AI models (a) Random Forest ensemble learning, (b) Support Vector Machine (SVM) classification, (c) XGBoost gradient boosting, (d) Convolutional Neural Network (CNN) for image analysis, and (e) Graph Neural Network (GNN).

classes in the feature space maximally and at the same time maximise the class-margin between them (Marjanović et al., 2011). The effective processing of small and imbalanced sets and fast detection of high accuracy are among the virtues of SVMs (Arabgol et al., 2015). The working nature of SVM is based on the Structural Risk Minimisation (SRM) scheme, that is, it minimises the generalisation errors rather than Empirical Risk Minimisation (ERM) (Arabgol et al., 2015). When the data can be linearly separable, it optimises the class-margin of SVM to find the best-separating hyperplane. However, the kernel trick is made for the non-linearly separable data, which maps the data into a highly dimensional space where a linear discriminator can be found (Marjanović et al., 2011). Radial Basis Function (RBF), polynomial, linear, and sigmoid functions are used in different RS data. The support vectors, the nearest data points to the hyperplane, are responsible for defining the decision boundary, and thus the stability and computational efficiency of the model (Sajedi-Hosseini et al., 2018).

SVMs are preferred compared to other classifiers due to their high accuracy, ability to generalise from a few training samples, and efficiency when operating in high-dimensional spaces, particularly when handling multispectral and hyperspectral RS applications (Marjanović et al., 2011). Model performance is greatly affected by hyperparameter selection. Some key parameters that could be used are C (regularisation parameter), which is the trade-off between minimising the mistake and maximising the margin, and γ (gamma parameter) in RBF kernels, which is the degree of effect that a training sample has on the model (Sajedi-Hosseini et al., 2018). Suitable hyper-tuning of the hyperparameters using methods similar to grid search and cross-validation leads to much improved performance of the SVM model when working in environmental monitoring applications.

8.2.1.3 Extreme Gradient Boosting

XGBoost is an optimal variant of the gradient boosting technique, which operates on the principle of ensemble learning. XGBoost constructs decision trees so that each next tree corrects the mistakes of the previous one and has regularisers to prevent overfitting. XGBoost performs both regression and classification, and is particularly effective at dealing with bulky, many original disparate, and possibly sparse data sets, which are commonly seen in environmental applications. It is the XGBoost that differs from others because of its ability to handle missing values, parallel processing and customer object function, and all of them are reasons for its increasing popularity with environmental applications covering landslide susceptibility mapping (Zhang et al., 2023), water quality modelling (Uddin et al., 2022) and RS soil moisture estimation (Zhang et al., 2022).

XGBoost uses a 'regularised objective' that combines the loss and a penalty to ensure the model balances accuracy and simplicity. It employs second-order derivatives (both gradient and Hessian) to steer the learning procedure, accelerating the convergence and the convergence stability of the model. The algorithm employs significant hyperparameters such as learning rate (eta), the maximum depth of the tree (max_depth), the minimal child weight (min_child_weight), and subsampling ratios (subsample, colsample_bytree), all of which affect the performance and capacity of the model to generalise (Niazkar et al., 2024). Explanation is one of the benefits of XGBoost when used in conjunction with methods such as SHAP, and this has been used to elucidate the spatial heterogeneity of environmental drivers of susceptibility (Zhang et al., 2023).

8.2.1.4 Convolutional Neural Networks

CNNs are one of the most effective architectures of DL in RS analysis of satellite and hyperspectral images. CNN aimed to learn and extract image data automatically by propagating an image through layers of convolution filters, activation, pooling, and fully connected layers. Compared to other highly manual feature-based ML models, CNN can recognise sophisticated spatial features, for example, urban structures, water bodies, and acceptable spectral variations in land cover. Therefore, they can be rather sturdy to use in flood alarm mapping (Vincent et al., 2023), groundwater potential mapping (Panahi et al., 2020), and the computation of soil characteristics (Ng et al., 2019). Besides this, CNNs are considered to be a versatile method as well. They apply to various types of input, such as 2D and 3D, and are efficient in spatial and spectral properties of hyperspectral imagery (Yang et al., 2018).

A conventional CNN optimises on training captured data by repeatedly slotting them on convolution layers where filters are utilised to recognise particular patterns, that is, edges and textures. The next ones are non-linear activation functions, such as ReLU, to ensure the model learns complex relations. The pooling layer reduces the dimensionality of the data, thereby reducing performance while preserving the characteristics. The final part consists of fully connected layers, where all the extracted features are utilised for classification or regression. Moreover, CNN performance may also be fine-tuned with the help of hyperparameter tuning, and the approaches to the hyperparameter tuning are grid search, random search, or by sets of kernel dimensions, learning rate, or the number of layers modification with the help of Bayesian optimisation or evolution strategy (Vincent et al., 2023). Therefore, CNNs are economical in dealing with spatial analysis of the problems of the contemporary environment.

8.2.1.5 Graph Neural Networks

GNNs are an advanced DL technique that has developed specialisation in processing data that is already graphically represented and augmented with nodes and edges to denote relations amongst spatial or functional objects. GNNs can describe asymmetrical and non-linear relations typical of geophysical and satellite-related applications compared to the existing problems of conveying data in a linear structure, namely image or time series information, based on DL models. Consider, for example, when sensing data or samples are spread over space, one obtains graphs wherein the information of its neighbours evaluates the information at each node. GNNs apply the message passing to the calculation, namely, connectivity is included through each iteration, with node features being updated according to the features of local nodes and the topology of the graph (Zhou et al., 2020). This renders GNNs especially helpful in every scope, including land cover classification, spatial interpolations, and hydrological process models across adjoining geographical boundaries.

In addition, spatial and temporal properties can be learnt with the newest iteration of Graph Convolutional Recurrent Neural Networks (GCRNNs), and, in turn, they apply to spatiotemporal forecasting models like groundwater level forecasting and water demand forecasting (Taccari et al., 2024; Zanfei et al., 2022). Modifiers include alteration of such critical hyperparameters as the number of layers in a graph, the size of a neighbourhood (K-hop), the selection of an aggregation function such as averaging or attention, and the learning rate. GNNs are even more preferable to ordinary ML/DL models in terms of the ability to analyse scarce and missing data in some contexts, explainability, and scale to support the more complex applications of RS (Wu et al., 2022; W. Zhang et al., 2023). As advancements in AI continue in the geospatial area, GNNs provide the next level of integration of the real spatial structure within environmental prediction.

8.3 AI TECHNIQUES FOR REMOTE SENSING DATA

Integrating AI and RS has transformed the processing and understanding of environmental monitoring data. Conventional analytical methods often face challenges while handling spatial-temporal variability, sensor heterogeneity (e.g., multispectral, hyperspectral, LiDAR, SAR), and high-dimensional RS data. In contrast, data-driven solutions offered by AI approaches, especially ML and DL, can extract useful information from complex inputs. They enable automated learning of features, pattern recognition, and predictive modelling, thereby enhancing the precision, processing capacity, and responsiveness of environmental monitoring through RS. Figure 8.7 shows a workflow for ecological monitoring and application using GIS and RS.

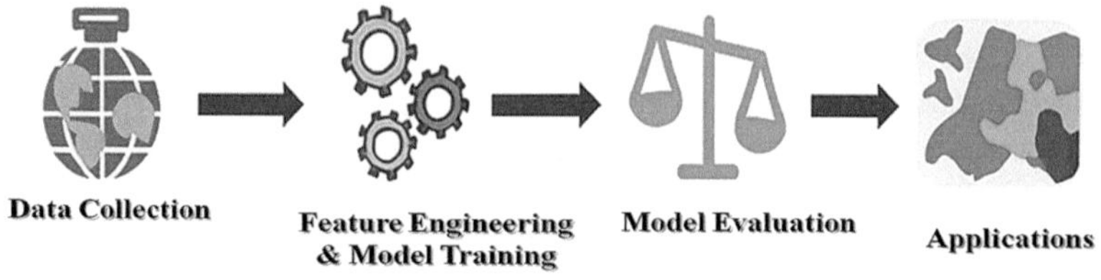

FIGURE 8.7 Generalised workflow for environmental monitoring.

A single ML model may not perform well for all applications; sometimes, combining models or an ensemble model may work well. Depending on the multidimensional data, select relevant features and conduct a performance comparison. For instance, Zhang et al. (2022) suggest that ensemble learning and ancillary RS environmental data perform well when mapping physical variables such as soil moisture. DL further enhances the possibilities of AI in RS as it allows the training of the model from the raw data without involving any preprocessing steps. CNNs, a primary architecture of DL, are specifically advantageous for RS imagery since they directly identify the hierarchy and texture at the pixel level. This patching is essential when utilising detailed images or intricate data sets, such as hyperspectral or Synthetic Aperture Radar (SAR) imagery.

As established by Zhu et al. (2017), DL models are beneficial in object detection, image segmentation, and multi-temporal change, in which, due to the complexity of the related features, it is challenging to design the features manually. In the case of CNN-based architectures, it has been ascertained that they can integrate multi-source data, thereby enabling the integrated consideration of environmental data available from optical, radar, and supplementary sensors (Panahi et al., 2020). Due to the development of optimisation methods that include AutoML and evolutionary algorithms, DL is incorporable into solution fields, including resource-limited modes (Vincent et al., 2023). Thus, AI has a crucial role in RS since it provides solutions to overcome the shortcomings of manual interpretation of RS images. AI also helps with real-time large-scale environmental monitoring. Unlike ML, which is more transparent and delivers better performance for precise, labelled data, DL can handle vast, unstructured, diverse RS data.

8.4 APPLICATIONS OF AI IN ENVIRONMENTAL MONITORING

Environmental monitoring is a critical process for addressing and controlling environmental issues such as climate change, pollution, loss of ecosystems, and natural disasters. In the past, it used observations from the ground, satellite data, and inference from experts to gather, represent, and analyse big data on the environment. ML and DL, as subcategories of AI, have proven to be at the forefront of environmental monitoring. These techniques provide the ability to automate feature extraction, classification, and regression, as well as pattern discovery, uncertainty assessment, and improved precision at various spatiotemporal scales. Analysing satellite imagery, spectral indices, and other atmospheric and topography layers, combined with other datasets, and AI models, has become critical to developing early warning, resource management, and dynamic ecosystem assessment. Moreover, their integration with platforms like GEE and cloud-based computing infrastructures has made near-real-time environmental intelligence more accessible than ever.

This section comprehensively analyses AI applications across key environmental domains, including air quality forecasting, water quality assessment, land-use change detection, agricultural monitoring, and disaster risk modelling. As summarised in Table 8.1, these applications reveal the breadth of AI model deployment, ranging from traditional algorithms like RF and SVM to hybrid ensembles and deep CNN architectures, and demonstrate their effectiveness in enhancing model accuracy, explainability, and operational readiness for sustainable environmental decision-making.

The key observation from the compiled application table is the diversity of AI model architectures from traditional ML models (e.g., SVM, RF, XGBoost) to DL approaches (e.g., CNNs, MLPs) tailored according to data complexity, spatial resolution, and prediction objectives. For instance, tree-based ensemble models like RF and XGBoost dominate water quality and flood susceptibility tasks due to their robustness, low computational cost, and suitability for tabular or mixed-type input datasets. On the other hand, CNNs are the preferred model for image-based tasks, such as soil classification, LST retrieval, and gully erosion detection, where spatial pattern recognition is crucial.

The table also reveals that model performance highly depends on the quality and integration of multi-source data. Many high-performing models, particularly in air quality and soil monitoring, leverage satellite-derived indices (e.g., AOD, NDVI, NDMI), meteorological parameters, and temporal trends to improve prediction granularity. Many studies used hybrid frameworks, combining classical models with optimisation techniques (e.g., SHAP for feature importance or AutoML for tuning), which consistently outperformed standalone methods. Regarding evaluation metrics, the preference for R^2, RMSE, AUC, and Kappa statistics shows a consistent effort across studies to validate model outputs rigorously. However, some works went further to include measures of uncertainty and model interpretability, such as SHAP and ALE, which are crucial in policy and decision-making, where accuracy is not the only thing that matters. In summary, the table demonstrates that AI techniques have emerged beyond the sphere of innovative pilots' projects and have become routine for environmental surveillance. The most relevant use cases include integrating first, domain knowledge; second, RS skills; and third, optimal AI architectures, which bring forward prospects for more innovative and explainable environmental management methods on a large scale.

8.5 CHALLENGES, FUTURE SCOPE, AND CONCLUSION

Several hurdles still must be overcome regarding using AI with RS for environmental purposes. RS data primarily originates from satellites, drones, or sensor networks, which are often constrained by spatial resolution, temporal frequency, or geographic coverage, particularly in under-monitored regions. Problems such as clouds, sensor faults, disturbances in weather conditions, and variation in format make these problems more complex, influencing the results of environmental studies and

TABLE 8.1
Applications in Environmental Monitoring Across Diverse Domains

Application Area	Study Area	Input Data	Model Used	Output	Accuracy assessment	Ref
Water Quality Evaluation and Treatment	Khalkhal region, Ardebil Province, Iran	DEM-derived factors, DR, DD, FD, LU, LT	SVM, RF, RFGA	Groundwater Potential Mapping	ROC Curve (AUC)	Naghibi et al. (2017)
Water Quality Evaluation and Treatment	Lake Loktak, Manipur, India	Water quality parameters	RF, GBM, DNN	WQI Prediction and Interpretation	R^2, RMSE, SHAP, PDP, ALE	Talukdar et al. (2023)
Water Quality Evaluation and Treatment	NA	Dye conc, adsorbent dose, pH, contact time	RF, XGB	Dye Removal Efficiency Prediction	RMSE, R^2, MSE, MAE	Shukla et al. (2024)
Water Quality Evaluation and Treatment	Cork Harbour, Ireland	Water quality parameters	DNN, SVM, GPR, KNN, LR, RF, XGB	Predicted Coastal Water Quality Index (WQI)	RMSE, MAE, MSE, MAPE, AIC, BIC, NSE, MEF, ASRE, Z-statistics	Uddin et al. (2022)
Air Pollution	Tehran, Iran	PM2.5, Meteorological Data, AOD (3 km, 10 km), Spatial Data	RF, XGB	PM2.5 Concentration Prediction	R2, MAE, RMSE	Joharestani et al. (2019)
Air Pollution	Shanghai, China	Hourly PM2.5, Meteorological Data, Temporal Series	Hybrid DL (DBN, LSTM, MLP)	PM2.5 Forecasting	MAE, MAPE, RMSE, r	Liu and Dong, (2020)
Air Pollution	USA (Urban)	PM2.5, PM10, wind speed (m/s), wind direction (degrees), precipitation levels (mm)	BT, L-SVM, NN	PM forecast and regression analysis for extreme climate condition	Performance metrics (RMSE, MAE, and MAPE)	Deters et al. (2017)
Air Pollution	Dhaka, Bangladesh	PM2.5, PM10, NOX, SO2, CO, O3, meteorological variables	L-SVM, M-SVM, RF, ANN, PHOPHET	Predicted concentrations of PM2.5 and PM10	R^2, RMSE, and MAE metrics	Shahriar et al. (2020)

Soil and Agricultural Monitoring	Hubei Province, China	Soil samples, MODIS reflectance and vegetation index, climate, terrain	GBRT, RF, DT, BDT	SOM content maps (2000–2017)	R^2, RMSE, ME, LCCC	Chen et al. (2019)
Soil and Agricultural Monitoring	Tiben Plateau	Landsat 8 data, 1,154 stations from the international soil moisture network, SMAP, ERA5, auxiliary datasets	XGB, RF	Soil moisture at 30m resolution	RMSE, R, 10-fold CV	Zhang et al. (2022)
Soil and Agricultural Monitoring	Not Specified	Spectral data from visible-near-infrared (vis-NIR) and mid-infrared (MIR) spectrometers, soil property	CNN, PLSR, CRTM,	Multiple Soil Properties	R^2, RMSE	Ng et al. (2019)
Soil and Agricultural Monitoring	Northern Iran	Atmospheric parameters, Crop factors, crop stress factors, soil properties, hydraulic parameters	HYDRUS-2D, MLR, ANFIS, SVM	Time series simulations of soil Water content (SWC) under different irrigation strategies	RMSE, MBE	Karandish and Šimůnek (2016)
Disaster Management	Kerala, India	DEM-derived features, R, DR, geology, soil, LU, NDVI, SPI, LT	AdaBoost, RF, GBM, and XGBoost	Flood susceptibility mapping	Accuracy, AUC, precision, recall, loss, kappa	Vincent et al. (2023)
Disaster Management	BâscaChiojdului River Basin, Romania	DEM-derived factor, LULC, ST, LT, SPI,	CART, RF, XGB, BRT	Flash flood susceptibility mapping	AUC	Abedi et al. (2022)
Disaster Management	Fruška Gora Mountain, Serbia	DEM-derived factor, LT, distance from fault, distance from geological boundary, NDVI	SVM, DT, LR, AHP	Landslide susceptibility mapping	AUC, RUC	Marjanović et al. (2011)

(Continued)

TABLE 8.1 (Continued)
Applications in Environmental Monitoring Across Diverse Domains

Application Area	Study Area	Input Data	Model Used	Output	Accuracy assessment	Ref
Disaster Management	Tongnan, Wushan, Fengjie counties, China	DEM-derived factors, DRD, DF SPI, DR, LT, R, LULC, NDVI, historical landslide records	SHAP-XGBoost	Landslide susceptibility map	AUC, RUC	Zhang et al. (2023)
Land Surface Dynamics and Urban Growth	Ontario & Quebec, Canada	LANDSAT 8 (Bands 4, 5, 10)	Algorithm in ERDAS IMAGINE 2014	Land Surface Temperature (LST)	Compared with meteorological station data	Avdan and Jovanovska (2016)
Land Surface Dynamics and Urban Growth	Pensacola, Florida, USA	Landsat TM/ETM+, High-Res DOQQ imagery	Multivariate Statistical Analysis	Landscape Imperviousness Index	Compared with DOQQ-based ground reference	Yang and Liu (2005)
Land Surface Dynamics and Urban Growth	Nanjing, China	Landsat TM (Red & TIR bands)	Normalised Difference Impervious Index (NDII)	Impervious Surface Area	Accuracy assessment with validation points	Wang et al. (2015)
Land Surface Dynamics and Urban Growth	Ganga River Basin, India	LULC maps (1985–2005), SL, EL, roads, water bodies	Land Change Modeler (LCM) with Markov + MLP	Predicted LULC for 2030, 2060, and 2090	Validation against 2005 LULC maps	Anand et al. (2018)

DR – Distance from Rivers, DD – Drainage Density, FD – Fault Density SL – Slope, EL – Elevation, LU – Land Use, LT – Lithology, RFGA – Genetic Algorithm Optimised Random Forest, ROC – Receiver Operating Characteristic, AUC – Area Under the Curve, TWI – Topographic Wetness Index, ST – Soil Type, GBM – Gradient Boosting Machine, DNN – Deep Neural Network, GPR – Gaussian Process Regression, KNN – K-Nearest Neighbours, LR – Logistic Regression, DBN – Deep Belief Network, LSTM – Long Short-Term Memory Network, MLP – Multilayer Perceptron, L-SVM – Linear-Support Vector Machine, M-SVM – Medium Gaussian-Support Vector Machine, ANN – Artificial Neural Network, TS – Time Series Model, BT – Boosted Trees, NN – Neural Network, GBRT – Gradient Boosting Regression Trees, LCCC – Lin's Concordance Correlation Coefficient, SMAP – Soil Moisture Active Passive, ERA5 – ECMWF Reanalysis v5, PLSR – Partial Least Squares Regression, CRTM – Cubist Regression Tree Model, MLR – Multiple Linear Regressions, ANFIS – Adaptive Neuro-Fuzzy Inference Systems, MBE – Mean Bias Errors, R – Rainfall, NDVI – Normalised Difference Vegetation Index, SPI – Stream Power Index, DT – Decision Trees, LR – Logistic Regression, AHP – Analytical Hierarchy Process, DOQQ – Digital Orthophoto Quarter Quadrangle, DRD – Distance to Roads, DF – Distance to faults.

forecasting (AghaKouchak et al., 2015). Additionally, data acquisition from satellite, aerial, and ground data sources poses challenges due to differences in resolution, format, and calibration, which can hinder the formation of effective datasets that support improved AI analysis (Janga et al., 2023). In the future, the application of AI for RS is expected to receive a positive prospect based on better computational capacity, enhanced models of AI, and a new way of integrating extensive data sets. The usage of bio-inspired AI algorithms and DL models, such as CNN and RNN, as well as the segregation of big data, enhancements of digital devices, and improvements of AI algorithms for more challenging tasks, such as image segmentation, object detection, and real-time monitoring, are continuously enhancing (Oche et al., 2021). The integration of Explainable AI (XAI) shall improve model interpretability and explainability, which is needed to foster trust and accountability. Additionally, establishing relevant connections with Internet of Things (IoT), Blockchain technology, and cloud computing facilities has paved the way for greater efficiency, scalability, and security in handling big RS data (Mukhamediev et al., 2022).

In conclusion, this chapter has focused on innovative advances in applying AI-integrated RS technologies in environmental monitoring in the current day. AI helps to improve accuracy, validity, and frequency of many analyses within different fields, such as air and water quality assessment, dynamics of land surfaces, agriculture, and disaster mapping and risk assessment. The chapter also provides value-added insights through many examples of the application of various ML models. As such, several types of issues remain open and require solutions to reach the full power of these technologies, including data and computation, ethical problems, and interpretability of the outcomes. Collaborating with researchers, industries, policymakers, and environmental managers is crucial to managing these issues and implementing sustainable and ethical use of AI-RS technologies. All these instances will be significant for effective environmental management on the global front. They will ensure that innovation on one side is in harmony with conservation and sound use of resources on the other side.

REFERENCES

Abedi, R., Costache, R., Shafizadeh-Moghadam, H., & Pham, Q. B. (2022). Flash-flood susceptibility mapping based on XGBoost, random forest and boosted regression trees. *Geocarto International*, *37*(19), 5479–5496.

AghaKouchak, A., Farahmand, A., Melton, F. S., Teixeira, J., Anderson, M. C., Wardlow, B. D., & Hain, C. R. (2015). Remote sensing of drought: Progress, challenges and opportunities. *Reviews of Geophysics*, *53*(2), 452–480. https://doi.org/10.1002/2014rg000456

Anand, J., Gosain, A., & Khosa, R. (2018). Prediction of land use changes based on the Land Change Modeller and attribution of changes in the water balance of the Ganga basin to land use change using the SWAT model. *The Science of the Total Environment*, *644*, 503–519. https://doi.org/10.1016/j.scitotenv.2018.07.017

Arabgol, R., Sartaj, M., & Asghari, K. (2015). Predicting nitrate concentration and its spatial distribution in groundwater resources using a Support Vector Machine (SVM) model. *Environmental Modeling & Assessment*, *21*(1), 71–82. https://doi.org/10.1007/s10666-015-9468-0

Avdan, U., & Jovanovska, G. (2016). Algorithm for automated mapping of land surface temperature using LANDSAT 8 satellite data. *Journal of Sensors*, *2016*, 1–8. https://doi.org/10.1155/2016/1480307

Balha, A., Mallick, J., Pandey, S., Gupta, S., & Singh, C. K. (2021). A comparative analysis of different pixel and object-based classification algorithms using multi-source high spatial resolution satellite data for LULC mapping. *Earth Science Informatics*, *14*(4), 2231–2247. https://doi.org/10.1007/s12145-021-00685-4

Bhagat, S. K., Pilario, K. E., Babalola, O. E., Tiyasha, T., Yaqub, M., Onu, C. E., Pyrgaki, K., Falah, M. W., Jawad, A. H., Yaseen, D. A., Barka, N., & Yaseen, Z. M. (2022). Comprehensive review of machine learning methodologies for modeling dye removal processes in wastewater. *Journal of Cleaner Production*, *385*, 135522. https://doi.org/10.1016/j.jclepro.2022.135522

Breiman, L. (2001). Random Forests. *Machine Learning*, *45*(1), 5–32. https://doi.org/10.1023/a:1010933404324

Chen, C., Zhang, H., Shi, W., Zhang, W., & Xue, Y. (2023). A novel paradigm for integrating physics-based numerical and machine learning models: A case study of an eco-hydrological model. *Environmental Modelling & Software*, *163*, 105669. https://doi.org/10.1016/j.envsoft.2023.105669

Chen, D., Chang, N., Xiao, J., Zhou, Q., & Wu, W. (2019). Mapping dynamics of soil organic matter in croplands with MODIS data and machine learning algorithms. *The Science of the Total Environment*, *669*, 844–855. https://doi.org/10.1016/j.scitotenv.2019.03.151

Deters, J. K., Zalakeviciute, R., Gonzalez, M., & Rybarczyk, Y. (2017). Modeling PM2.5 urban pollution using machine learning and selected meteorological parameters. *Journal of Electrical and Computer Engineering*, *2017*, 1–14. https://doi.org/10.1155/2017/5106045

Golkarian, A., Naghibi, S. A., Kalantar, B., & Pradhan, B. (2018). Groundwater potential mapping using C5.0, random forest, and multivariate adaptive regression spline models in GIS. *Environmental Monitoring and Assessment*, *190*(3). https://doi.org/10.1007/s10661-018-6507-8

Greifeneder, F., Notarnicola, C., & Wagner, W. (2021). A machine learning-based approach for surface soil moisture estimations with Google Earth Engine. *Remote Sensing*, *13*(11), 2099. https://doi.org/10.3390/rs13112099

Haggerty, R., Sun, J., Yu, H., & Li, Y. (2023). Application of machine learning in groundwater quality modeling - A comprehensive review. *Water Research*, *233*, 119745. https://doi.org/10.1016/j.watres.2023.119745

Hoyos, I. C. P., Krakauer, N., & Khanbilvardi, R. (2015). Random forest for identification and characterisation of groundwater-dependent ecosystems. *WIT Transactions on Ecology and the Environment*, *1*, 89–100. https://doi.org/10.2495/wrm150081

Janga, B., Asamani, G., Sun, Z., & Cristea, N. (2023). A review of practical AI for remote sensing in Earth Sciences. *Remote Sensing*, *15*(16), 4112. https://doi.org/10.3390/rs15164112

Jasmin, I., & Mallikarjuna, P. (2011). Review: Satellite-based remote sensing and geographic information systems and their application in the assessment of groundwater potential, with particular reference to India. *Hydrogeology Journal*, *19*(4), 729–740. https://doi.org/10.1007/s10040-011-0712-7

Joharestani, M. Z., Cao, C., Ni, X., Bashir, B., & Talebiesfandarani, S. (2019). PM2.5 prediction based on Random Forest, XGBoost, and deep learning using multi-source remote sensing data. *Atmosphere*, *10*(7), 373. https://doi.org/10.3390/atmos10070373

Karandish, F., & Šimůnek, J. (2016). A comparison of numerical and machine-learning modeling of soil water content with limited input data. *Journal of Hydrology*, *543*, 892–909. https://doi.org/10.1016/j.jhydrol.2016.11.007

Liu, H., & Dong, S. (2020). A novel hybrid ensemble model for hourly PM2.5 forecasting using multiple neural networks: A case study in China. *Air Quality Atmosphere & Health*, *13*(12), 1411–1420. https://doi.org/10.1007/s11869-020-00895-7

Luo, Z., Huang, F., & Liu, H. (2020). PM2.5 concentration estimation using a convolutional neural network and a gradient boosting machine. *Journal of Environmental Sciences*, *98*, 85–93. https://doi.org/10.1016/j.jes.2020.04.042

Marjanović, M., Kovačević, M., Bajat, B., & Voženílek, V. (2011). Landslide susceptibility assessment using the SVM machine learning algorithm. *Engineering Geology*, *123*(3), 225–234. https://doi.org/10.1016/j.enggeo.2011.09.006

Mukhamediev, R. I., Popova, Y., Kuchin, Y., Zaitseva, E., Kalimoldayev, A., Symagulov, A., Levashenko, V., Abdoldina, F., Gopejenko, V., Yakunin, K., Muhamedijeva, E., & Yelis, M. (2022). Review of artificial intelligence and machine learning technologies: Classification, restrictions, opportunities, and challenges. *Mathematics*, *10*(15), 2552. https://doi.org/10.3390/math10152552

Naghibi, S. A., Ahmadi, K., & Daneshi, A. (2017). Application of support vector machine, random forest, and genetic algorithm optimised random forest models in groundwater potential mapping. *Water Resources Management*, *31*(9), 2761–2775. https://doi.org/10.1007/s11269-017-1660-3

Ng, W., Minasny, B., Montazerolghaem, M., Padarian, J., Ferguson, R., Bailey, S., & McBratney, A. B. (2019). Convolutional neural network for simultaneous prediction of several soil properties using visible/near-infrared, mid-infrared, and their combined spectra. *Geoderma*, *352*, 251–267. https://doi.org/10.1016/j.geoderma.2019.06.016

Niazkar, M., Menapace, A., Brentan, B., Piraei, R., Jimenez, D., Dhawan, P., & Righetti, M. (2024). Applications of XGBoost in water resources engineering: A systematic literature review (Dec 2018–May 2023). *Environmental Modelling & Software*, *174*, 105971. https://doi.org/10.1016/j.envsoft.2024.105971

Oche, P. A., Ewa, G. A., & Ibekwe, N. (2021). Applications and challenges of artificial intelligence in space missions. *IEEE Access*, *12*, 44481–44509. https://doi.org/10.1109/access.2021.3132500

Panahi, M., Sadhasivam, N., Pourghasemi, H. R., Rezaie, F., & Lee, S. (2020). Spatial prediction of groundwater potential mapping based on a convolutional neural network (CNN) and support vector regression (SVR). *Journal of Hydrology*, *588*, 125033. https://doi.org/10.1016/j.jhydrol.2020.125033

Sajedi-Hosseini, F., Malekian, A., Choubin, B., Rahmati, O., Cipullo, S., Coulon, F., & Pradhan, B. (2018). A novel machine learning-based approach for the risk assessment of nitrate groundwater contamination. *The Science of the Total Environment*, *644*, 954–962. https://doi.org/10.1016/j.scitotenv.2018.07.054

Shahriar, S. A., Kayes, I., Hasan, K., Salam, M. A., & Chowdhury, S. (2020). Applicability of machine learning in the modeling of atmospheric particle pollution in Bangladesh. *Air Quality Atmosphere & Health*, *13*(10), 1247–1256. https://doi.org/10.1007/s11869-020-00878-8

Shukla, A. K., Alam, J., Mallik, S., Ruokolainen, J., Kesari, K. K., & Alhoshan, M. (2024). Optimisation and prediction of dye adsorption utilising cross-linked chitosan-activated charcoal: Response Surface Methodology and machine learning. *Journal of Molecular Liquids*, *411*, 125745. https://doi.org/10.1016/j.molliq.2024.125745

Taccari, M. L., Wang, H., Nuttall, J., Chen, X., & Jimack, P. K. (2024). Spatial-temporal graph neural networks for groundwater data. *Scientific Reports*, *14*(1). https://doi.org/10.1038/s41598-024-75385-2

Talukdar, S., Shahfahad, N., Bera, S., Naikoo, M. W., Ramana, G. V., Mallik, S., Kumar, P. A., & Rahman, A. (2023). Optimisation and interpretation of machine and deep learning models for improved water quality management in Lake Loktak. *Journal of Environmental Management*, *351*, 119866. https://doi.org/10.1016/j.jenvman.2023.119866

Uddin, M. G., Nash, S., Rahman, A., & Olbert, A. I. (2022). Performance analysis of the water quality index model for predicting water state using machine learning techniques. *Process Safety and Environmental Protection*, *169*, 808–828. https://doi.org/10.1016/j.psep.2022.11.073

Vincent, A. M., Parthasarathy, K.S.S., & Jidesh, P. (2023). Flood susceptibility mapping using AutoML and a deep learning framework with evolutionary algorithms for hyperparameter optimisation. *Applied Soft Computing*, *148*, 110846. https://doi.org/10.1016/j.asoc.2023.110846

Wang, Z., Gang, C., Li, X., Chen, Y., & Li, J. (2015). Application of a normalised difference impervious index (NDII) to extract urban impervious surface features based on Landsat TM images. *International Journal of Remote Sensing*, *36*(4), 1055–1069. https://doi.org/10.1080/01431161.2015.1007250

Weng, Q. (2011). Remote sensing of impervious surfaces in the urban areas: Requirements, methods, and trends. *Remote Sensing of Environment*, *117*, 34–49. https://doi.org/10.1016/j.rse.2011.02.030

Wu, L., Cui, P., Pei, J., Zhao, L., & Song, L. (2022). Graph neural networks. In *Graph Neural Networks: Foundations, Frontiers, and Applications* (pp. 27–37). Springer Nature Singapore. https://doi.org/10.1007/978-981-16-6054-2_3

Yang, X., & Liu, Z. (2005). Use of satellite-derived landscape imperviousness index to characterise urban spatial growth. *Computers Environment and Urban Systems*, *29*(5), 524–540. https://doi.org/10.1016/j.compenvurbsys.2005.01.005

Yang, X., Ye, Y., Li, X., Lau, R. Y. K., Zhang, X., & Huang, X. (2018). Hyperspectral image classification with deep learning models. *IEEE Transactions on Geoscience and Remote Sensing*, *56*(9), 5408–5423. https://doi.org/10.1109/tgrs.2018.2815613

Zanfei, A., Brentan, B. M., Menapace, A., Righetti, M., & Herrera, M. (2022). Graph Convolutional recurrent neural networks for water demand forecasting. *Water Resources Research*, *58*(7). https://doi.org/10.1029/2022wr032299

Zhang, J., Ma, X., Zhang, J., Sun, D., Zhou, X., Mi, C., & Wen, H. (2023). Insights into geospatial heterogeneity of landslide susceptibility based on the SHAP-XGBoost model. *Journal of Environmental Management*, *332*, 117357. https://doi.org/10.1016/j.jenvman.2023.117357

Zhang, Y., Liang, S., Zhu, Z., Ma, H., & He, T. (2022). Soil moisture content retrieval from Landsat 8 data using ensemble learning. *ISPRS Journal of Photogrammetry and Remote Sensing*, *185*, 32–47. https://doi.org/10.1016/j.isprsjprs.2022.01.005

Zhou, J., Cui, G., Hu, S., Zhang, Z., Yang, C., Liu, Z., Wang, L., Li, C., & Sun, M. (2020). Graph neural networks: A review of methods and applications. *AI Open*, *1*, 57–81. https://doi.org/10.1016/j.aiopen.2021.01.001

Zhu, X. X., Tuia, D., Mou, L., Xia, G., Zhang, L., Xu, F., & Fraundorfer, F. (2017). Deep learning in remote sensing: A comprehensive review and list of resources. *IEEE Geoscience and Remote Sensing Magazine*, *5*(4), 8–36. https://doi.org/10.1109/mgrs.2017.2762307

9 Implementing Artificial Intelligence-Based Algorithms for Sustainable Environmental Monitoring

Karim Ennouri, Mohamed Ali Triki, Fathi Ben Amar, and Monia Ennouri

9.1 INTRODUCTION TO AI IN ENVIRONMENTAL MONITORING

9.1.1 Overview of Machine Learning and Artificial Intelligence in Environmental Observing

Artificial intelligence (AI) and machine learning (ML) have become the fundamental technologies in the field of environmental observing, with the aptitude to convert beyond records collection to sophisticated analysis and prediction. Here, AI is predestined to be computer-based systems that imitate human perceptive processes, comprising learning, problem-solving, and management, in a way that is specially designed to resolve difficult ecological problems. These systems are made more competent, which is simplified by the usage of ML, a division of AI, and is typically accomplished by developing patterns in huge datasets without clear programming.

AI and ML have an extensive range of applications in environmental monitoring that include air and H_2O quality, biodiversity defense, climate-change modeling, and disaster management. The field has traditionally had an important growth trajectory, as the number of academic publications on AI-driven environmental monitoring has grown considerably, particularly since 2010 (Alotaibi & Nassif, 2024). The reason for this surge can be explicated by the cumulative availability of high-resolution databases, such as satellite pictures and sensor networks, and the growing influence of computers that can be employed to train and arrange complex prototypes at scale (Ogunwumi & Matindike, 2025). The rising form of literature specifies that there is a multidisciplinary conjunction of computer skill, environmental discipline, and data

DOI: 10.1201/9781003616207-11

engineering, which highlights the significance of AI in confirming that sustainability objectives are met.

The application of AI in environmental situations has an important impact on sustainable development, as it permits gaining a better perception of ecological systems, providing responses to environmental threats in time, and encouraging resource-efficient management. With the current anthropogenic activities threatening environments in techniques never perceived before, AI has delivered explanations that are complementary and additional to the usual monitoring approaches due to its innovative real-time and predictive analytics systems (Abubakar et al., 2024). Simultaneously, the varying algorithms of AI open new prospects in the explanation of environmental records, proposing the perceptions to guide the sound policymaking and sustainability determinations worldwide (Saleh & Battseren, 2023).

9.1.2 Objectives and Benefits of AI-Based Environmental Monitoring

The central aim of implementing artificial-intelligence algorithms in environmental observing systems is to influence advanced computational techniques to obtain more precise and appropriate assessments, as well as a more comprehensive one. One of the most important benefits is the capability to offer real-time surveillance and data-driven decision-making, thus providing environmental supervisors with real-time access to actionable intelligence. This aptitude raises the receptiveness to quickly varying environmental situations like emerging natural disasters, thus decreasing the damaging environmental and social consequences (Mobo et al., 2024).

AI significantly increases the precision of predictive prototypes and environmental impact investigations by applying machine-learning methods that can combine different data streams that can comprise Internet-of-Things sensor records with satellite pictures. The prototypes allow projections of environmental adjustments at a higher resolution, thus supporting proactive contributions aimed at checking or adapting to dangerous actions like environmental degradation or climatic variability (Cukjati et al., 2022). Besides, AI algorithms will manage the properties in a sustainable way by exploiting the use of natural resources, such as water, soil, and forest resources, and, consequently, minimize waste and improve protection efforts through knowledgeable allocation and conservation strategies (Farooq et al., 2025).

Introducing AI into the process of environmental observing, the investors will be capable of acquiring not only an advanced level of operational efficiency but also encouraging the modernization in the management of ecological structures. These benefits are aligned with global sustainability priorities, such as the United Nations Sustainable Development Goals (SDGs), and contribute to the aptitude of sustainable development at the local, regional, and global stages.

9.1.3 Challenges and Considerations

Despite the fact that AI promises innovative changes in improving environmental observation, numerous ethical and methodological apprehensions should undergo strict inspection to promise answerable and effective employment. First is the data quality, since the precision and reliability of the AI-driven forecast depend on whether it has high-fidelity, illustrative, and prudently maintained datasets (Koldasbayeva et al., 2024).

The transparency of algorithms is an important challenge since numerous AI patterns, in particular deep learning networks, are "black box" arrangements that offer imperfect information about how they make their internal deductions. Such an unusual complication undermines assurance among end-users and supervisory organizations, a sharp subject in high-stakes settings like ecological risk assessment or public safety (Singh et al., 2024). The new development of explainable AI (XAI) structures offers a means of addressing these insufficiencies, endeavoring to deliver understandable model outputs without surrendering (and in fact improving) predictive performance.

Algorithm bias is an additional concern and can arise due to inadequate or unsatisfactory training data, and can result in different outcomes for different groups of people or even geographic zones. Equal access to AI tools has a significant impact, as the unbalanced distribution of infrastructural and technical capacities can further increase existing environmental dissimilarities on a global scale (Turyasingura et al., 2024). Ethical codes and authority structures should be formulated and executed to protect equality, responsibility, and inclusivity in the AI practice. Such contexts should address questions of privacy, data rights, and liable AI use that raise positive interaction between technologists, legislators, and local populations.

9.2 CORE AI ALGORITHMS AND METHODOLOGIES FOR ENVIRONMENTAL MONITORING

9.2.1 Machine Learning Techniques and Their Applications

ML comprises various algorithms, which are broadly classified into supervised, unsupervised, and reinforcement paradigms, all of which have been used in the application field of environmental monitoring. The training of predictive models using supervised learning methods, where labeled data is used to teach machines, is widely used in classification and regression applications like soil property mapping, species recognition, and estimation of pollutant concentrations. Unsupervised learning techniques, which include clustering and dimensionality reduction algorithms, can be used to identify intrinsic patterns and anomalies in environmental data and thus improve exploratory analyses in cases where there are few labeled data. Reinforcement learning, with agent-environment interactions and agent-feedback optimization, is becoming popular in dynamic environments such as adaptive irrigation systems or optimization of disaster responses in real time (Hua, 2021).

The application of machine-learning methods to environmental issues has proven to be effective in specific use cases. An example is the use of soil mapping, which is based on models to combine environmental covariates and remote-sensing data to forecast soil carbon or moisture content with a high level of accuracy. ML algorithms are used in species identification to classify images or acoustic recordings in biodiversity measurements. Supervised-learning models are used to detect pollution by examining data provided by chemical sensors to detect the presence and concentration of contaminants. The progress in deep learning, such as convolutional neural networks (CNNs) and generative adversarial networks (GANs), has even improved the performance of manipulating dense environmental data like imagery and time-series signals. Instead, GANs have been applied in data augmentation and missing-data imputation, which improves the resistance of models to data constraints (Madeshwaran & Govindarajan, 2024; Sadegh et al., 2025).

9.2.2 Integration of Convolutional Neural Networks and Image Analysis

CNNs have become an enabler technology in the remote sensing and environmental image analysis field due to their unprecedented performance in multispectral and hyperspectral data in terms of hierarchical spatial feature extraction. CNNs enable the classification of land cover, detecting deforestation, or mapping susceptibility to floods by absorbing large pixel-level information when considered atop satellite images. Similarly, ML of camera-trap videos (typically with the help of CNNs) automates camera tracking and species identification, which, in turn, contributes to the increased efficiency of biodiversity surveillance and anti-poaching programs (Mamapule et al., 2025).

The combination of AI and hyperspectral imaging is synergistic and serves to enhance a more careful observation of ecosystems due to a growing capacity to identify minor spectral trends. CNNs are employed to interpret high-dimensional hyperspectral streams of data and are appropriate in an assortment of hyperspectral applications, such as crop-health sensing, malady, and invasive-species detection. The approach has also been applied to natural-disaster management, where early-warning instruments based on high-resolution image records are more efficiently recognized (Lanke & Chandak, 2024).

In forestry ecology, unmanned aerial vehicles that come with imaging sensors joined with AI algorithms are changing the forest-management practices. They support mapping the tree species, their healthiness, and the structural diversity of forests with great accuracy in huge areas of space. These self-sustaining systems are also enabled to detect wildfires in real time using thermal imaging, which provides real-time alerts, which are essential in preventing and controlling disasters (Buchelt et al., 2024).

9.2.3 Advanced AI Models for Real-Time Environmental Data

The implementation of advanced AI-based models adapted to handle real-time environmental data is thus reported to have grown tremendously with the usage of schemes like federated learning models and multi-agent systems. Federated learning enables training of distributed models over heterogeneous data sources without necessarily growing central collections of data and, thus, protects privacy and supports data security. This approach has been shown to be useful in holistic and air-pollution surveillance and health-impact analyses, where privacy issues are particularly relevant. Federated AI models allow the definition of the risks of pollutant exposure and the dynamic forecast of the Air Quality Index (AQI) by synthesizing socio-economic and environmental factors and adhering to regional data-protection requirements (Neo et al., 2022). Table 9.1 shows the principal AI-based environmental monitoring.

TABLE 9.1
Main AI-Based Environmental Monitoring

Challenge	Description
Data Quality and Reliability Issues	AI-based environmental checking systems face trials related to information quality and consistency, with issues such as record degradation, sensor drift, and measurement mistakes. Low-cost sensors produce noisy and unreliable data, affecting AI model performance.
Algorithmic Bias and Environmental Integrity	AI algorithms can perpetuate environmental inequities if training data reflects biases or if equity considerations are not included. There is a risk of less accurate coverage in underserved communities.
Scalability and Resource Constraints	Scaling AI-based systems presents technical challenges due to diverse environmental conditions. Specialized AI models may be needed, and computational resource requirements can be substantial, raising concerns about energy consumption and environmental impact.

Table 9.1 outlines three major challenges facing AI-based environmental monitoring systems, emphasizing the technical and ethical hurdles that must be addressed for reliable deployment. First, data quality and reliability issues arise from sensor drift, measurement errors, and the use of low-cost devices that often generate noisy or inconsistent data, which undermines AI model accuracy. A combination of statistical measures and a set of structured means of evaluation was used to validate that the model accuracy and performance were validated to achieve reliability as well as generalizability. K-fold cross-validation and train/test splits were used to divide the dataset, which minimized bias of one partition and emphasized consistency within subsets. In regression applications like prediction of continuous Water Quality Index (WQI) values, a coefficient of determination (R2), root mean square error (RMSE), mean absolute error (MAE), and mean absolute percentage error (MAPE) were used to measure the variance accounted and the magnitude of the error. In the case of classification, the general accuracy, precision, recall, F1-score, and the area under the Receiver Operating Characteristic (ROC) curve (AUC-ROC) were created as measures. The other validation techniques were residual analysis to identify systematic errors, sensitivity analysis to assess the effect of critical parameters (e.g., pH, dissolved oxygen, biochemical oxygen demand), and comparisons with traditional statistical models to indicate the value added by AI techniques. Collectively, the approaches were very strong techniques for establishing predictive efficiency, reducing overfitting, and practical applicability in real-life monitoring of water quality.

Second, algorithmic bias and environmental integrity present risks when training data reflects inequities or lacks representation, potentially reinforcing disparities in underserved communities. Finally, scalability and resource constraints highlight the difficulty of adapting AI models to diverse environmental conditions while managing the heavy computational demands, which themselves raise concerns about energy use and environmental impact.

Multi-agent AI models provide parallel, domain-specialized sub-agents that have specific programs that run discrete research or monitoring tasks within a collaborative architecture. The designs increase qualitative data analysis and decision support capabilities within environmental inquiries and thus enhance scalability and interpretive accuracy in large textual or sensor-generated data sets (Oprea et al., 2011).

Sensors with embedded AI algorithms can be used in the field of environmental health to enable continuous measurement of individual physiological conditions and the surrounding temperature and humidity. These sensors facilitate early disease detection and decision-making and accelerate the identification of environmental risks and hazards, thus optimizing the protection of the citizen population by increasing situational awareness (Shobana, 2025).

9.3 DATA SOURCES AND SENSOR TECHNOLOGIES SUPPORTING AI MODELS

9.3.1 Remote Sensing and Satellite Imagery Data

Remote sensing systems, especially satellite systems like Landsat and Sentinel, are key sources of data used to monitor environmental conditions by getting systematic multispectral and radar observations of the environment of a region or the entire globe. Such datasets allow modeling forest structure, biodiversity indicators, and changes in land use over time, with continuous time showing time series, which is possible with tools like Google Earth Engine to run large time-series analyses. An example of this includes deep learning models that have been utilized on Sentinel imagery, which have been able to predict forest species diversity and structural heterogeneity, which is a critical proxy of ecosystem health and management (Hoffmann et al., 2022).

The ability to produce high-resolution maps with a temporal viewpoint as a result of nationwide land-use/land-cover reconstruction exercises using Landsat archives indicates that it is possible to support better application of environmental policy. Through such comprehensive datasets, analyses extending to the identification of deforestation and habitat fragmentation, and the determination of the restoration progress are possible. However, satellite data are limited by both spatial and temporal resolution limitations, interference with cloud cover, and complexity of data processing, which need a synergistic approach with complementary methods of sensing to provide a comprehensive environmental analysis (Park & Park, 2024).

9.3.2 Internet of Things (IoT) and Sensor Networks

By combining IoT and AI, the possibilities of environmental monitoring become even more pronounced through the active use of sensor networks that can be deployed extensively and retrieve various physical and chemical indicators continuously. IoT models can be used to gather water-quality measures, climate parameters, and pollutant levels. The algorithms in the form of AI that are installed in such networks make it possible to predict models, identify anomalies, and initiate automatic reactions, therefore making operations more efficient and improving timely decision-making (Aramide, 2025).

Expensive sensors embedded on platforms like Raspberry Pi offer low-cost and customizable explanations for the health tracking of the environment. They obtain environmental records and process them through cloud services to track their health patterns and inform users of possible threats. These distributed sensor networks can make environmental monitoring more accessible and scalable, particularly when they are combined with smart-city infrastructures designed to utilize resources sustainably and easily safeguard the public (Ouni & Saleem, 2022).

Smart-city projects take advantage of the convergence of the IoT and AI to maintain consistent environmental data records and optimize the city services, such as air-quality management, waste management, and energy consumption. Such systems are examples of how emergent technologies can work together in aiding environmental sustainability in dynamic, living cities.

9.3.3 Emerging Sensing Techniques: Raman Spectroscopy and Electrochemical Sensors

State-of-the-art sensing strategies, for example, laser-induced deep UV Raman spectroscopy and AI-based processing architectures, have opened up new paths of real-time aquatic pollution detection and quantification of pollutants such as nitrates, pharmaceuticals, and microplastics. CNNs are AI models trained to identify spectral signatures with noise as well as interferences to increase the sensitivity and selectivity of pollutant measurements. Some of the chronic problems are the detection of contaminants at trace levels and the modulation of a heterogeneous distribution of particles, which current research activities are working toward solving (Babu & Malarkodi, 2024).

Cellulose-based electrochemical sensors are very innovative as biogenic sensors that can be applied in different environmental studies. Such sensors have inherent features of biocompatibility and heterogeneity, which offer a sensitive response to various analytics using sustainable materials. A combination of these sensors with AI and Internet of Things (IoT) technologies promises potential directions of high-performance and environmentally friendly environmental monitoring. However, challenges that relate to the question of scalability, selectivity, and long-term stability highlight issues that demand thorough research to bring these technologies out of laboratory demonstration to field implementation (Rajesh et al., 2019).

9.4 AI Applications in Air and Water Quality Monitoring

9.4.1 AI to Predict Air Pollution and Assess Its Health Impact

AI models are used more to forecast AQI and analyze the correlated health effects because of the complexity of environmental and socio-economic data. Federated learning models can facilitate the creation of models between multiple institutions and maintain data privacy, thus enhancing predictive accuracy of pollutant concentrations, including particulate matter and gaseous emissions. These models help recognize hotspots and susceptible populations and therefore inform mitigation-focused strategies and health service operations. The sophisticated algorithms help in the detection of pollution sources and prediction of negative health outcomes, such as respiratory diseases, which subsequently guide the interventions of the population health. The implementation of AI in air pollution monitoring can help in compliance with the regulations and improve the community's resilience to environmental health hazards (Rathi & Vasal, 2025).

AI technologies have become the key solutions in the process of predicting air pollution and estimating health effects. Through ML and computational approaches, scientists would be capable of predicting the level of pollution and health dangers using complex data. Different AI techniques, such as hybrid models, have demonstrated a higher level of excellence in the prediction of air pollutants such as PM2.5, PM10, and NO2, with parameters such as R^2 and RMSE with high accuracy (Yadav et al., 2023). The integration of the records enables the evaluation of the effects of pollution, enabling real-time monitoring and response actions (Rathi & Vasal, 2025).

9.4.2 Water Management AI-Based and Best Resource Quality Control

The use of ML methodologies in water resource management is wide-ranging and enables optimizing irrigation systems, monitoring water quality, and predicting flooding. To promote sustainability in water use of agriculture and urban-related water, AI systems prepare a wide range of sensor data sets, such as hydrological and climatic indicators, and forecast water demand and its allocation efficiency, which makes these systems more efficient (Drogkoula et al., 2023).

AI-enhanced analytical methods can determine chemical pollution and microplastics in water in real time and enhance the sensitivity and specificity of both methods compared to traditional ones. The monitoring with AI assistance makes operational decisions within water-treatment facilities, which will secure the desired quality of drinking water (Jogee et al., 2024).

The AI models are vital with WQI since they help in improving the accuracy of prediction and real-time monitoring. Artificial Neural Network (ANN) models have shown that they are highly accurate in the predictions and can achieve performance metrics of 0.98114 in WQI prediction (Murivhami et al., 2023). Integrating mechanisms such as Particle Swarm Optimization (PSO) and Support Vector Machines (SVM) has resulted in flexible models that increase predictive potential (Naeem et al., 2023). Long Short-Term Memory (LSTM) network models and Nonlinear Autoregressive Neural Networks (NARNET) have proven to be more robust when used to predict WQI with an accuracy up to 97.01% (Aldhyani et al., 2020).

9.4.3 Artificial Intelligence to Manage Water Resources

Land use and groundwater levels must be observed and correctly modelled to promise a supportable interaction of geothermal and groundwater assets. Optimization of parameters and comparative studies support the efficiency of such algorithms in the design of sophisticated land-cover processes that are reliable for sustainability purposes (Yousefi et al., 2022). Figure 9.1 shows the algorithm used for monitoring.

The figure illustrates a process that transforms environmental conditions into knowledge. This data is processed by AI algorithms, which detect patterns, predict risks, and generate interpretations that are impossible to extract manually.

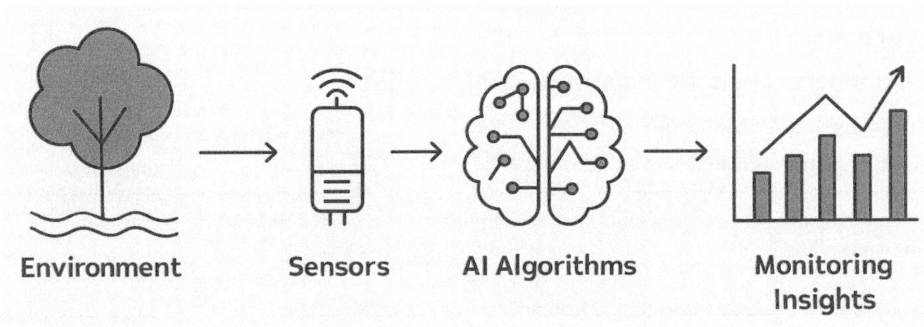

FIGURE 9.1 AI algorithms in environmental monitoring.

Deep-learning prototypes, such as CNNs and LSTM networks, overcome the problem of missing records by using imputation networks, like GANs. These methods increase the precision of measurements (Chen et al., 2023).

9.5 AI-BASED CONSERVATION AND ECOSYSTEM MONITORING OF BIODIVERSITY

9.5.1 Wildlife Tracking and Species Recognition

Algorithms used on camera-trap data, audio data, and environmental DNA (eDNA) data are tools of biodiversity observation. These methods permit the measurement of biodiversity on a mass scale (Thomsen & Willerslev, 2014). AI-based territory assessment also uses picture recognition to assess ecosystem health, deforestation, and restoration. With the combination of numerous data streams, AI assists with reliable conservation plans to allow species interaction and the effects of human activities (Stowell and Sueur, 2020).

9.5.2 Forest Ecology and Forest Land Cover

Deep learning applied to satellite remote sensing descriptions is used to model the biodiversity of plantations. Identification of invasive species and deforestation patterns using such techniques enhances the abilities of early intervention and decision-making within the management strategies (Ali et al., 2025). Table 9.2 summarizes technologies used for environmental modeling.

The table highlights three cutting-edge technologies. Edge and neuromorphic systems enable real-time environmental monitoring by processing data. Quantum computing offers tools for data analysis, using specialized algorithms to tackle complex optimization problems and improve the accuracy of climate projections. Blockchain integration ensures environmental data management is secure, transparent, and tamper-proof, supporting trust and collaboration across stakeholders, even if governance and integration remain challenges.

AI-based imaging technologies deployed to drones allow the individual tree species identification, health evaluation, and wildfire detection in the forest inventory

TABLE 9.2
Technologies Used for Environmental Modeling

Technology	Application	Key Features	Benefits	Challenges
Edge Computing and Neuromorphic Systems	Environmental Monitoring	Real-time data processing, neuromorphic computing architectures	Increased data throughput, reduced power consumption	Requires less centralized processing
Quantum Computing	Environmental Modeling and Data Analysis	Quantum algorithms, QAOA	Solving complex optimization problems, more accurate climate modeling	Current limitations require hybrid approaches
Blockchain Integration	Environmental Data Management	Secure, transparent, tamper-proof systems	Ensures data integrity, enables secure data sharing	Challenges with integration and stakeholder governance

and monitoring process. These automated systems also offer the ability to scan large and intricate wooded regions in a short time, hence assisting decision-making in the conservation and management of resources in real time (Buchelt et al., 2023).

9.5.3 Biodiversity Monitoring with Ecoacoustics and Acoustic Sensors on the Large-Scale Level

Ecoacoustics involves passive acoustic sensing to detect the biodiversity and ecosystem health on scales beyond those of conventional visual surveys. ML techniques are applied to soundscape data in order to identify the presence of species, patterns of behavior, and disturbances of the environment. This can be used to effectively record both time and space fluctuations in biodiversity, which provides additional evidence compared to other surveillance tools like satellite imagery and eDNA analyses. By combining ecoacoustic data with the others, it becomes possible to assess ecosystems holistically by incorporating information on other ecological measures, and this information is very useful in conservation planning and biodiversity management. The standardization of data collection procedures and automated evaluations are progressing the relevance of ecoacoustics into actual conservation programs (Robinson et al., 2024).

9.6 DISASTER MANAGEMENT AND MITIGATION OF CLIMATE CHANGE USING AI

9.6.1 Prediction of Floods and Other Natural Catastrophes Using AI

Early warning systems used to predict natural disasters and floods are AI-based and utilize remote sensing data and IoT network sensors to measure risks in time and assist in evacuation planning. They are systems that simulate the dynamics of hydrological and meteorological variables and thus increase resilience and preparedness among vulnerable communities, as these systems help preempt disaster events more accurately. Fusion of real-time environmental records with AI analytics can be connected with the general monitoring of disasters that permits dynamic management and diminishes human and economic damages. The practical management method enabled by AI predictive prototypes is approximately that which improves the enhancement of society's resilience against ever-increasing and natural disasters (Venkadesh et al., 2024). Figure 9.2 presents remote sensing as a central hub that connects multiple technologies and applications, showing how data from satellites, unmanned aerial vehicles (UAVs), and AI-driven systems is transformed into actionable insights.

9.6.2 Machine Learning-Based Carbon Capture and Reduction of Emissions

Algorithms based on AI can also help minimize factors in carbon capture plants, hence exploiting resource effectiveness by utilizing large datasets. Additionally, the generative AI prototypes can be used to accelerate selection of new materials that could be critical in efficient CO_2 capture (Manikandan et al., 2024). Combining AI with renewable energy systems and smart grid technologies promotes more advanced carbon capture strategies because it will allow better control. These innovations use AI assistance in achieving International Sustainability Goals and SDGs and addressing climate action. Figure 9.3 resumes ecological environmental observing.

9.6.3 AI Application Risk and Resilience and AI Governance: Systemic Risks, Resilience, and Governance

Despite its advantages, AI implementation in sectors that are of vital sustainability concern should consider the systemic risks of using AI, including algorithmic bias, cascading failure, and trade-offs between efficiency and resilience, which are complicated. The consequences of algorithmic bias might be allocative harms, since it affects the allocation of benefits and subjects certain groups or regions to a higher risk. Besides, the system interruptions arising due to the failure of the AI system can be propagated through interrelated environment and economic systems, thus posing challenges to governance systems that have been traditionally organized to account for the traditional technologies (Schmidt & Biessmann, 2020).

Existing mechanisms of governance are sometimes falling behind changes in technology without the strength to deal with such emerging risks in a holistic manner. To overcome systemic risks, computer scientists, environmental researchers, policymakers, and ethicists must work across disciplines to ensure the development of models of adaptive governance that would guarantee AI will produce sustainable, equitable results without causing more vulnerabilities.

9.7 CONCLUSION

Unlike any other application, the use of AI algorithms in the study of the environment offers chances to achieve unprecedented sustainability, reliability, and efficiency levels of work. The most recent substantive developments in AI technology have had an

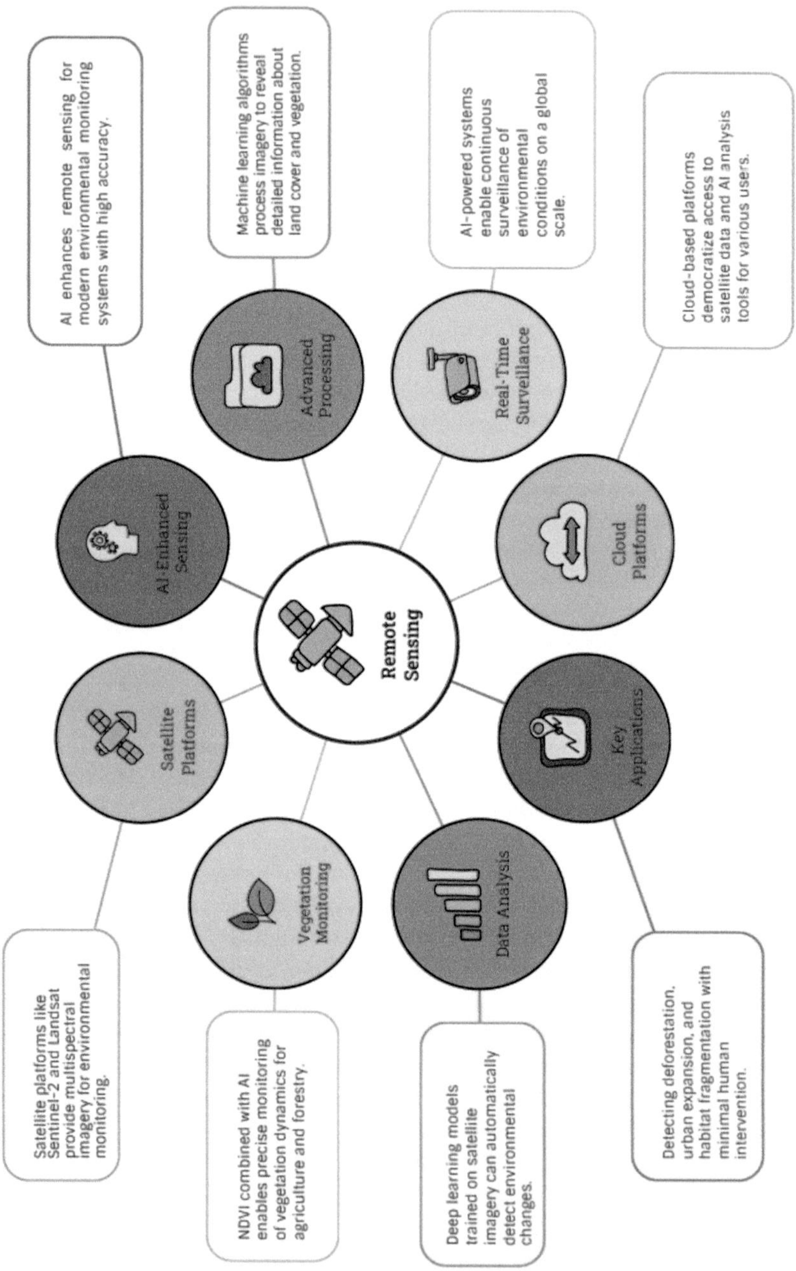

FIGURE 9.2 The uses of remote sensing for environmental monitoring.

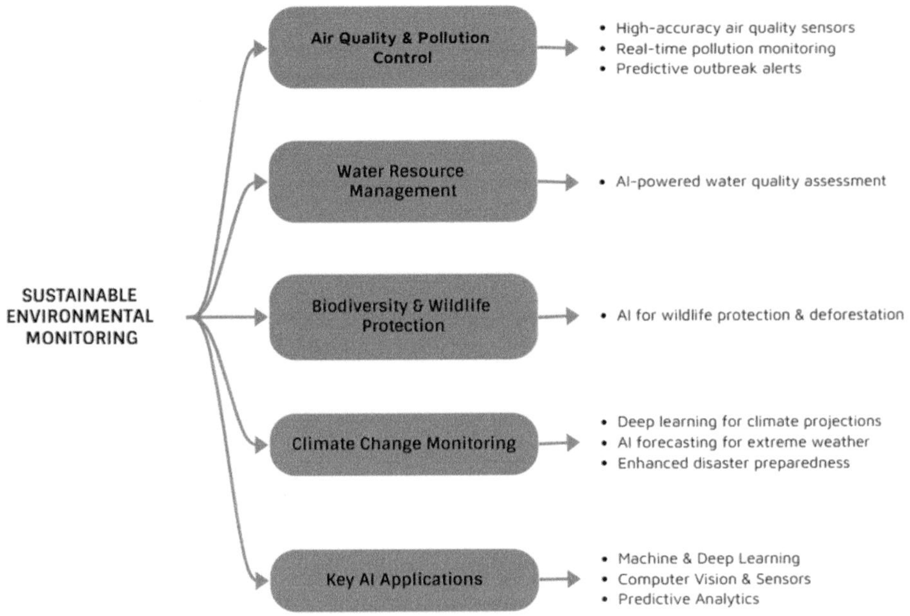

FIGURE 9.3 Sustainable environmental monitoring diagram.

enormous impact on the conservation of biodiversity, monitoring disasters, and mitigating climate change through facilitating the real-time processing of data, effective predictive models, and the availability of management responses. Once the operational issues and the ethical problems are managed in a systematic way, AI is set to completely reinvent the approaches to environmental safety and resource usage.

AI technologies offer scalable and innovative solutions compatible with international sustainability commitments. The high investment in the data infrastructure is necessary to ensure the successful implementation of AI in environmental monitoring to ensure better quality and availability of data, which will improve model training and validation accuracy and reliability.

AI will be highly significant in sustainable development by optimizing transactions of smart cities, precision agriculture, industrial methods, and other trials to develop sustainable ecosystems. Interdisciplinary cooperation, application of new technologies, and thematic policy synergy will all help to create an ecologically sustainable and socially fair future by implementing AI-based surveillance and control systems.

REFERENCES

Abubakar, A. M., Zakarya, I. A., Hasnain, M., Sarkinbaka, Z. M., Mukwana, K. C., & Abdo, A. I. (2024). Potential breakthroughs in environmental monitoring and management. In *Harnessing AI in Geospatial Technology for Environmental Monitoring and Management*, Mobo, F. D. (ed.) (pp. 239–282). IGI Global Scientific Publishing. https://doi.org/10.4018/979-8-3693-8104-5.ch011

Aldhyani, T. H. H., Al-Yaari, M., Alkahtani, H., & Maashi, M. S. (2020). Water quality prediction using artificial intelligence algorithms. *Applied Bionics and Biomechanics, 2020*, 6659314. https://doi.org/10.1155/2020/6659314

Ali, G., Mijwil, M. M., Adamopoulos, I., & Ayad, J. (2025). Leveraging the internet of things, remote sensing, and artificial intelligence for sustainable forest management. *Babylonian Journal of Internet of Things*, 2025, 1–65. https://doi.org/10.58496/bjiot/2025/001

Alotaibi, E., & Nassif, N. (2024). Artificial intelligence in environmental monitoring: In-depth analysis. *Discover Artificial Intelligence, 4*(1). https://doi.org/10.1007/s44163-024-00198-1

Aramide, O. O. (2025). Predictive network maintenance and anomaly detection with AI. *International Journal of Technology, Management and Humanities, 11*(02), 1–11. https://doi.org/10.21590/ijtmh.11.02.08

Babu, V. K., & Malarkodi, K. (2024). Comparative analysis between enhanced convolution neural network compared over support vector machine in predicting PM2.5 air pollutants with improved accuracy. In *15th International Conference on Computing Communication and Networking Technologies (ICCCNT), Kamand, India*, 1–5. https://doi.org/10.1109/ICCCNT61001.2024.10724339

Buchelt, A., Adrowitzer, A., Kieseberg, P., Gollob, C., Nothdurft, A., Eresheim, S., Tschiatschek, S., Stampfer, K., & Holzinger, A. (2023). Exploring artificial intelligence for applications of drones in forest ecology and management. *Forest Ecology and Management, 551*, 121530. https://doi.org/10.1016/j.foreco.2023.121530

Buchelt, A., Adrowitzer, A. C., Kieseberg, P., Gollob, C., Nothdurft, A., Eresheim, S., Tschiatschek, S., Stampfer, K., & Holzinger, A. (2024). Exploring artificial intelligence for applications of drones in forest ecology and management. *Forest Ecology and Management*. https://doi.org/10.1016/j.foreco.2023.121530

Chen, H.-Y., Vojinović, Z., Lo, W., & Lee, J.-W. (2023). Groundwater level prediction with deep learning methods. *Water, 15*(17), 3118. https://doi.org/10.3390/w15173118

Cukjati, J., Mongus, D., Rizman Žalik, K., & Žalik, B. (2022). IoT and satellite sensor data integration for assessment of environmental variables: A case study on NO2. *Sensors, 22*(15), 5660. https://doi.org/10.3390/s22155660

Drogkoula, M., Kokkinos, K., & Samaras, N. (2023). A comprehensive survey of machine learning methodologies with emphasis in water resources management. *Applied Sciences*. https://doi.org/10.3390/app132212147

Farooq, M. U., Murtaza, G., Mahmood, M. H., Iqbal, S., Tanveer, H., Haq, E. U., Gatti, G., & Sabir, R. M. (2025). AI-powered solutions for sustainable water conservation. *Advances in Computational Intelligence and Robotics Book Series*, 473–512. https://doi.org/10.4018/979-8-3373-0680-3.ch016

Hoffmann, J., Muro, J., & Dubovyk, O. (2022). Predicting species and structural diversity of temperate forests with satellite remote sensing and deep learning. *Remote Sensing, 14*(7), 1631. https://doi.org/10.3390/rs14071631

Hua, C. (2021). *Reinforcement Learning and Feedback Control* (pp. 27–57). Springer Vieweg, Wiesbaden. https://doi.org/10.1007/978-3-658-33034-7_3

Jogee, D., Nowbuth, M. D., Proag, V., & Probst, J. (2024). *Artificial Intelligence–Based Clean Water and Sanitation Monitoring* (pp. 69–80). Emerald (MCB UP). https://doi.org/10.1108/978-1-83753-540-820241006

Koldasbayeva, D., Tregubova, P., Gasanov, M., Zaytsev, A., Petrovskaia, A., & Burnaev, E. (2024). Challenges in data-driven geospatial modeling for environmental research and practice. *Nature Communications, 15*(1). https://doi.org/10.1038/s41467-024-55240-8

Lanke, N. P., & Chandak, M. (2024). Role of convolutional neural networks in hyperspectral imaging applications: A review. In *Fourth Congress on Intelligent Systems. CIS 2023. Lecture Notes in Networks and Systems*, Kumar, S., Balachandran, K., Kim, J. H., Bansal, J. C. (eds.), Vol. 869. Springer, Singapore. https://doi.org/10.1007/978-981-99-9040-5_14

Madeshwaran, S., & Govindarajan, M. (2024). Deep learning architectures for image processing including convolutional neural networks and generative adversarial networks. In *Image Processing and it's Applications in Computer Vision and Artificia Intelligence* (pp. 215–243). https://doi.org/10.71443/9788197933660-09

Mamapule, S., Esiefarienrhe, B. M., & Obagbuwa, I. C. (2025). Automatic identification and counting of South African animal species in camera traps using deep learning. *International Journal of Intelligent Systems, 2025*(1). https://doi.org/10.1155/int/1561380

Manikandan, S., Karmegam, N., Subbaiya, R., Kaviya, R. S., Govarthanan, M., Shreeharan, D. H., Vickram, S., & Kim, W. (2024). Artificial intelligence-driven sustainability: Enhancing carbon capture for sustainable development goals – A review. *Sustainable Development*, 33(2), 2004–2029. https://doi.org/10.1002/sd.3222

Mobo, F. D., García, A. C. B., & Miłek, K. (2024). Leveraging AI for real-time environmental monitoring. *Advances in Geospatial Technologies Book Series*, 201–212. https://doi.org/10.4018/979-8-3693-8104-5.ch009

Murivhami, T., Tartibu, L. K., & Olayode, I. O. (2023). Applications of artificial neural network in the prediction of water quality index. In *2023 14th International Conference on Mechanical and Intelligent Manufacturing Technologies, ICMIMT 2023* (pp. 100–104). Institute of Electrical and Electronics Engineers Inc. https://doi.org/10.1109/ICMIMT59138.2023.10199388

Naeem, H., Mokharzadeh, A. A., & Khan, S. (2023). Leveraging artificial intelligent model for water quality indices assessment: a comprehensive study and framework. In *2023 International Conference on the Cognitive Computing and Complex Data (ICCD), Huaian, China* (pp. 290–295). https://doi.org/10.1109/ICCD59681.2023.10420685

Neo, E. X., Hasikin, K., Mokhtar, M. I., Lai, K. W., Azizan, M. M., Razak, S. A., & Hizaddin, H. F. (2022). Towards integrated air pollution monitoring and health impact assessment using federated learning: A systematic review. *Frontiers in Public Health, 10*. https://doi.org/10.3389/fpubh.2022.851553

Ogunwumi, O. T., & Matindike, R. (2025). AI applications in environmental monitoring. *Advances in Computational Intelligence and Robotics Book Series*, 65–100. https://doi.org/10.4018/979-8-3373-0680-3.ch003

Oprea, M., Dragomir, E. G., & Carbureanu, M. (2011). On the use of collaborative intelligence in an agent-based environmental monitoring and analysis system. *International Conference on System Theory, Control and Computing* (pp. 1–6).

Ouni, R., & Saleem, K. (2022). Framework for sustainable wireless sensor network based environmental monitoring. *Sustainability, 14*(14), 8356. https://doi.org/10.3390/su14148356

Park, S., & Park, N.-W. (2024). Generation of cloud-free high spatial resolution optical images using spatio-temporal fusion. In *IGARSS 2024 - 2024 IEEE International Geoscience and Remote Sensing Symposium, Athens, Greece* (pp. 9084–9087). https://doi.org/10.1109/igarss53475.2024.10640912

Rajesh, G., Gomadhi, G., Malathi, G., Nehul, J. N., & Krishnaveni, A. (2019). Innovative pathways in environmental monitoring and advanced technologies for sustainable resource management. *Environmental Reports, 1*(1), 17–20. https://doi.org/10.51470/er.2019.1.1.17

Rathi, P., & Vasal, A. (2025). Predictive modeling of air pollution and its impact on public health: A review. In *2025 2nd International Conference On Multidisciplinary Research and Innovations in Engineering (MRIE), Gurugram, India* (pp. 300–304). https://doi.org/10.1109/MRIE66930.2025.11156661

Robinson, J. M., Annells, A., Cavagnaro, T. R., Liddicoat, C., Rogers, H., Taylor, A., & Breed, M. F. (2024). Monitoring soil fauna with ecoacoustics. *Proceedings of the Royal Society B: Biological Sciences, 291*(2030). https://doi.org/10.1098/rspb.2024.1595

Sadegh, R., Moctar, A. O. M., Salihi, M. L., & Nanne, M. F. (2025). A survey of missing data imputation techniques: statistical methods, machine learning models, and GAN-based approaches. *IAES International Journal of Artificial Intelligence*, *14*(4), 2876. https://doi.org/10.11591/ijai.v14.i4.pp2876-2888

Saleh, S., & Battseren, B. (2023). AI-driven solutions for sustainable environment monitoring. *Embedded Selforganising Systems*, *10*(8). https://doi.org/10.14464/ess.v10i8.615

Schmidt, P., & Biessmann, F. (2020). Calibrating human-AI collaboration: impact of risk, *Ambiguity and Transparency on Algorithmic Bias* (pp. 431–449). Springer. https://doi.org/10.1007/978-3-030-57321-8_24

Shobana, D. (2025). AI-powered biosensors for real-time physiological monitoring and response. In *The Role of Artificial Intelligence in Advanced Prosthetics and Implantable Devices* (pp. 373–403). Rademics Reseach Institute. https://doi.org/10.71443/9789349552975-13

Singh, J., Rani, S., & Srilakshmi, G. (2024). Towards explainable AI: Interpretable models for complex decision-making. In *2024 International Conference on Knowledge Engineering and Communication Systems (ICKECS), Chikkaballapur, India* (pp. 1–5). https://doi.org/10.1109/ICKECS61492.2024.10616500

Stowell, D. & Sueur, J. (2020). *Ecoacoustics: Acoustic Sensing for Biodiversity Monitoring at Scale*. Wiley. https://doi.org/10.1002/rse2.174

Thomsen, P. F. & Willerslev, E. (2014). *Environmental DNA: An Emerging Tool in Conservation for Monitoring Past and Present Biodiversity*. Elsevier BV. https://doi.org/10.1016/j.biocon.2014.11.019

Turyasingura, B., Ayiga, N., Kayusi, F., & Tumuhimbise, M. (2024). Application of artificial intelligence (AI) in environment and societal trends: Challenges and opportunities. *Babylonian Journal of Machine Learning*, *2024*, 177–182. https://doi.org/10.58496/bjml/2024/018

Venkadesh, P., Divya, S. V., Marymariyal, P., & Keerthana, S. (2024). *Predicting Natural Disasters with AI and Machine Learning* (pp. 39–64). IGI Global. https://doi.org/10.4018/979-8-3693-3362-4.ch003

Yadav, N., Singh, S. K., & Sharma, D. (2023). Forecasting air pollution for environment and good health using artificial intelligence. In *2023 3rd International Conference on Innovative Sustainable Computational Technologies (CISCT), Dehradun, India* (pp. 1–5). https://doi.org/10.1109/CISCT57197.2023.10351334

Yousefi, S., Mirzaee, S., Almohamad, H., Al Dughairi, A. A., Gomez, C., Siamian, N., Alrasheedi, M., & Abdo, H. G. (2022). Image classification and land cover mapping using Sentinel-2 imagery: Optimization of SVM parameters. *Land*, *11*(7), 993. https://doi.org/10.3390/land11070993

10 Data Preparation, Collecting, Cleaning, and Managing Datasets in Generative AI

Vishal Dutt, Sartajvir Singh, and Ganesh Kumar Sethi

10.1 INTRODUCTION

Generative AI, comprising models like generative adversarial networks (GANs), virtual assistant models (VAEs), and large language models, creates fresh content by discovering patterns from large datasets (Maranghi et al., 2022; Sancricca et al., 2024). Since such models can only perform to the level of trained data, the success of this undertaking relies on the addition of good-quality data. Poor data quality can lead to biased results, poor performance from the model, or ethical concerns regarding the process of modeling (Sancricca et al., 2024). Data preparation, comprising data collection, data cleansing, and data administration, is therefore a critical step in the process of obtaining strong results from generative AI (see Figure 10.1). A good data preparation process makes sure that the collected data sets are correct, representative, and free from any discrepancies. The data collection process entails obtaining a variety of pertinent data to capture the target domain adequately. For data issues such as missing values, duplication, or noise, which might interfere with model learning (Doorly et al., 2025), cleaning is performed. For effective iteration and model training, data organization and structuring to make it accessible and scalable might be identified as data management. These procedures, in general, result in a boost to the reliability, generalizability, and fairness of the model. This chapter presents an overall roadmap for preparing and managing datasets for generative artificial AI applications. Methods, best practices, and tools developed to manage typical data processing challenges are studied in this research paper (Dong & Wang, 2024; Duan, 2024; Gogoi & Valan, 2024; Parida, 2024). These materials should be utilized by artificial AI professionals, developers, and researchers who desire to maximize artificial AI processes. Stakeholders can realize full generative artificial AI capacity by gaining an understanding of and putting in place good data preparation protocols. Innovation and responsible application in a wide range of disciplines will be promoted by this action (Balan et al., 2025; Gebru et al., 2021; Kalluri, 2020).

DOI: 10.1201/9781003616207-12

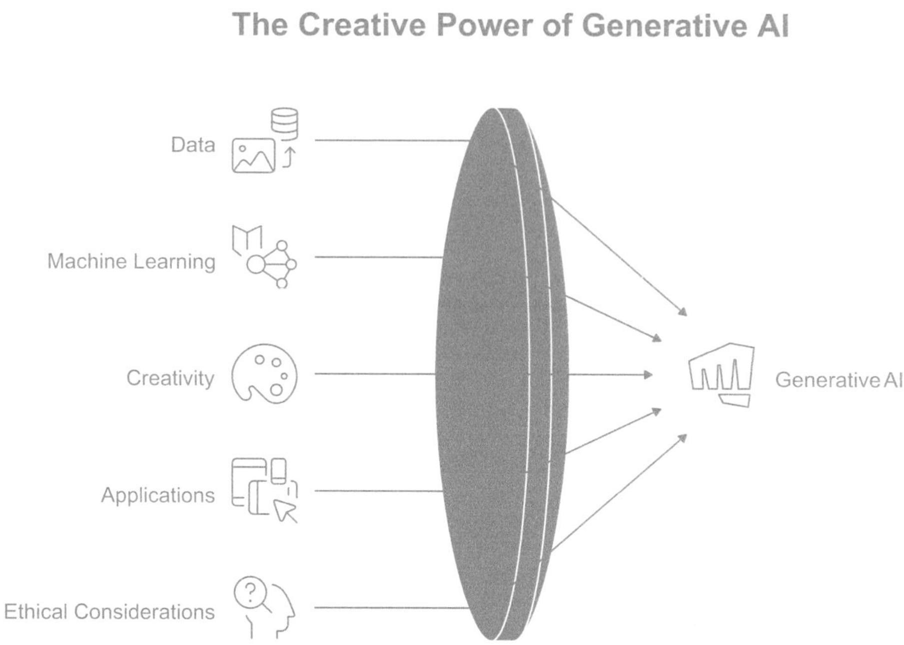

FIGURE 10.1 The concept of generative AI.

10.2 BACKGROUND AND CONTEXT

Differing from classical AI systems, which classify or predict from inputs, generative systems create entirely new information—images, language, music, etc. It is a creative task, so the training data must be varied with much diversity to get a good representation of the target domain. For example, a language generation model with biased or deficient language data will be able to produce offensive or misleading output. In the same manner, image synthesis with a GAN will not learn to generalize true-world variation with training data being specific or repetitive (Radford et al., 2023; Roh et al., 2021). Preparation of this sector poses challenges. Volume is the first—the deep generative networks need large sets to learn tough distributions. Diversity is second—the training sets need to span diverse sets of examples to prevent overfitting or unwanted biases. Quality is third—errors, invalid values, and inconsistencies can produce mis-shaped outputs or non-operationalized networks. These issues take place in the practical world within multiple industries (Ying et al., 2024). For a chatbot program to produce natural speech, it takes billions of text samples from a variety of places. For a computer vision image generation program to produce legitimate pictures, it must be trained with varied, adequately labeled vision sets. Similarly, for sound generation technologies to produce human speech or music sounds, high-fidelity, pure recordings on multiple speakers and settings are needed. Understanding the role of data for these applications is paramount. Data is not fuel for generative AI—it determines its scope, capacity, and potential risks. Good-quality data preparation makes machine creativity viable, reliable, and accountable (Qiu et al., 2024; Yoon et al., 2024).

10.3 DATA COLLECTION FOR GENERATIVE AI

To build a generative AI system, data collection is the most critical step. It is essential to maintain the quality of data, data relevance to the system objectives, and variety of the data during the data collection phase, as it can directly impact the system's performance (Sancricca et al., 2024). The generative AI systems learn and behave based on the provided data, making data collection the most sophisticated task. There are different data sources available to build a robust generative AI system (Baack, 2024), as shown in Figure 10.2.

10.3.1 Data Sources

- **Public datasets**: These are publicly available and widely used for research purposes. Some of these include Common Crawl for webpages, ImageNet for images, and LibriSpeech for sound files. Public datasets are useful for benchmarking as well as for prototyping purposes (Song et al., 2024).
- **Proprietary data**: Companies rely on internal data gathered from within the platform by the company (Garudasu et al., 2024). For instance, a company with images created by users trains a generative image model using the same data. Such data tends to be domain-specific but poses a privacy and consent problem.

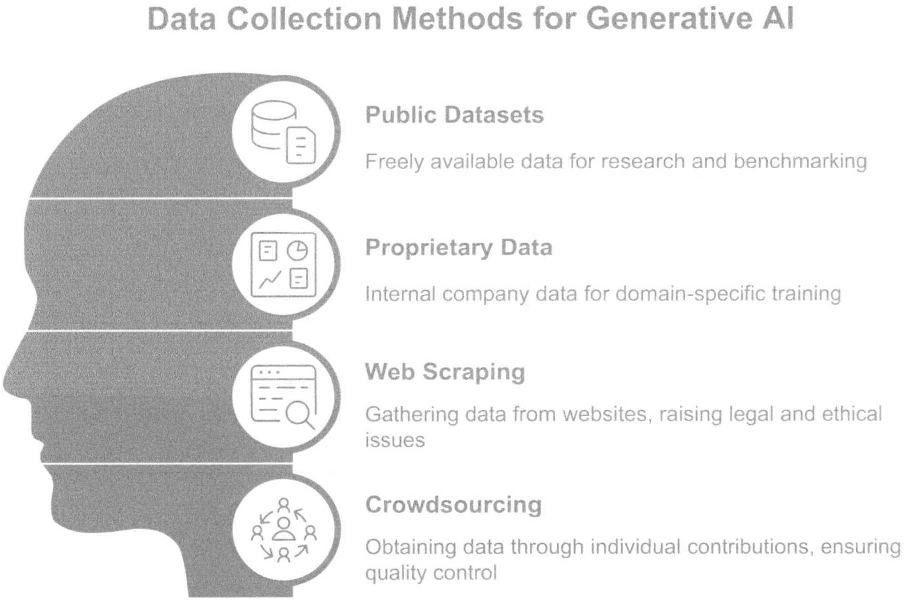

FIGURE 10.2 Different data sources for generative AI.

- **Web scraping**: Some projects gather large-scale data from websites. While scraping allows access to vast content, it can create legal and ethical problems, especially if the data is copyrighted or contains personal information (Rathod et al., 2024).
- **Crowdsourcing**: Data can also be obtained by having individuals create or annotate content. Such customized data is typically obtained through the use of sites like Prolific or Amazon Mechanical Turk. The approach offers more structure and content quality control but is costly (Christoforou et al., 2024).

10.3.2 Ethical Considerations

Data collection must follow ethical guidelines. First, data must be collected with informed consent when it involves individuals. Users should know how their data will be used. Second, privacy must be protected. Personally identifiable information (PII) should be removed or anonymized. Third, bias in data collection must be minimized. If the model is trained with the same data, then the bias is clearly visible in the results of the model. If the model is trained with the same data (same patterns), then the problem of class imbalance occurs, and due to this problem, bias can occur in the performance and output of the model (García-Peñalvo et al., n.d.). For example, if a face detection model is trained only on images of the front profile of a human face, then this model will not be able to identify any other profile, even if it is the face of the same human. All collected data may not be useful in some cases. To build a strong generative model, data should meet several criteria (Bandarapu, 2025):

- **Relevance**: The information has to be relevant to the task. For a law advisor chatbot, information from the internet or blogs is not suitable as a feed. The validated literatures of laws are required.
- **Diversity**: There should be a diverse dataset for training. For example, to build a language translator, it cannot be trained on a single language. The training must be in different languages.
- **Representativeness**: The training data must be balanced for all the available target domains. Skewed data would create biased or shallow models.

10.3.3 Tools and Platforms

There are several instruments (see Figure 10.3) for collecting data efficiently:

- **APIs**: Platforms like Twitter, Reddit, or YouTube offer APIs to gather structured data (Padarha & Vijaylakshmi, 2024; Shivam Hajong, 2024).
- **Open-source repositories**: Hugging Face, Kaggle, and Google Dataset Search host thousands of datasets (Song et al., 2024).
- **Scraping**: Various scraping tools like Scrapy or BeautifulSoup for scraping the data from the web sources (Rathod et al., 2024).

The best example of such a system is ChatGPT, which boasts a vast collection of data across various domains. Due to its extensive data gathering, it can provide

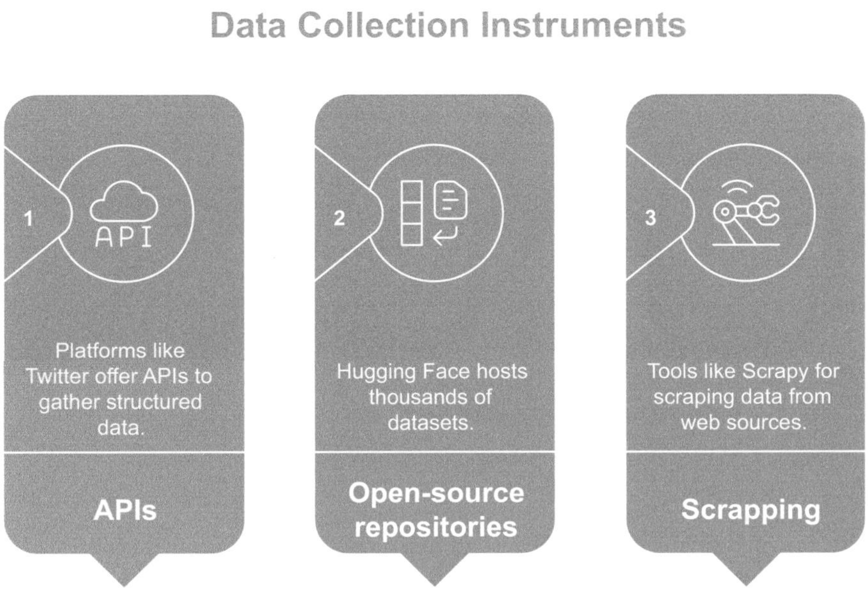

FIGURE 10.3 Various platforms for data collection.

information on any field. Another example is NVIDIA's StyleGAN (Karras et al., 2019). This system is built on an extensive collection of images. Its image repository includes a diverse array of FFHQ (Flicker Faces HQ) (Matuzeviius, 2024), enhancing the system's diversity and scalability.

10.4 DATA CLEANING AND PREPROCESSING

Data cleaning and preprocessing are essential tasks that should be done before training a generative AI model. The collected data may be in its raw form, and it may contain impurities. If impure data is supplied to the model for training, this may lead to poor performance of the model (Alijoyo et al., 2024; Ebert et al., 2025; Elhersh & Alqawasmeh, 2024; Jilani et al., 2022). The following are some common issues, as visualized in Figure 10.4, that may arise during the generative tasks for AI:

- Noise: The random or irrelevant information might be included during the data collection process, which may mislead the learning process of the system.
- Missing values: Collected data may have missing values or incomplete records.
- Duplicates: The redundant data might be there, which may lead to overfitting of the model.
- Inconsistencies: Although the collected data is relevant and appropriate for the system's objectives, it may have inconsistent formats due to the diverse sources from which it was obtained.

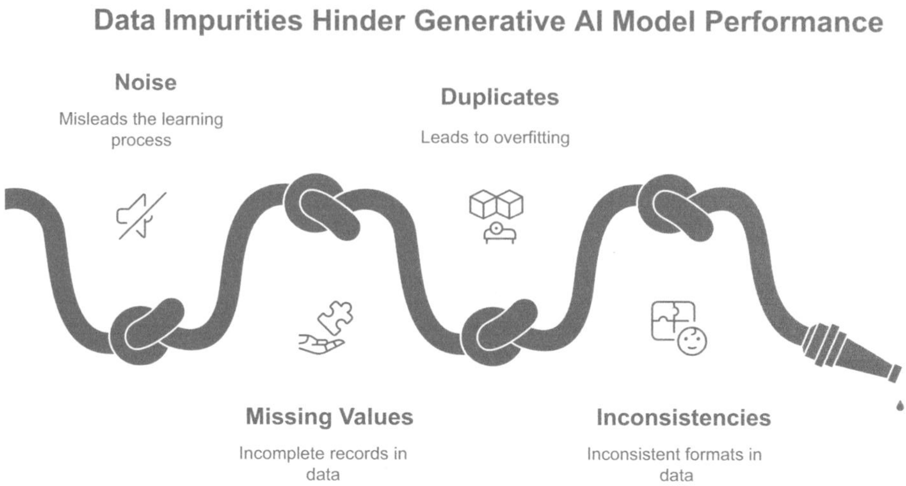

FIGURE 10.4 Various data impurities.

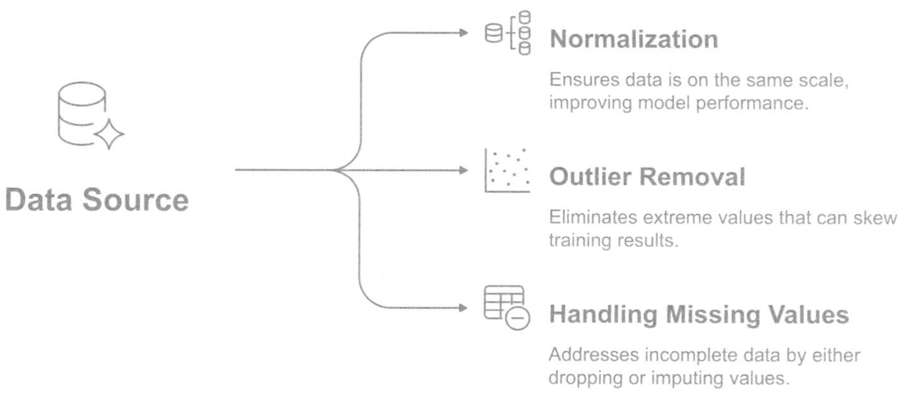

FIGURE 10.5 Data cleaning techniques.

These kinds of data problems are addressed in the data cleaning and preprocessing phase, which plays a critical role in model performance by removing errors, impurities, noise, etc.

10.4.1 Techniques for Data Cleaning

There are several techniques available for data cleaning (Dai et al., 2024; Danendra et al., 2023; Gheni, 2024; Mashhadi et al., 2025; Rather & Kumar, 2024; Sujakumari & Dassan, 2023). The most common techniques (refer to Figure 10.5) are as follows:

- Normalization: In this process, the data is distributed on the same scale. For example, if the marks are from 0 to 100, then scale them between 0 and 10.

- Outlier removal: In this process, values that are extremely low or high on the scale are eliminated so that they cannot impact the training.
- Handling missing values: In this process, those records that have missing values are dropped. Apart from this, if it is not appropriate to drop the records, then the missing values are filled with the help of a mean or model-based imputation method.

In the development of a generative AI system, various types of data (text, images, audio) require different data preprocessing. There are different data preprocessing techniques given according to the data type (Alipour & Gallegos, 2025; Becerra et al., 2024; Jin et al., 2024; Liu et al., 2025; Pushpalatha & Shree, 2025), as given in Figure 10.6.

Text

- Tokenization: To analyze the text, the complete text is split into small chunks (words or sub-words).
- Stemming: In this process, the suffixes are removed to get the simplest form of the text.
- Lemmatization: Converting the word to the root word by applying language rules.
- Stop-word removal: The process of removing stop-words, for example, "and" or "the" that contribute minimally, decreases vocabulary size and enhances learning by the model.

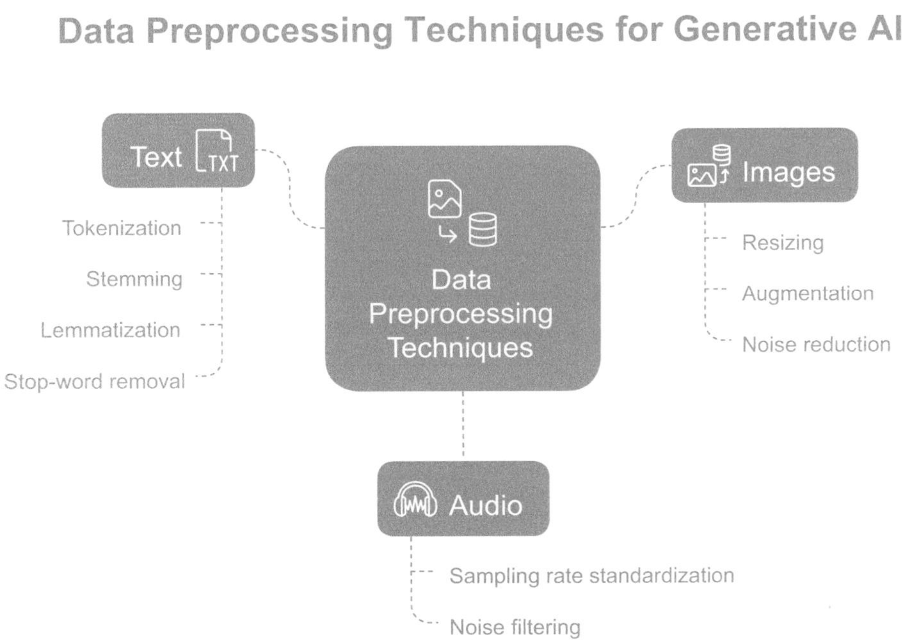

FIGURE 10.6 Data types and suitable preprocessing techniques.

Images

- Resizing: Preparing all pictures to be the same size to enable batch processing.
- Augmentation: This process involves creating new training samples from a single image by performing various operations like flipping, color changes, rotation, scaling, etc.
- Noise reduction: The extraneous pixels that may confuse the model are removed.

Audio

- Sampling rate standardization: It is performed to ensure the sampling rate of each audio file.
- Noise filtering: The extra ambient or static noise embedded in the voice/audio may make the speech synthesis harder, so the removal or filtering of these noises is compulsory.

10.4.2 Tools and Libraries

Python is the most powerful language for dealing with data cleaning and preprocessing, as it contains an extensive collection of tools and libraries for such tasks (Ebsen et al., 2024; Mahendar, 2025; Vetluzhskikh et al., 2024), as depicted in Figure 10.7.

- Pandas: It is a Python library that provides robust and is well-suited for structured data. It is very helpful in identifying, handling, and normalizing missing values.

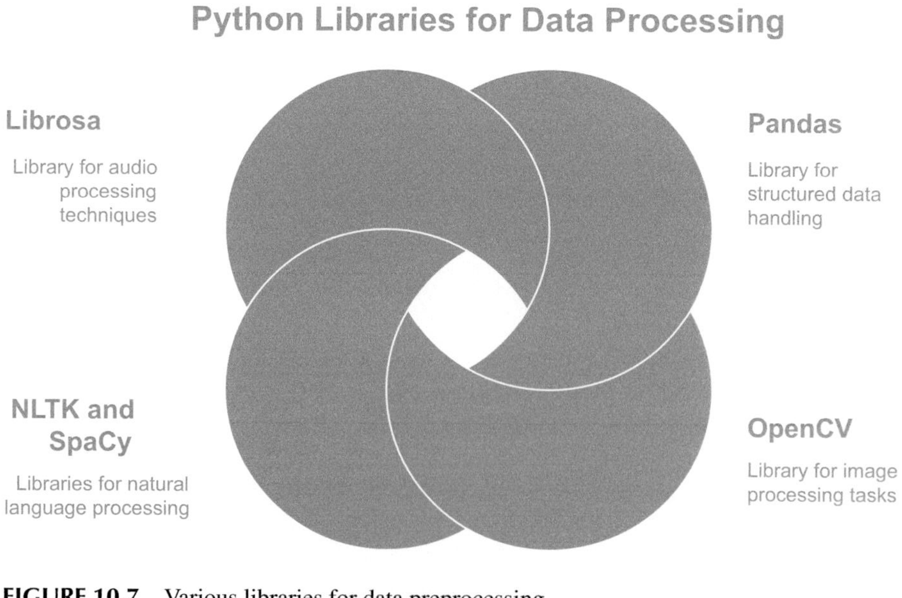

FIGURE 10.7 Various libraries for data preprocessing.

- OpenCV: This library is primarily used for image processing tasks such as resizing, cropping, and filtering.
- NLTK and SpaCy: These are Python libraries that are primarily used for Natural Language Processing (NLP). These libraries offer tokenization, stop-word removal, and lemmatization for the text.
- Librosa: To process the audio, Librosa provides several techniques like sampling, noise reduction, etc.

10.5 DATA MANAGEMENT AND ORGANIZATION

How good an AI generative model has less to do with the algorithm and more to do with the quality of data management. A poorly tracked, unstructured dataset may result in wasted effort, inconsistent output, and regulatory problems. Good data management helps lay the groundwork for efficient development and future reliability.

A. Structuring Datasets

Well-structured data is easier to browse, edit, and verify. The choice of file format begins the process. CSV works well for small tabular data, but JSON is more flexible if the data are nested or hierarchical. For larger or more complex data, HDF5 or Parquet are better choices because of their improved read/write times and better compressibility. Besides format, the directory structure has an impact as well. A flat hierarchy—by data type, data origin, or date—enables collaborators to instantly access the data they require. A single, dedicated metadata file per directory ensures every dataset has clear descriptions of contents, source, and data preprocessing (Varma et al., 2024).

B. Versioning and Reproducibility

Data changes with time, and if version control is not appropriately utilized, the results of the past cannot be reproduced. Systems such as DVC (Data Version Control) are embedded with Git to track dataset versions as well as the code, and teams are thus allowed to roll back to an older version if needed. This prevents confusion when results are unanticipated from the experiment and guarantees that results presented later are reproducible. Reproducibility entails recording the concrete preprocessing steps used on every version of the dataset (Ceri et al., 2013; Daniotti et al., 2025). This could be achieved through the use of automated scripts or a pipeline that logs data detailing every manipulation of the input data, from raw to final training data.

As data volumes scale, storage locally becomes unfeasible. Flexible cloud storage provides a way to scale capacity without hardware upgrades. Services such as AWS S3, Google Cloud Storage, and Azure Blob Storage even come with integration with huge databases to enable fast queries and retrieval. For huge data volumes, consider storage and database engines suited to the workload of big data. Techniques such as partitioning, caching, and indexing ensure performance remains good even when handling billions of rows.

C. Data Security and Compliance

Generative AI projects generally involve sensitive or restricted data. Security must be embedded into data storage and transfer. Encryption, either at rest or in transit, protects data from unauthorized access. Complying with regulatory requirements such as GDPR or CCPA entails more than encryption. Data has to be pseudonymized or anonymized to remove identifiable information. Accurate records of the consent and of the data processing serve to comply with legal requisites and to engender trust among users (Siddappa & Researcher, 2021).

D. Workflow Management

A working workflow ensures data moves steadily from source to model. This generally means automated pipelines for ingesting, cleaning, and integrating data. A good pipeline won't just ingest data but will also validate data at every step of the way in order to catch problems early. In handling ingestion, preprocessing, and integration in a single system, teams avoid human errors and ensure datasets are always up to date and in sync with model demands. This further ensures that scaling or replacing data sources becomes easier (Bhagat, 2025).

Data management for generative AI is a mix of structure, discipline, and adaptability. Clear organization, strong version control, scalable storage, rigorous security, and automated workflows together form the backbone of a sustainable AI development process. When these elements are in place, teams can focus on innovation instead of firefighting data problems.

10.6 CHALLENGES AND SOLUTIONS IN DATA PREPARATION FOR GENERATIVE AI

Preparing data for generative AI models involves more than gathering information—it involves making sure the data enables good, fair, and efficient model behavior. At each stage, there are unique obstacles that, if overlooked, can compromise model performance and credibility. Here are four common pitfalls and practical solutions (Javaid et al., 2024; Liu et al., 2024; Muthusamy & Lee, 2024; Patel & Lim, 2024).

10.6.1 Bias and Fairness in Datasets

Generative AI models are trained on patterns from the data they are provided with. When such data is biased socially, culturally, or demographically, these models may unwittingly reproduce or even exacerbate biases. This may give rise to unfair results, particularly in such sensitive uses as recruitment, lending, or healthcare. To overcome this issue, the first step is detection. Tools that measure representation balance, sentiment differences, or correlation with sensitive attributes can help identify bias before training begins. Once identified, data teams should source material from multiple, diverse channels, ensuring coverage across demographic, geographic, and cultural dimensions. Fairness-aware preprocessing techniques—such as reweighting samples, anonymizing sensitive attributes, or balancing classes—can further reduce bias without distorting the dataset's core meaning (Zhao et al., 2025). Figure 10.8 visualizes the concept of bias and fairness in datasets.

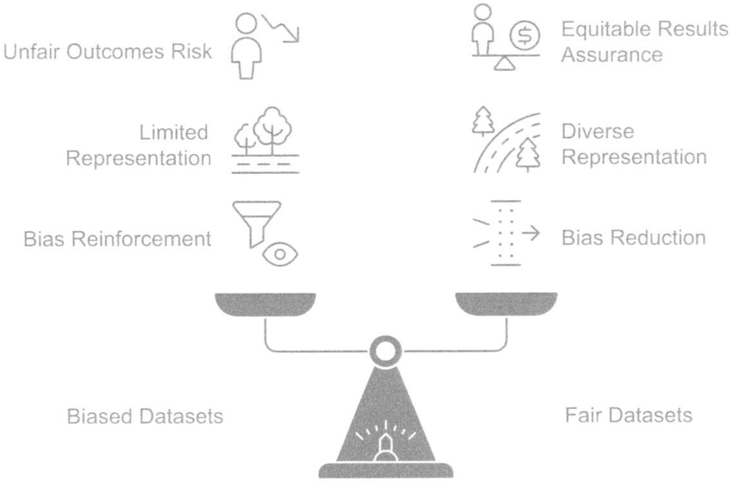

FIGURE 10.8 Balancing bias for equitable AI.

10.6.2 Scalability for Big Data Sets

As generative models become more complex, they demand enormous volumes of data. Storing, accessing, and processing billions of records can strain infrastructure and budgets. Poorly managed scaling often leads to slow training cycles and high operational costs. Distributed computing frameworks (such as Apache Spark or Ray) enable parallel computing across multiple nodes, and training time decreases drastically. Web-based environments have elastic scaling, allowing storage and compute to expand or shrink on demand. Data must also be stored in compact formats—like Parquet or ORC—that compress nicely and enable quick read/write operations. Keeping everything together makes large-scale computing practical and cost-effective (Li et al., 2024; Vu et al., 2024). The practical solutions for the scalability issue for large datasets are suggested graphically in Figure 10.9.

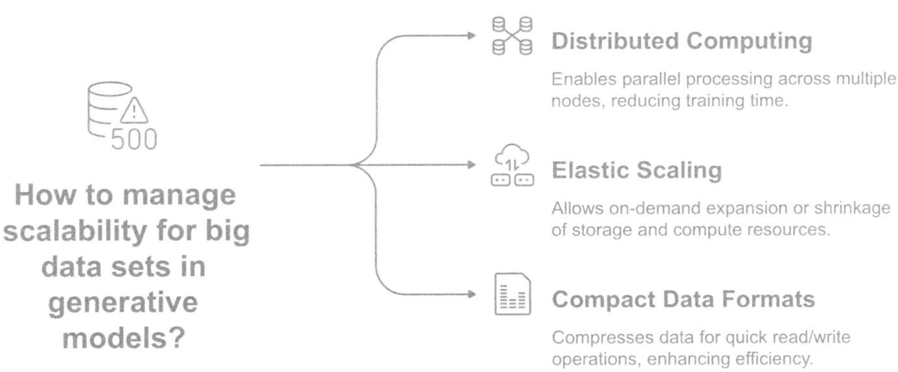

FIGURE 10.9 Scalability issue and solution.

10.6.3 Lack of Data for Niche Markets

In highly specialized domains—like rare disease medicine or low-resource languages—getting sufficient quality data tends to be the primary barrier. Scraping or publicly available data collections don't tend to give us sufficient instances to train effectively on. Synthetic data generation fills the gap, employing controlled simulations or small generative models to generate more instances corresponding to the attributes of the initial dataset. Transfer learning provides another avenue: models learned on more comprehensive, related datasets are fine-tuned using small niche datasets, minimizing data demands. Data augmentation—by means of transformation, paraphrasing, or domain-specific manipulation—also boosts dataset variety without adding noise (Andreoni et al., 2024). See Figure 10.10 for a better understanding of data insufficiency.

10.6.4 Preserving Data Quality When Training Models

Even a carefully curated dataset may deteriorate in quality during training—due to corruption, mislabeling, or missed anomalies. The deterioration may cause model drift or unanticipated behavior. The solution to this issue is to set up constant monitoring pipelines to report abnormalities in real time. Add validation checks in between training epochs to catch things early. Incremental cleaning—re-evaluation and correction of data throughout the project—keeps the dataset consistent with quality expectations. This step should have automated checks and human verification of nuanced situations (David et al., 2024; Javaid et al., 2024); see Figure 10.11.

Data preparation is not a one-time task but an ongoing process. Addressing bias, scaling efficiently, handling scarcity, and maintaining quality are all essential to building generative AI models that perform well and produce trustworthy results.

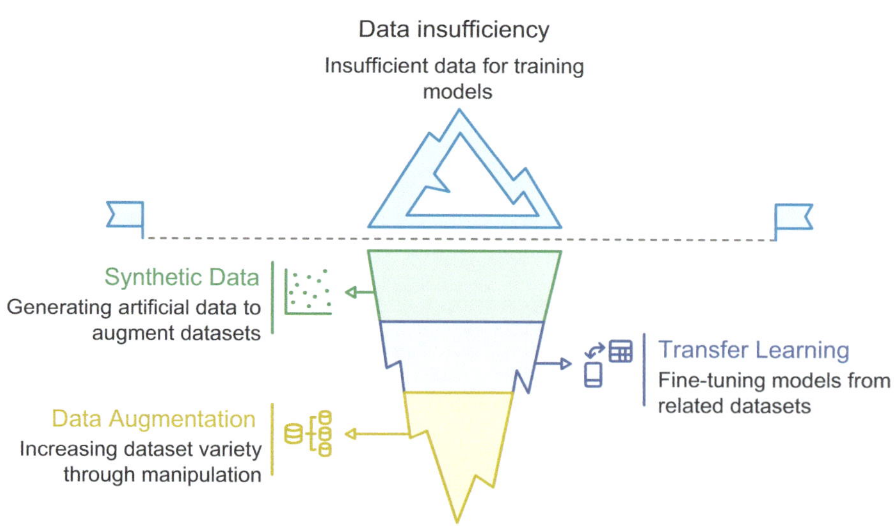

FIGURE 10.10 Overcoming data insufficiency.

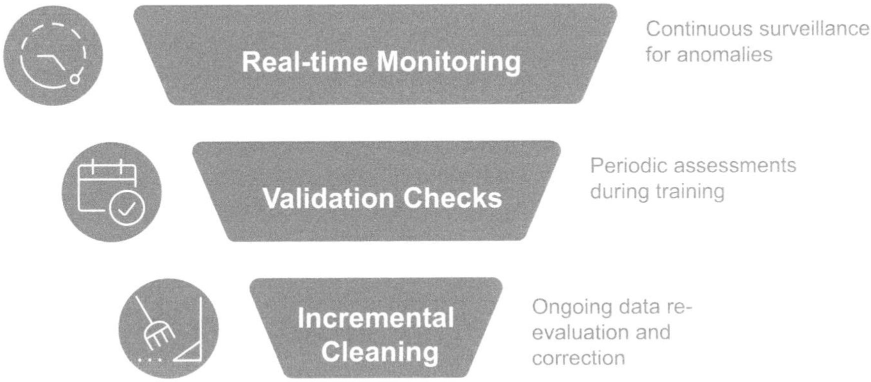

FIGURE 10.11 Maintaining data quality in model training.

By combining technical strategies with careful oversight, teams can create datasets that truly support the goals of their AI systems.

10.7 CASE STUDIES AND PRACTICAL EXAMPLES

- **Case Study 1:** Preparing Wikipedia Text for Fine-Tuning of LLMs
  Cleaned Wikipedia dumps are used regularly by OpenAI and scientists to fine-tune deep language models. A tutorial on fine-tuning using Wikipedia data details the processes of extracting, cleaning, and formatting specified topic updates (such as the last hurricane data updates), and then feeding them into a fine-tuning pipeline to enhance model relevance (Javaid et al., 2024). Another paper examines the more extensive approaches to preparing Wikimedia data for NLP applications—pointing to processes such as cleaning, filtering according to language, and normalizing article content in an effort to create trustworthy data sets to train models on (Johnson et al., 2024).
- **Case Study 2:** Curating and Augmenting Face Image Datasets for GANs
  Experiments explored why GANs create artificial face data to supplement sparse real-world data—and showed the potential of such artificial instances to enhance model performance on image tasks (Sáez Trigueros et al., 2021). The StyleGAN-Human project trained a GAN data-first. The team brought together and labeled over 230,000 images of humans—over varying poses and textures—and systematically varied dataset size, balance, and alignment, and corresponding effects on generation quality (Fu et al., 2022). Transfer learning with small datasets of images is another relevant case, where pre-trained GANs are fine-tuned given small, new datasets like scraped images from the internet, to credibly generate images in new environments (Jilani et al., 2022).
- **Case Study 3:** Managing Text-and-Image Multimodal Datasets
  In text- and image-data combined applications, authors stress important preparation steps: A recently published article reviews how multimodal AI

systems—such as Contrastive Language-Image Pre-training (CLIP) and image-text generation tools—depend on well-aligned image-text pairs. It outlines the importance of metadata linkage and preprocessing to establish consistent connections between modalities. Another piece on multimodal Retrieval-Augmented Generation (RAG) systems shows how embedding both text and images into a vector database enables efficient retrieval in creative tools—demonstrating real-world methods of aligning and integrating multimodal datasets (Garudasu et al., 2024).

10.8 FUTURE TRENDS IN DATA PREPARATION FOR GENERATIVE AI

Data preparation of generative AI is shifting toward more automation, integration, and compliance. Breakthroughs in automated cleaning and preprocessing, often fueled by AutoML methods, are minimizing the need for labor-intensive intervention by identifying anomalies, filling in missing information, and streamlining data format with minimal intervention. This is being supplemented by the generation of synthetic data combined with real-world data, enabling models to transcend scarcity in niche applications while maintaining diversity and realistic appearance (Garudasu et al., 2024; Luo et al., 2020). Federated learning is becoming an important method of decentralized data preparation, permitting organizations to jointly train models on dispersed data without disclosure of sensitive information—a capability ever more important in healthcare, finance, and other privacy-conscious industries (Cui et al., 2024). At the same time, maturing regulations like GDPR, CCPA, and mooted AI-centric legislation are defining permissible ways data may be harvested, stored, and processed, compelling adoption of privacy-by-design principles and more stringent governance processes. In the next five to ten years, generative AI data pipelines are set to become more autonomous, adaptive, and regulation-sensitive, with end-to-end mechanisms capable of ingesting raw multi-format data, validating in real time, supplementing with synthetic sources, and orchestrating decentralized workflows safely. These trends foreshadow a future in which high-quality, compliant data assets may be prepared more quickly and at scale hitherto unaffordable (Kandulna, n.d.).

10.9 CONCLUSION

Vigorous data preparation is the cornerstone of any successful generative AI effort. From the outset of data gathering to thorough cleaning, thoughtful organization, and continued management, every phase has a direct impact on the resulting models' reliability, fairness, and creative potential. High-quality data not only enhances accuracy but also decreases bias, boosts scalability, and helps ensure compliance with shifting regulations. Best practices for researchers and practitioners are to create a dataset structure with explicit formats and metadata, use versioning for reproducibility, take advantage of scalable storage offerings, and include validation checks throughout the pipeline. The use of multiple diverse real-world sources combined with synthetic augmentation has the potential to fill gaps, and automating workflow at scale ensures compound improvements in performance and trustworthy results.

The message is simple: make data quality an equal priority—maybe even greater—than model architecture to realize generative AI's full potential. Teams devoted to proper data preparation at a high level will have an advantage in performance and trustworthiness. For more information, some sources to explore are the *Journal of Machine Learning Research*, the arXiv repository, and technology guides by sites such as TensorFlow and Hugging Face, which provide extensive strategies, case studies, and tools. Ultimately, however, the journey to revolutionary generative AI starts not with code but with disciplined, thoughtful data stewardship.

REFERENCES

Alijoyo, F. A., Reddy, L. C. S., Selvi, V., Murugan, R., Kakad, S., & Balakumar, A. (2024). Transforming business with Generative AI models applications trends. *2024 International Conference on Artificial Intelligence and Quantum Computation-Based Sensor Application (ICAIQSA)*, 1–5. https://api.semanticscholar.org/CorpusID:276510597

Alipour, P., & Gallegos, E. (2025). Leveraging generative AI, synthetic, and social media data for content generalizability to overcome data constraints in vision deep learning. *Artificial Intelligence Review*, *58*, 145. https://api.semanticscholar.org/CorpusID:276590187

Andreoni, M., Lunardi, W. T., Lawton, G., & Thakkar, S. P. (2024). Enhancing autonomous system security and resilience with Generative AI: A comprehensive survey. *IEEE Access*, *12*, 109470–109493. https://api.semanticscholar.org/CorpusID:271763176

Baack, S. (2024). A critical analysis of the largest source for Generative AI training data: Common Crawl. *Proceedings of the 2024 ACM Conference on Fairness, Accountability, and Transparency*. https://api.semanticscholar.org/CorpusID:270286602

Balan, K., Gilbert, A., & Collomosse, J. P. (2025). Content ARCs: Decentralized content rights in the age of Generative AI. *ArXiv, abs/2503.14519*. https://api.semanticscholar.org/CorpusID:277113230

Bandarapu, S. R. (2025). Enhancing transparency in Generative AI: Strategies for ethical and accountable deployment. *The Edge Review*. https://api.semanticscholar.org/CorpusID:277233218

Becerra, Á., Mohseni, Z. (Artemis), Sanz, J., & Pérez, R. C. (2024). A Generative AI-based personalized guidance tool for enhancing the feedback to MOOC learners. *2024 IEEE Global Engineering Education Conference (EDUCON)*, 1–8. https://api.semanticscholar.org/CorpusID:271068959

Bhagat, S. K. (2025). Enterprise architecture in the age of Generative AI: Adapting ERP systems for next-generation automation. *International Journal of Scientific Research in Computer Science, Engineering and Information Technology*. https://api.semanticscholar.org/CorpusID:276849904

Ceri, S., Bozzon, A., Brambilla, M., Valle, E. Della, Fraternali, P., & Quarteroni, S. (2013). *An Introduction to Information Retrieval*. https://api.semanticscholar.org/CorpusID:262553219

Christoforou, E., Demartini, G., & Otterbacher, J. (2024). Generative AI in crowdwork for web and social media research: A survey of workers at three platforms. *International Conference on Web and Social Media*. https://api.semanticscholar.org/CorpusID:270193469

Cui, L., Li, H., Chen, K., Shou, L., & Chen, G. (2024). Tabular data augmentation for machine learning: Progress and prospects of embracing Generative AI. *ArXiv, abs/2407.21523*. https://api.semanticscholar.org/CorpusID:271571270

Dai, S., Liu, M.-Y., Valle, R., & Gururani, S. (2024). ExpressiveSinger: Multilingual and multi-style score-based singing voice synthesis with expressive performance control. *Proceedings of the 32nd ACM International Conference on Multimedia*. https://api.semanticscholar.org/CorpusID:273641653

Danendra, A. D., Kaurow, J. E., Osvaldo, N., Pratama, G. D., & Hasana, S. (2023). Traffic congestion classification using convolutional neural networks: A simplified approach to traffic management. *2023 6th International Conference on Information and Communications Technology (ICOIACT)*, 171–176. https://api.semanticscholar.org/CorpusID:268254790

Daniotti, S., Wachs, J., Feng, X., & Neffke, F. M. H. (2025). Who is using AI to code? Global diffusion and impact of generative AI. *ArXiv, abs/2506.08945*. https://api.semanticscholar.org/CorpusID:279261182

David, T., Muhammad, K., Nassisid, K., & Farus, B. (2024). Enhancing vehicular networks with Generative AI: Opportunities and challenges. *ArXiv, abs/2407.11020*. https://api.semanticscholar.org/CorpusID:271218297

Dong, Z. Y., & Wang, T. (2024). Artificial intelligence driving perception, cognition, decision-making, and deduction in energy systems: State-of-the-art and potential directions. *Energy Internet*. https://api.semanticscholar.org/CorpusID:272170039

Doorly, R., Ong, J., Waisberg, E., Sarker, P., Zaman, N., Tavakkoli, A., & Lee, A. G. (2025). Applications of generative adversarial networks in the diagnosis, prognosis, and treatment of ophthalmic diseases. *Graefe's Archive for Clinical and Experimental Ophthalmology = Albrecht von Graefes Archiv Fur Klinische Und Experimentelle Ophthalmologie*. https://api.semanticscholar.org/CorpusID:277996327

Duan, J. (2024). Deep learning anomaly detection in AI-powered intelligent power distribution systems. *Frontiers in Energy Research*. https://api.semanticscholar.org/CorpusID:268431141

Ebert, C., Nawalramka, P., & Ebert, C. (2025). Generative AI for data science. *IEEE Software*, *42*, 20–24. https://api.semanticscholar.org/CorpusID:279127401

Ebsen, T., Segall, R. S., Aboudja, H., & Berleant, D. (2024). A customer service chatbot using Python, machine learning, and artificial intelligence. *Proceedings of the 15th International Multi-Conference on Complexity, Informatics and Cybernetics: IMCIC 2024*. https://api.semanticscholar.org/CorpusID:268518806

Elhersh, G. A., & Alqawasmeh, H. K. (2024). Public sentiment and ethical considerations of ChatGPT in higher education: Insights from data analytics of conversations on platform X. *Journal of Infrastructure, Policy and Development*. https://api.semanticscholar.org/CorpusID:273685391

Fu, J., Li, S., Jiang, Y., Lin, K.-Y., Qian, C., Loy, C. C., Wu, W., & Liu, Z. (2022). StyleGAN-Human: A data-centric odyssey of human generation. *ArXiv Preprint ArXiv:2204.11823*.

García-Peñalvo, P. F.-J., Casañ-Guerrero, M.-J., Alier-Forment, M., & Pereira-Varela, J.-A. (n.d.). The ethics of generative artificial intelligence in education are under debate. A perspective from the development of a theoretical-practical case study. https://api.semanticscholar.org/CorpusID:279014036

Garudasu, S., Mehrotra, T., Banoth, D. N., Bhat, S. R., Akisetty, A. S. V. V., & Jain, A. (2024). Generative AI model development for enhancement of real-time medical image-based training datasets. *2024 International Conference on Advances in Computing, Communication and Materials (ICACCM)*, 1–7. IEEE. https://doi.org/10.1109/icaccm61117.2024.11059126

Gebru, T., Morgenstern, J., Vecchione, B., Vaughan, J. W., Wallach, H., Daumé, H. III, & Crawford, K. (2021). Datasheets for datasets. *Communications of the ACM*, *64*(12), 86–92. https://doi.org/10.1145/3458723

Gheni, A. A. (2024). Urban traffic congestion prediction using GTFS data and advanced machine learning models. *International Research Journal of Innovations in Engineering and Technology*. https://api.semanticscholar.org/CorpusID:273385017

Gogoi, P., & Valan, J. A. (2024). Machine learning approaches for predicting and diagnosing chronic kidney disease: Current trends, challenges, solutions, and future directions. *International Urology and Nephrology*. https://api.semanticscholar.org/CorpusID:274139333

Javaid, S., Fahim, H., He, B., & Saeed, N. (2024). Large language models for UAVs: Current state and pathways to the future. *IEEE Open Journal of Vehicular Technology*, *5*, 1166–1192. https://doi.org/10.1109/ojvt.2024.3446799

Jilani, U., Asif, M., Rashid, M., Siddique, A. A., Talha, S. M. U., & Aamir, M. (2022). Traffic congestion classification using GAN-based synthetic data augmentation and a novel 5-layer convolutional neural network model. *Electronics*, *11*(15), 2290. https://doi.org/10.3390/electronics11152290

Jin, Y., Carrasco-Revilla, A., & Chen, M. (2024). iGAiVA: Integrated Generative AI and visual analytics in a machine learning workflow for text classification. *ArXiv, abs/2409.15848*. https://api.semanticscholar.org/CorpusID:272831935

Johnson, I., Kaffee, L.-A., & Redi, M. (2024). Wikimedia data for AI: A review of Wikimedia datasets for NLP tasks and AI-assisted editing. *ArXiv Preprint ArXiv:2410.08918*.

Kalluri, P. (2020). Don't ask if artificial intelligence is good or fair, ask how it shifts power. *Nature*, *583*(7815), 169–169. https://doi.org/10.1038/d41586-020-02003-2

Kandulna, B. (n.d.). *A study of the prospects of higher education*. https://api.semanticscholar.org/CorpusID:274435958

Karras, T., Laine, S., Aittala, M., Hellsten, J., Lehtinen, J., & Aila, T. (2019). Analyzing and improving the image quality of StyleGAN. *2020 IEEE/CVF Conference on Computer Vision and Pattern Recognition (CVPR)*, 8107–8116. https://api.semanticscholar.org/CorpusID:209202273

Li, B., Tripathi, S., Hosain, S., Zhang, R., Xie, J., & Wang, M. (2024). When learning meets dynamics: Distributed user connectivity maximization in UAV-based communication networks. *ArXiv, abs/2409.06010*. https://api.semanticscholar.org/CorpusID:272550401

Liu, G., Van Huynh, N., Du, H., Hoang, D. T., Niyato, D. T., Zhu, K., Kang, J., Xiong, Z., Jamalipour, A., & Kim, D. I. (2024). Generative AI for unmanned vehicle swarms: Challenges, applications and opportunities. *ArXiv, abs/2402.18062*. https://api.semanticscholar.org/CorpusID:268041856

Liu, R., Bin Ismail, A. I., & Lu, Y. (2025). Research on enhancing the generalization ability of Generative AI based on K-means clustering algorithm: Text-to-image task for wind-responsive passive building facade. *Proceedings of the 2025 International Conference on Generative Artificial Intelligence and Digital Media*. https://api.semanticscholar.org/CorpusID:279453528

Luo, Y., Zhang, Z., Chen, Y., Li, Z., & Tao, F. (2020). ChinaCropPhen1km: a high-resolution crop phenological dataset for three staple crops in China during 2000–2015 based on leaf area index (LAI) products. *Earth System Science Data*, *12*, 197–214. https://api.semanticscholar.org/CorpusID:212701553

Mahendar, M. (2025). A survey on AI powered personalized learning platform. *International Journal of Scientific Research in Engineering and Management*. https://api.semanticscholar.org/CorpusID:279137698

Maranghi, M., Anagnostopoulos, A., Cannistraci, I., Chatzigiannakis, I., Croce, F., Di Teodoro, G., Gentili, M., Grani, G., Lenzerini, M., Leonardi, S., Mastropietro, A., Palagi, L., Pappa, M., Rosati, R., Valentini, R., & Velardi, P. (2022). AI-based data preparation and data analytics in healthcare: The case of diabetes. *ArXiv, abs/2206.06182*. https://api.semanticscholar.org/CorpusID:249626222

Mashhadi, A. H., Rashidi, A., Hamedi, M., & Marković, N. (2025). Traffic estimation in work zones using a custom regression model and data augmentation. *Computer-Aided Civil and Infrastructure Engineering*, *40*, 1422–1444. https://api.semanticscholar.org/CorpusID:275373054

Matuzeviius, D. (2024). Diverse dataset for eyeglasses detection: Extending the Flickr-Faces-HQ (FFHQ) dataset. *Sensors (Basel, Switzerland)*, *24*. https://api.semanticscholar.org/CorpusID:274710973

Muthusamy, M., & Lee, S. I. (2024). Abiotic stress-induced secondary metabolite production in Brassica: opportunities and challenges. *Frontiers in Plant Science, 14*. https://api.semanticscholar.org/CorpusID:266833585

Padarha, S., & Vijaylakshmi, S. (2024). A multifaceted approach at discerning Redditors feelings towards ChatGPT. *EAI Endorsed Transactions on Internet of Things*. https://api.semanticscholar.org/CorpusID:270829839

Parida, S. (2024). Multiscalar brain adaptability in AI systems. *Journal of Multiscale Neuroscience*. https://api.semanticscholar.org/CorpusID:275147602

Patel, N. S., & Lim, J. T. (2024). Critical design futures thinking and GenerativeAI: a Foresight 3.0 approach in higher education to design preferred futures for the industry. *Foresight*. https://api.semanticscholar.org/CorpusID:273679921

Pushpalatha, M., & Shree, M. (2025). Comparative analysis of web scraping methodologies using Generative AI. *2025 6th International Conference on Recent Advances in Information Technology (RAIT)*, 1–6. https://api.semanticscholar.org/CorpusID:280394194

Qiu, Y., Duvvuri, V., Yadavalli, P., & Prasad, N. (2024). Evaluation of Generative AI Q&A Chatbot chained to optical character recognition models for financial documents. *Proceedings of the 2024 8th International Conference on Machine Learning and Soft Computing*. https://api.semanticscholar.org/CorpusID:269114943

Radford, A., Kim, J. W., Xu, T., Brockman, G., Mcleavey, C., & Sutskever, I. (2023). Robust speech recognition via large-scale weak supervision. In A. Krause, E. Brunskill, K. Cho, B. Engelhardt, S. Sabato, & J. Scarlett (Eds.), *Proceedings of the 40th International Conference on Machine Learning* (Vol. 202, 28492–28518). PMLR. https://proceedings.mlr.press/v202/radford23a.html

Rather, I. H., & Kumar, S. (2024). Generative adversarial network based synthetic data training model for lightweight convolutional neural networks. *Multimedia Tools and Applications, 83*(2), 6249–6271. https://doi.org/10.1007/s11042-023-15747-6

Rathod, M., Yadav, A., Phutane, S., & Bhise, S. (2024). Comment compass: An approach to analyze YouTube comments through web scraping, sentiment analysis, and Generative AI for actionable insights. *2024 First International Conference on Pioneering Developments in Computer Science & Digital Technologies (IC2SDT)*, 1–5. https://api.semanticscholar.org/CorpusID:273103951

Roh, Y., Heo, G., & Whang, S. E. (2021). A survey on data collection for machine learning: A Big Data - AI integration perspective. *IEEE Transactions on Knowledge and Data Engineering, 33*(4), 1328–1347. https://doi.org/10.1109/TKDE.2019.2946162

Sáez Trigueros, D., Meng, L., & Hartnett, M. (2021). Generating photo-realistic training data to improve face recognition accuracy. *Neural Networks, 134*, 86–94. https://doi.org/10.1016/j.neunet.2020.11.008

Sancricca, C., Siracusa, G., & Cappiello, C. (2024). Enhancing data preparation: Insights from a time series case study. *The Journal of Intelligent Information Systems, 62*, 1503–1530. https://api.semanticscholar.org/CorpusID:271470236

Shivam, H. (2024). Enhancing voice assistants with Generative AI language models for interactive conversations. *International Scientific Journal of Engineering and Management*. https://api.semanticscholar.org/CorpusID:269764212

Siddappa, N. B. R., & Researcher, I. (2021). AI-driven document management systems: revolutionizing information retrieval and workflow automation. *World Journal of Advanced Research and Reviews*. https://api.semanticscholar.org/CorpusID:275599054

Song, F., Agarwal, A., & Wen, W. (2024). The impact of Generative AI on collaborative open-source software development: evidence from GitHub Copilot. *ArXiv, abs/2410.02091*. https://api.semanticscholar.org/CorpusID:271136080

Sujakumari, P. D., & Dassan, P. (2023). Generative adversarial networks (GAN) and HDFS-based realtime traffic forecasting system using CCTV surveillance. *Symmetry, 15*, 779. https://api.semanticscholar.org/CorpusID:257735626

Varma, S., Shivam, S., Ray, B., & Biswas, S. (2024). Reimagining enterprise data management using generative artificial intelligence. *2024 11th IEEE Swiss Conference on Data Science (SDS)*, 107–114. https://api.semanticscholar.org/CorpusID:272722808

Vetluzhskikh, M., Gnanasekaran, R. K., & Marciano, R. (2024). Can Generative AI uncover hidden patterns in historical domestic traffic ads through data analysis? A ChatLoS-DTA exploration. *2024 IEEE International Conference on Big Data (BigData)*, 2543–2548. https://api.semanticscholar.org/CorpusID:275586570

Vu, T.-H., Jagatheesaperumal, S. K., Nguyen, M.-D., Van Huynh, N., Kim, S., & Pham, V. Q. (2024). Applications of Generative AI (GAI) for mobile and wireless networking: A survey. *IEEE Internet of Things Journal*, *12*, 1266–1290. https://api.semanticscholar.org/CorpusID:270123870

Ying, W., Wang, D., Hu, X., Qiu, J., Park, J., & Fu, Y. (2024). Revolutionizing biomarker discovery: Leveraging Generative AI for bio-knowledge-embedded continuous space exploration. *Proceedings of the 33rd ACM International Conference on Information and Knowledge Management*. https://api.semanticscholar.org/CorpusID:272832460

Yoon, S.-W., Cho, Y., & Ahn, T. I. (2024). Melon fruit detection and quality assessment using generative AI-based image data augmentation. *ArXiv, abs/2407.10413*. https://api.semanticscholar.org/CorpusID:271213664

Zhao, C., Wang, J., Zhang, R., Niyato, D., Sun, G., Du, H., Kim, D. I., & Jamalipour, A. (2025). Generative AI-enabled wireless communications for robust low-altitude economy networking. *IEEE Communications Magazine*. 10.1109/MWC.2025.3597910

11 High-Resolution Soil Erosion Mapping for the Narmada Basin

A RUSLE-Google Earth Engine Synergy

Akash Deep, Rakesh Singh, Shalini Kumari, and Sandeep Kunwar

11.1 INTRODUCTION

Soils perform essential tasks in the general functioning of Earth's environment (Allafta and Opp, 2022). However, soil erosion has emerged as a significant environmental threat globally, driven by both natural forces and anthropogenic actions such as deforestation, overgrazing, and unsustainable land-use practices (Hazbavi et al., 2019). River basins and watersheds in regions experiencing rapid land-use changes are highly vulnerable to accelerated erosion, which leads to land degradation, sedimentation, soil fertility loss, and reduced crop yields. (Almouctar et al., 2021; Ganasri and Ramesh, 2016). The Narmada Basin in central India exemplifies such a landscape, where increasing sediment loads threaten reservoir storage, flood regulation, and hydropower generation, thereby affecting ecological stability and human livelihoods (Sharma et al., 2023; Sharma et al., 1991).

Quantifying and monitoring soil erosion over vast extents is essential for designing efficient land cover management practices. For basin-scale erosion assessment, Revised Universal Soil Loss Equation (RUSLE) remains the preferred data-driven model because it systematically integrates key factors—including precipitation intensity, soil vulnerability, terrain steepness, vegetation coverage, and anthropogenic management measures (Bezak et al., 2024; Kayet et al., 2018). Although RUSLE is effective, its practical implementation over extensive areas like the Narmada Basin has historically been limited by challenges related to data availability, spatial resolution, and processing capabilities. Recent developments in geospatial technologies, particularly Google Earth Engine (GEE), have helped overcome many of these limitations by facilitating rapid access to multi-source satellite data and enabling scalable environmental analyses (Chen et al., 2020). The GEE platform allows researchers to efficiently model spatial patterns of soil loss using consistent

DOI: 10.1201/9781003616207-13

datasets for rainfall (e.g., CHIRPS), vegetation (e.g., NDVI), terrain (e.g., SRTM), and land use, all within a unified computational environment.

Integrating RUSLE with GEE has demonstrated promising results in soil erosion studies across diverse regions (Avand et al., 2023; Madhukar et al., 2023). This combination offers a powerful framework for high-resolution erosion assessment, enabling the identification of critical erosion-prone zones and informing targeted conservation planning. For example, the application of GEE-based RUSLE modeling in the Narmada Basin facilitates the creation of thematic erosion intensity maps that support region-specific recommendations for land management interventions. Furthermore, few studies have explored the integration of machine learning techniques to enhance model accuracy or assessed the long-term sustainability of erosion control strategies. This study contributes to filling these gaps by leveraging the synergy between RUSLE and GEE to deliver an efficient, scalable, and reproducible approach for assessing soil erosion in the Narmada Basin—thereby laying the groundwork for informed environmental planning and policy development.

11.2 STUDY AREA

The Narmada Basin, formed by the Narmada River, is a defining geographical dimension of Central India, covering an area of approximately 98,796km^2. Geographically, it is bounded from 72°32' to 81°45' E, Latitudes from 21°20' to 23°45' N, and is dispersed throughout the states of MP, Gujarat, and Maharashtra. Physiographically, the Narmada Basin lies between the Vindhya Range to the north and the Satpura Range to the south, with elevations ranging from –12 m in the alluvial lowlands to 1,333 m in the highlands. Narmada River rises on the Amarkantak Plateau of Madhya Pradesh, which flows west for a length of about 1312 km into the Arabian Sea of Gujarat. (Figure 11.1).

The basin has a wide variety of topographic features, including mountainous areas, hilly country, high plateaus, and fertile plains. Forest-covered highlands and sloping land characterize the upper part, and the middle and lower parts form alluvial plains and major valleys. Inside the basin, the climatic zoning is strong, from arid plains to more humid forest climates at higher elevations. Rainfall is 800 mm–1200 mm a year, mostly around the southwest monsoon, June–September. The Narmada Basin has a well-established tributary system, such as the Tawa, Hiran, and Barna Rivers, which make significant contributions to hydrological roles. The extensive alluvial

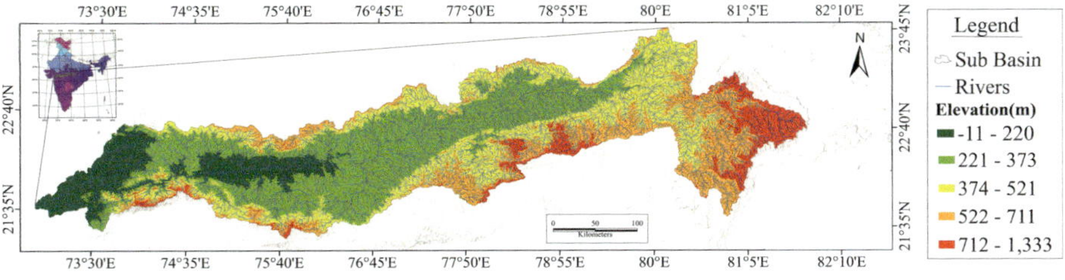

FIGURE 11.1 Spatial distribution of slope angles across the Narmada Basin, generated from SRTM DEM.

plains of the productive river are substantial contributors to agricultural use, supporting the production of wheat, soybeans, and cotton crops, as well as the local economies through irrigation systems. The basin, however, is characterized by various problems, including variable rainfall, heterogeneous soils, and variable land-use patterns, which make soil erosion and degradation of the environment more frequent. Human activities, including urbanization, deforestation, and agriculture, have significantly disrupted the hydrological and ecological balance of the basin. To avoid such problems, significant research with state-of-the-art methods like cloud-based geospatial analysis and the RUSLE model is necessary. A work of such a nature provides an understanding of soil erosion processes and suggests viable land-use management options. Proper management of the Narmada Basin requires coordinated conservation programs to protect ecological health and water balance of the basin, which then guarantees long-term sustainability of this natural resource.

11.3 DATASET AND METHODOLOGY

Using the computational framework of GEE, we have applied RUSLE models to estimate the average Annual Soil Loss (A) of the Narmada Basin. This analysis integrated key factors—rainfall-runoff erosivity (R), soil erodibility (K), slope length and steepness (LS), crop management (C), and support practices (P), each derived from thematic layers derived from satellite imagery and ancillary data. Long-term precipitation records available on GEE were employed to compute the R factor, while soil characteristics of the Narmada basin informed the K factor estimation. The LS, C, and P factors were also derived using remote sensing data processed within the GEE environment. Notably, NDVI was integrated into the calculation of the C factor, effectively accounting for vegetation dynamics and seasonal variation. Among all variables, terrain slope emerged as a significant driver of erosion potential. The cloud-based processing capabilities of GEE facilitated a spatially detailed analysis of soil erosion potential zones in the basin, highlighting the platform's strengths in scalability, accuracy, and handling of large geospatial datasets. These insights inform sustainable land-use planning in erosion-prone regions. A graphic representation of the methodology is shown in Figure 11.2.

11.3.1 RUSLE Thematic Maps

In the current GEE-based study, we employed RUSLE to calculate the A, which the RUSLE Equation represents:

$$A = R * K * L * S * C * P \tag{11.1}$$

A: Mean yearly soil loss per hectare (t ha^{-1} yr^{-1})
R: Rainfall's erosive power (in MJ·mm/ha·h·yr).
K: Soil's susceptibility to erosion (in tons·ha·h/MJ·ha·mm).
L & **S**: dimensionless and influence how terrain affects erosion.
C: from 0 to 0.5, which reflects how different vegetation or land uses affect erosion.
P: values 0 or 1 with respect to effectiveness of soil conservation practices.

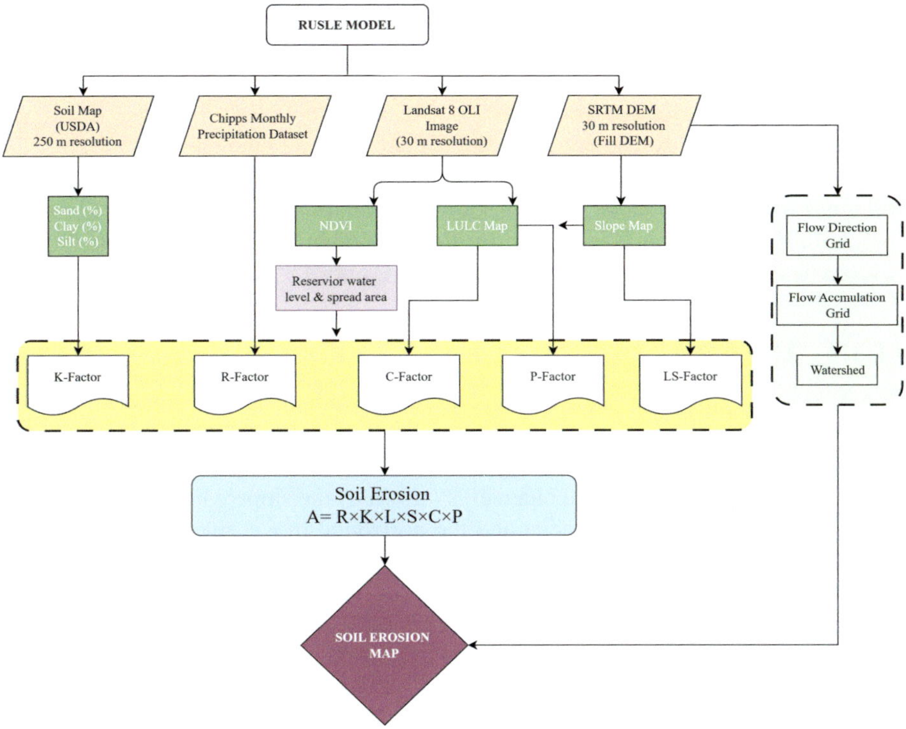

FIGURE 11.2 An illustration of the RUSLE model's workflow that shows the input datasets and erosion factors derivation.

11.3.2 R Factor

The R factor, reflecting rainfall erosivity, is a crucial indicator of erosion vulnerability in our GEE-based study. Heavy rainfall detaches and moves soil particles, driving sheet and rill erosion (Nam et al., 2014). High-intensity storms produce larger raindrops and increased runoff, intensifying erosion (Shen et al., 2016). We calculated the R factor from 2001 to 2024 monthly rainfall using CHIRPS data to ensure precise rainfall-runoff erosivity estimates.

The following formula, which is widely used by researchers worldwide (Parveen and Kumar, 2012), was employed:

$$R = 79 + 0.363 * P \tag{11.2}$$

where P- rainfall in a year (mm).

11.3.3 K Factor

One of the main factors influencing the soil's vulnerability to erosion from natural rainfall or surface runoff is the K factor (Parveen and Kumar, 2012). K is established by variability of soil properties, including mineral composition, structure, and chemistry. The K factor is equal to the mean amount of soil lost on an annual basis along a standard given unit of erosive energy due to rainfalls that are under standard conditions—clean tilled soils under bare soil and a slope length of 22 m with a slope rate of 9% (Wischmeier and Smith, 1978). K-factor values were computed from soil-type records, as well as results of relevant research on soil type using soil type maps and unique to Narmada Basin texture maps (Parveen and Kumar, 2012). Based on soil types found in the study area, four K-factor values were determined. The soil's susceptibility to erosion is indicated by these numbers, which range from 0 to 1. A value nearer 1 denotes greater susceptibility (Figure 11.3).

11.3.4 LS Factor

Gravity forces acting on sediment and water movement are the main drivers of topography's effect on soil erosion. L factor and S Factor variables are the two parts that make up the LS factor. The length of overland water flow from the source of the overland runoff to its outfall channel is known as the slope's length, whereas the ratio between the downslope move and the horizontal move gives slope steepness. The combination of LS factors was calculated in this study using DEM maps derived from the SRTM satellite, which enabled the precise mapping of the length of the flow route and the slope gradients in the Narmada Basin. This calculation procedure involved depression filling of the DEM and using the following Equations (11.3) and (11.4) to compute the LS factor (Ganasri and Ramesh, 2016; Niyonsenga et al., 2020):

$$\mathrm{LS} = \left(\frac{\lambda}{22.1}\right)^{\mathrm{m}} \times \left(0.065 + 0.045S_{\mathrm{g}} + 0.0065S_{\mathrm{g}}^{2}\right) \tag{11.3}$$

$$\lambda = \left[\text{flow accumulation} * \text{cell size}\right]; \mathrm{S_g} = \frac{\sin\left(0.01745 \times \theta\right)}{0.09} \tag{11.4}$$

where θ represents the slope in ° and λ is its length in meters, and m is an empirical constant, usually set at 0.5, for moderate slopes. The elevation imagery was used to calculate filled sinks, and hence the construction of flow direction and accumulation grids. This method emphasizes the incorporation of topographic factors into soil erosion analysis, thus allowing the delineation of risk-prone areas and targeted conservation measures to counter erosion and conserve the soil.

11.3.5 C Factor

A higher NDVI, which indicates more vegetation, is associated with a lower chance of soil erosion. Specifically, higher NDVI values, indicating greater vegetative cover, are linked to smaller C values. Normalized C values between 0 and 1 were found by determining the maximum and minimum values from the exponential output (C4) by using the reduce regional operations. Normalized C was determined by subtracting the minimum value from the image and dividing the difference by the range. The resultant C-factor image is an expression of spatial variation in the vegetation

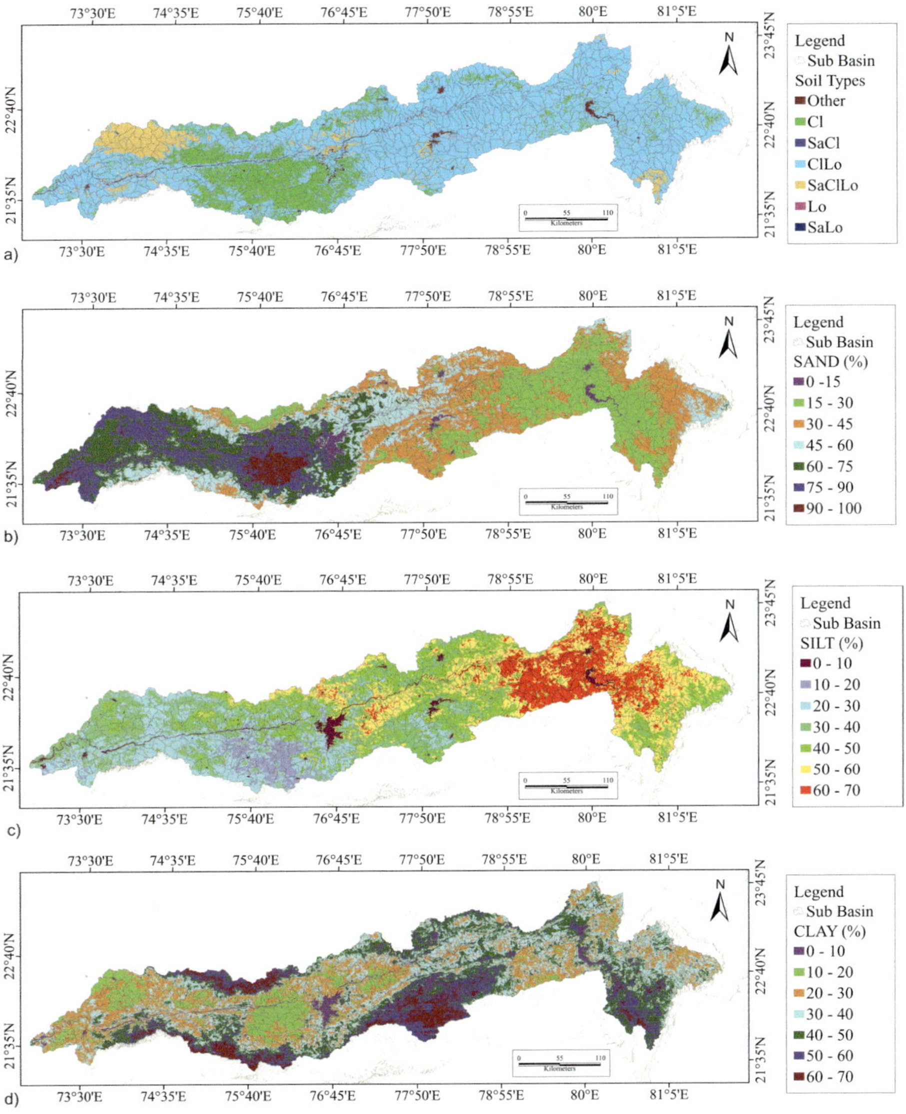

FIGURE 11.3 Narmada Basin soil data. (a) Soil types; (b) percentage of sand; (c) percentage of silt; (d) percentage of clay.

cover; sparse or no vegetation has high C values (close to 1), indicating a high erosion potential; and vegetative cover has low C values (close to 0), thus possessing a protective cover against soil erosion.

NDVI was then converted to C values using an exponential function:

$$\text{where } \alpha \text{ is } -2. \quad C = \exp\left(\frac{(\alpha \times \text{NDVI})}{(1-\text{NDVI})}\right) \tag{11.5}$$

This method utilizes the geospatial analytic abilities associated with GEE to provide scalable and efficient land cover classifications and C-factor assignments in the Narmada Basin. This work offers an excellent perspective to evaluate how land covers impact soil erosion mechanisms from the use of high-resolution satellite images as well as sophisticated classification approaches. GEE, in turn, is a powerful instrument for a watershed-scale study because it also allows analyzing large stocks of information and conducting complex geographical analysis.

11.3.6 P Factor

Soil erosion modeling is incomplete without the inclusion of P factor (between 0 and 1) that expresses the effectiveness of soil loss mitigation techniques (Ahmed et al., 2013). It considers the effect of different interventions on rate attenuation for erosion, such as plant cover, biomass retention, and runoff management. The RUSLE, which provides the overall impact of LULC support methods (Figure 11.4), slope length, and site-specific parameters as inputs that control the processes of soil erosion, was updated to include the P component. The P factor of the RUSLE model, which is the supporting practice factor, refers to the impact of low runoff and erosion conservation tillage practices. It depends on LULC and slope. Rule-based classification was used to deliver P values as a function of these factors.

Areas of LULC < 11, usually croplands with conservation practice, received a P-value of 0.8. Forest, wetlands, and water bodies (LULC = 11, 13, or >14) received a value of 1.0, with no assumed erosion control practice. For cropland classes (LULC = 12 or 14), slope was utilized to adjust the P factor: on slopes <2%, the value was 0.6, indicating effective support practice. On slopes <5% and <8%, a value of 0.5 was used, indicating high effectiveness of contouring or strip cropping. As slope increased, the P factor increased progressively: 0.6 (slope <12%), 0.7 (slope <16%), 0.8 (slope <20%), and 0.9 (slope >20%), reflecting decreasing conservation effectiveness. In all other instances, a default value of 1.0 was employed (Wischmeier, 1965). This strategy improves accuracy in estimating erosion by linking topography and land use with conservation practice applied.

In this study, we used the GEE's scalability and remote sensing capabilities to accurately compute the P factor over the Narmada Basin. Using GEE, we develop a function that extracts the appropriate bands or properties of the main elements, such as C factor, M factor, L factor, and S factor, and any changes made to them. The factors were combined pixel for pixel using the RUSLE technique to form a P-factor

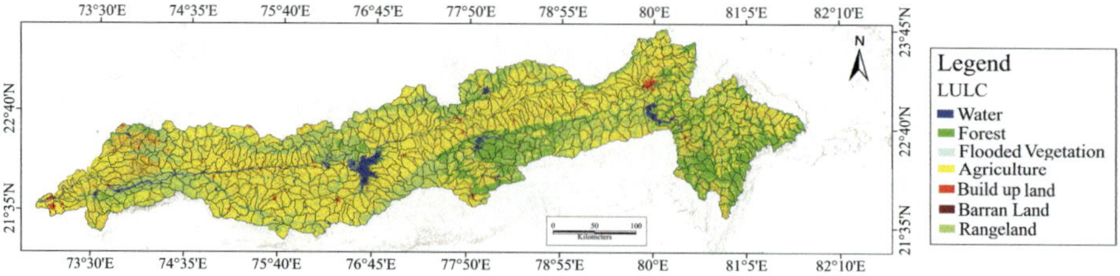

FIGURE 11.4 LULC map of Narmada Basin.

picture. The map on the GEE platform was then superimposed with the picture, delivering a spatially explicit depiction of conservation strategies for preventing soil erosion.

11.3.7 NDVI

The analysis of differences of red-light reflectance and near-infrared (NIR) is a standard vegetation index, NDVI, to estimate vegetation cover and health (Chen et al., 2006). NDVI was used as a reliable estimator of vegetation cover in the soil loss estimation with direct impact on the cover and management (C) factor of USLE (Almagro et al., 2019). The NDVI method used in soil loss estimation was performed using GEE. Vegetation cover seasonal fluctuations are recorded using NDVI (Figure 11.5), which is calculated using different intervals with this formula:

$$\text{NDVI} = (\text{NIR} - \text{Red}) / (\text{NIR} + \text{Red}) \tag{11.6}$$

Seasonal change often has regular NDVI patterns that contribute much to the total C factor. To include NDVI in the C factor of the RUSLE estimation, the following modified equation is used.

$$\text{CNDVI} = \text{Cbase} * (1 - \text{NDVI}) \tag{11.7}$$

In this case, the C base and base C factor are unadjusted C factors, whereas CNDVI and the adjusted C factor are modified versions that account for vegetation cover. Locations with high NDVI values indicate a strong cover of vegetation and a lower C factor, which minimizes erosion possibilities. This C component, as NDVI, is then used in the standard USLE framework alongside the support practice, slope length, slope steepness, and site-specific adjustments.

$$\text{P} = \left(\text{CNDVI} * \text{M} * \text{L} * \text{S} * \frac{\text{PCustom}}{100} \right) \tag{11.8}$$

Beyond the standard parameter (C, M, L, and S), P Custom refers to customized changes made for particular circumstances or soil erosion variables in Equation (11.8).

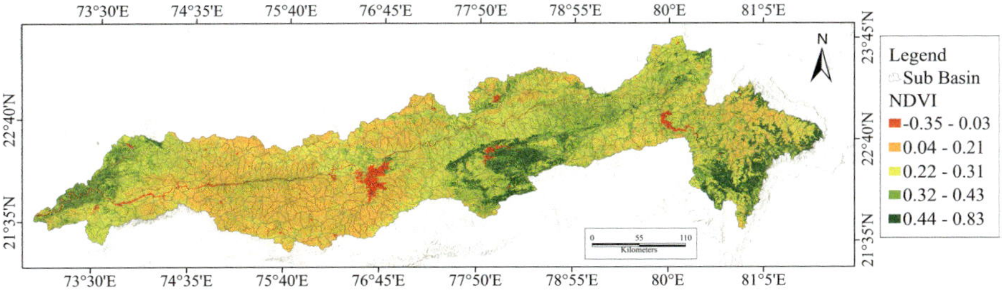

FIGURE 11.5 NDVI map of Narmada Basin.

P Custom and the NDVI-derived C factor enabled the USLE equation to be more adaptable to site-specific factors, such as local environmental circumstances, soil conservation strategies, and land management techniques. Such versatility dramatically enhances the accuracy and usability of the model across different geographical locations. NDVI-based C factor was added in the total P-factor computation. Adding NDVI to the model for predicting soil erosion rendered the model more sensitive to changes in vegetation cover.

11.3.8 Slope

Mapping land slope is essential for effective soil loss management and plays a key role in the RUSLE model. Steeper slopes accelerate runoff, leading to faster soil detachment and greater transport before deposition. In RUSLE, this effect is captured through the LS factor, which combines slope steepness and length to estimate their influence on erosion potential. Higher LS values indicate a greater likelihood of soil erosion as a result of the intensified impact of surface runoff. Slope maps represent the gradients of the terrain and therefore mark the zones that are likely subject to erosion and also indicate where conservation should be focused. When combined with RUSLE factors, slope information aids in the overall evaluation of soil erosion (Cardoso et al., 2020). Knowledge of slope processes allows specialized techniques to be used to minimize the velocity of runoff and erosion on steep slopes, including contour farming or terracing, or afforestation. Moreover, slope analysis provides details upon which land-use planning decisions can be based sustainably, for example, which areas of land are more suited to agriculture, development, or preservation in terms of erosion risk.

11.4 RESULT AND DISCUSSION

11.4.1 Soil Loss Assessment

The combination of GEE and the RUSLE model enables large-scale soil erosion assessment. Geospatial data is processed rapidly by GEE, and RUSLE calculates soil loss using five parameters. GEE retrieves satellite imagery, DEMs, and climate data to map these parameters. Multiplying these layers in RUSLE's equation, the approach measures soil loss amount and locates high-risk areas. The method uses cloud computing for scalable, current erosion analysis for supporting targeted land conservation planning (Figure 11.6).

High-intensity rainfall significantly increases soil erosion hazards through increased kinetic energy and runoff. In this study, the index of rain erosive potential ranged from 389.066 to 1163.55 $MJ \cdot mm \cdot ha^{-1} \cdot h^{-1} \cdot y^{-1}$ (Figure 11.6a). Precision was ensured using CHIRPS estimates of R factor, high-resolution (0.05°), long-term (30+ years) satellite and ground station merged rainfall data. CHIRPS enhances reliability through the merging of infrared satellite data with in-situ data, enabling precise trend analysis and drought monitoring. Merging these ensures robust erosion modeling, which is critical for the identification of regions prone to extreme rainfall-driven soil loss and planning of mitigation interventions.

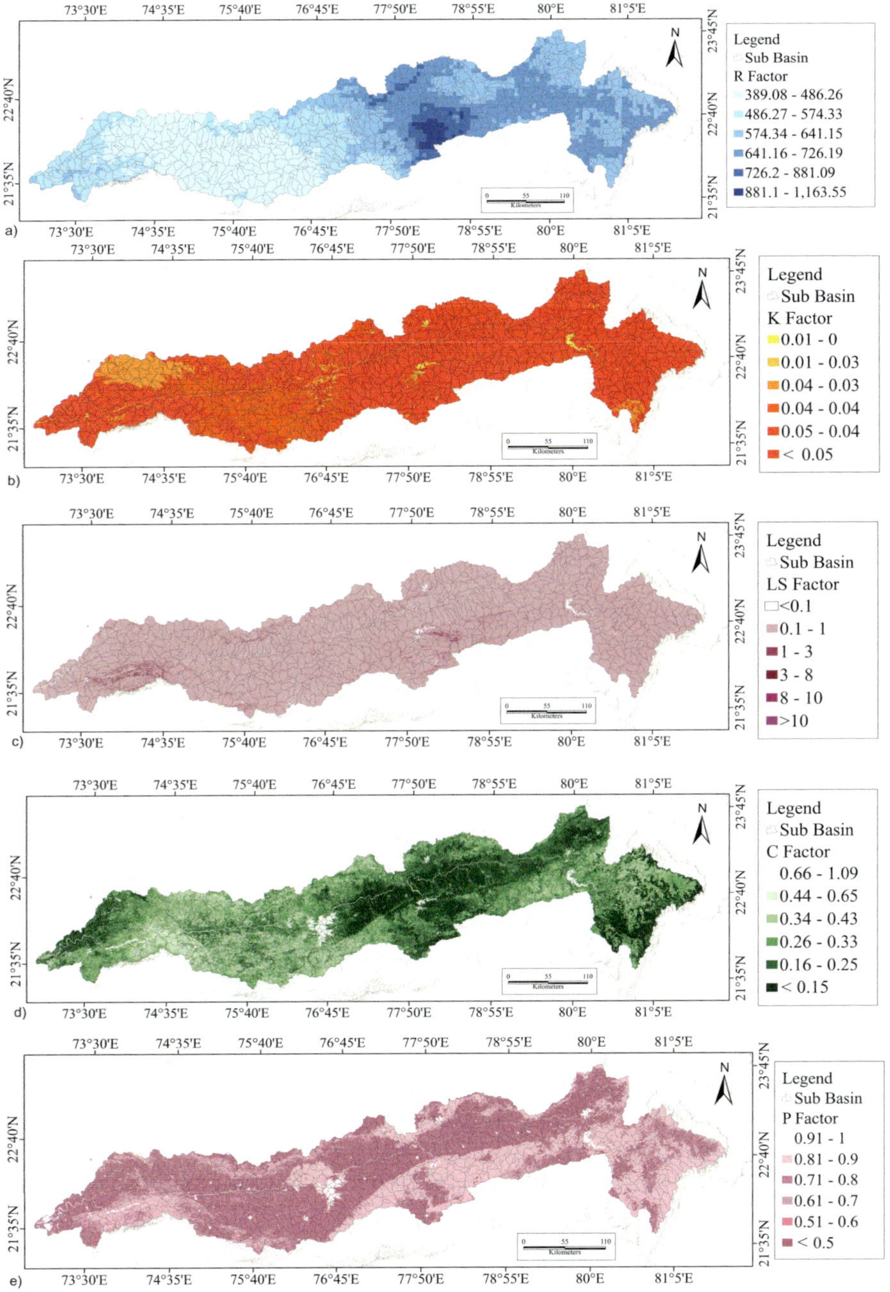

FIGURE 11.6 The RUSLE factors map: (a) R, (b) K, (c) LS, (d) C, (e) P factor.

The K factor measures soil erodibility as an index of a soil type's susceptibility to rain and runoff erosion. Large values of K (e.g., 0–0.05 $t \cdot ha \cdot h \cdot MJ^{-1} \cdot mm^{-1} \cdot ha^{-1}$ in the Narmada Basin) correspond to higher detachment probability, with texture such as sand, silt, and clay ratio and organic carbon concentration varying accordingly (Figure 11.6b). Soils with greater silt or organic matter loss are more erodible. Non-erodible land units such as settlements, glaciers, and water bodies are assigned K = 0. This attribute, derived from soil composition data, spatially maps the area susceptible to erosion, allowing for effective conservation planning. The range reflects heterogeneity of soil resistance in the basin, essential for realistic RUSLE-based erosion modeling.

The LS factor reflects the impact of topography on soil particle transport and surface runoff. LS values in the Basin increased gradually from the northwest (NW) to southeast (SE), with a mean value of 1.69 (Figure 11.6c). In mountainous subregions, values often exceed 5 at ridge crests. Non-topographic areas (e.g., water bodies) were excluded and assigned zero values.

The conservation practice factor, or P factor, ranges from 0 to 1, with 0 denoting effective erosion resistance driven by humans' practices. (e.g., terraces, contour farming) and 1 denotes the absence of erosion control measures (Renard et al., 1997). Settled land and grasslands within the study area possessed low P values (0.1), indicating robust erosion-resistant infrastructure. Riparian areas, which were mainly farmland, possessed moderate vulnerability (P = 0.5). Rocky/sandy high-altitude areas and forested downstream canyons (>79.2% cover) possessed P = 1, signifying negligible conservation practices. Water bodies possessed P = 0, as erosion resistance does not apply (Figure 11.6e).

11.4.2 Soil Erosion in the Narmada Basin

The annual soil erosion estimate for the Narmada Basin is divided into five classes: mild, moderate, high, very high, and severe. Each class has a different level of vulnerability. The results, summarized in Table 11.1, indicate that soil erosion is a significant concern in the Narmada Basin. Table 11.1 presents the distribution of soil loss across different classes in terms of affected area and percentage. As seen in Figures 11.7 and 11.8, the moderate soil loss category (10–20 $t\ ha^{-1}\ y^{-1}$) dominates, covering approximately 28,123 ha (30%) of the total area. The high soil loss category (20–30 $t\ ha^{-1}\ y^{-1}$) follows, affecting 21,812 ha (23%). Notably, slight soil loss (<10 $t\ ha^{-1}\ y^{-1}$) is observed in 17,220 ha (18%), whereas very high (30–40 $t\ ha^{-1}\ y^{-1}$) and severe (>40 $t\ ha^{-1}\ y^{-1}$) erosion levels impact 12,971 ha (14%) and 14,155 ha (15%), respectively.

Table 11.1 depicts the area and proportion of catchment under each soil loss severity category in the Narmada Basin.

The findings indicate that moderate to high soil loss (10–30 t/ha/yr) accounts for 53% of the total study area, emphasizing the urgent need for focused soil conservation interventions. Severe erosion levels in 15% of the region highlight areas that require immediate attention to mitigate further land degradation. The widespread moderate to high erosion suggests that soil erosion is caused by both natural effects (topography) and anthropogenic activities, including land-use patterns and agricultural activity.

TABLE 11.1
Area and Proportion of Catchment under Each Soil Loss Severity Category in the Narmada Basin

Sr. No	Class	Area (Ha)	Area (%)
1	Slight (<10)	17,220	18
2	Moderate (10–20)	28,123	30
3	High (20–30)	21,812	23
4	Very high (30–40)	12,971	14
5	Severe (>40)	14,155	15

11.4.3 Normalized Distribution of RUSLE Factors

This boxplots presents a simplified view of how different environmental and management factors influence soil erosion, using normalized data from the RUSLE model, which has been normalized to a scale from 0 to 1 to allow direct comparison (Figure 11.9). Among the five factors, K factor stands out with consistently high values that reflect the soil's natural susceptibility to erosion in many areas. In contrast, the LS factor is generally low, indicating that most of the land has gentle slopes, which by themselves don't contribute heavily to erosion.

R factor is widespread, with some regions experiencing much more intense rainfall, which can significantly accelerate erosion. The C factor and P factor show mixed patterns, reflecting differences in how land is managed—some areas have good coverage and conservation measures, while others don't. Altogether, this suggests that even in relatively flat places, erosion can be a serious issue if the soil is weak and rainfall is heavy, especially when protective vegetation or practices are lacking.

11.4.4 Correlation between LULC Classes and Soil Loss

Figure 11.10 illustrates the correlation coefficients between LULC categories and soil loss, indicating the direction and magnitude of each relationship. The rangeland class is most strongly correlated with soil loss, implying that rangeland areas experience more erosion (Figure 11.10). This may be a result of open vegetation or grazing

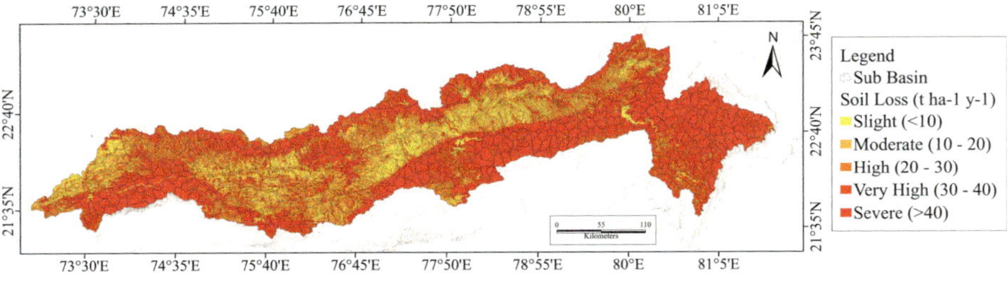

FIGURE 11.7 Soil loss erosion map in Narmada Basin.

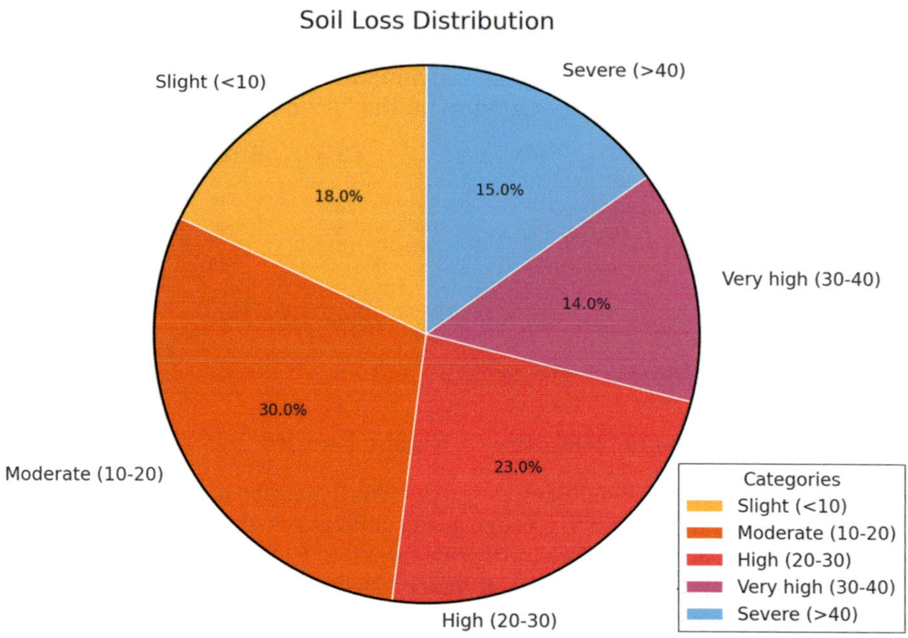

FIGURE 11.8 Distribution of soil loss severity classes in the Narmada River Basin, categorized by average annual soil loss (t ha^{-1} yr^{-1}).

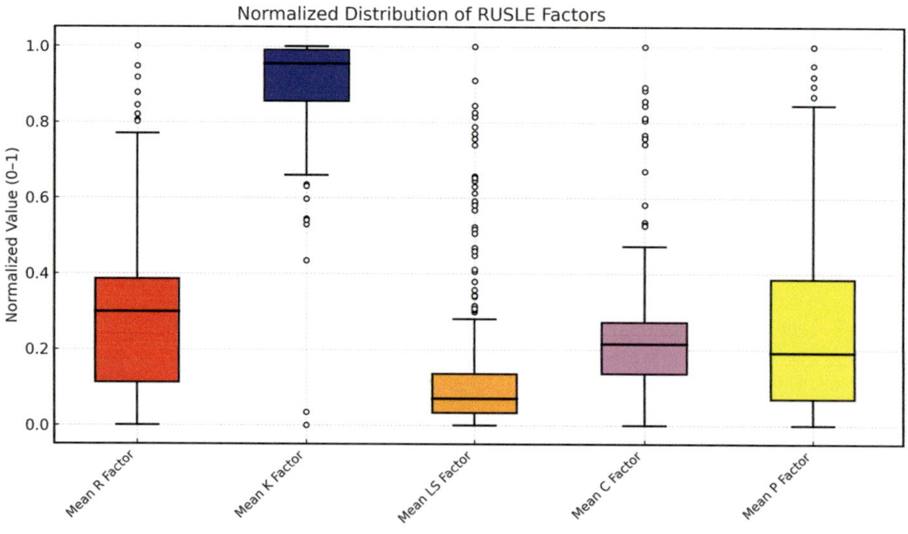

FIGURE 11.9 Spatial distribution of normalized RUSLE factors, indicating relative contributions to soil erosion across the study area.

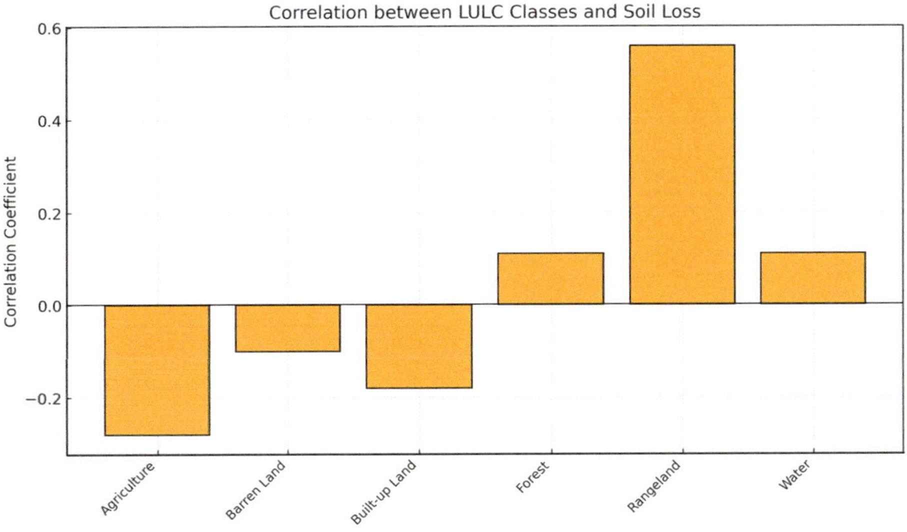

FIGURE 11.10 Map showing spatial correlation between LULC classes and estimated soil erosion, highlighting high-risk zones.

pressure that exposes the soil more. Forest and water classes show weak positive correlations, suggesting minimal to moderate erosion in these areas.

In contrast, agriculture and built-up land have negative correlations with soil loss, with agriculture showing the most noticeable negative relationship. This might reflect better soil conservation practices or more stable land cover in managed agricultural areas. Barren land, surprisingly, shows a mild negative correlation as well, possibly due to limited surface runoff or the presence of already degraded but stabilized surfaces. Overall, the chart highlights that rangelands are particularly vulnerable to soil erosion, while managed or less exposed land types like agriculture and urban areas show reduced erosion tendencies.

11.4.5 PCA Biplot of RUSLE Factors

This PCA biplot provides a simplified representation of the relationships among the RUSLE factors (R, K, LS, C, and P) and their contributions to the variability in the dataset (Figure 11.11). The plot breaks the data down into two main dimensions—PC1 and PC2—which together explain about 74.6% of the total variance (40.2% from PC1 and 34.4% from PC2).

Each yellow dot represents a data point, and the arrows show the direction and strength of each RUSLE factor's influence. From this plot, we can see that R and K factors have strong contributions along PC1, indicating that these two factors are closely related and largely explain the main variation in the data. C factor, however, points in a different direction—mostly along PC2, suggesting it captures a different aspect of variation, possibly related to land use or vegetation cover. P and LS factors also have distinct directions, suggesting they each contribute unique, independent information.

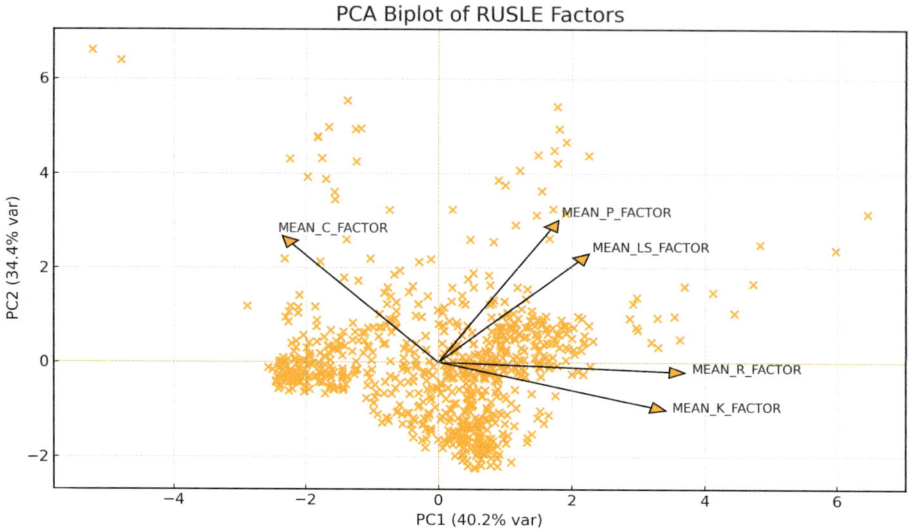

FIGURE 11.11 PCA biplot of RUSLE factors showing the loadings of R, K, LS, C, and P on PC1 (40.2% variance) and PC2 (34.4% variance), summarizing their interrelationships and contribution to overall variability.

11.4.6 Correlation Matrix

This correlation matrix provides an overview of the relationships among the different RUSLE factors, using color intensity to show the strength and direction of those relationships. Strong positive correlations are indicated by values around 1 in yellow color, whereas strong negative correlations are indicated by values near -1 in dark purple color (see Figure 11.12).

From the matrix, we see that most of the RUSLE factors are weakly to moderately correlated. For example, the R and K factors show a mild positive correlation, indicating that areas with more intense rainfall tend to have more erodible soils. Interestingly, the C factor (which reflects vegetation cover) shows a weak to moderate negative correlation with most of the other factors, especially R and K. This makes sense because dense vegetation usually reduces soil loss even in places with high rainfall or erodible soil. The P factor appears to have a modest positive relationship with other factors, suggesting a complementary role in managing erosion in conjunction with natural conditions.

11.5 CONCLUSION

This work successfully illustrates the implementation of the RUSLE model GEE for a large-scale, cloud-based assessment of soil erosion in the Narmada Basin. When geospatial analysis and empirical modeling are used together, erosion-prone areas can be accurately identified, and targeted land management techniques can be developed. The findings prove that over 50% of the basin reveals mild to severe levels of soil degradation, and it is imperative to take conservation actions based on the

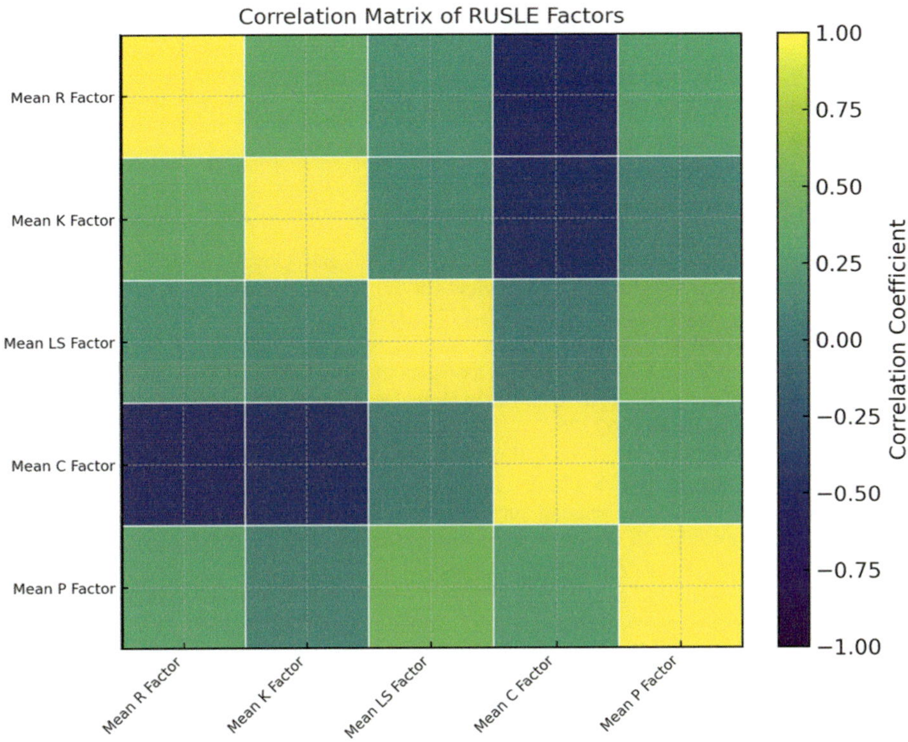

FIGURE 11.12 Heatmap of pairwise correlation coefficients among RUSLE factors (R, K, LS, C, and P), highlighting their interrelationships.

location. Topography and land cover were both found to have an undeniable weight; steep slopes and heavily worked lands were the most vulnerable to erosion, with forested and vegetated regions experiencing considerably lower rates.

The study also improved the erosion evaluation through the insertion of a rule-based classification to provide P-factor values based on slope and land-use designation, thus increasing the spatial accuracy of the conservation effectiveness analysis. RUSLE combined with GEE allowed performing an analysis of the erosion parameters on a pixel-wise basis over the whole scale of the basin using an efficient knowledge and process-poor approach, which demonstrated the scalability and adaptability of the framework to environmental monitoring efforts. The strategy not only develops the ability to measure and estimate soil erosion promptly but also provides a duplicate strategy that can be applied to other areas with comparable environmental concerns.

ACKNOWLEDGMENT

The authors would like to express their sincere gratitude to the Vice-Chancellors of Graphic Era (Deemed to be University) and SGRR University, Dehradun. The authors acknowledge the US National Aeronautics and Space Administration (NASA) for providing access to the extensive satellite data archive from the Landsat mission.

DECLARATION OF COMPETING INTEREST

The authors declare that they have no known competing financial interests or personal relationships that could have appeared to influence the work reported in this chapter.

REFERENCES

Ahmed, M., Asif, M., Hirani, A. H., Akram, M. N., and Goyal, A. (2013). Modeling for agricultural sustainability: A review. In N. K. Bhullar and G. S. Bhullar (Ed.), *Agricultural sustainability* (pp. 127–147). Academic Press. https://doi.org/10.1016/B978-0-12-404560-6.00007-1

Allafta H., and Opp, C. (2022). Soil erosion assessment using the RUSLE model, remote sensing, and GIS in the Shatt Al-Arab Basin (Iraq-Iran). *Applied Sciences*, *12*, 7776. https://doi.org/10.3390/app12157776

Almagro, A., Thomé, T. C., Barbosa Colman, C., Pereira, R. B., Junior, J. M., Rodrigues, D. B. B., and Oliveira, P. T. S. (2019). Improving cover and management factor (C-factor) estimation using remote sensing approaches for tropical regions. *International Soil and Water Conservation Research*, *7*, 325–334. https://doi.org/10.1016/j.iswcr.2019.08.005

Almouctar, M. A. S., Wu, Y., Zhao, F., and Dossou, J. F. (2021). Soil erosion assessment using the RUSLE model and geospatial techniques (remote sensing and GIS) in South-Central Niger (Maradi Region). *Water*, *13*, 3511. https://doi.org/10.3390/w13243511

Avand, M., Mohammadi, M., Mirchooli, F., Kavian, A., and Tiefenbacher, J. P. (2023). A new approach for smart soil erosion modeling: Integration of empirical and machine-learning models. *Environmental Modeling & Assessment*, *28*, 145–160. https://doi.org/10.21203/rs.3.rs-809330/v1

Bezak, N., Borrelli, P., Mikoš, M., Auflič, M. J., and Panagos, P. (2024). Towards multi-model soil erosion modelling: An evaluation of the erosion potential method (EPM) for global soil erosion assessments. *CATENA*, *234*, 107596. https://doi.org/10.1016/j.catena.2023.107596

Cardoso D. P. and Silva, E. M., Avanzi, J. C., Augusto Muniz, J., Ferreira, D. F., Silva, M. L. N., Acuña-Guzman, S. F., Curi, N. (2020). Rainfall erosivity factor: An R package for rainfall erosivity (R-factor) determination. *CATENA*, *189*, 104509. https://doi.org/10.1016/j.catena.2020.104509

Chen, C.-N., Tfwala, S. S., and Tsai, C.-H. (2020). Climate change impacts on soil erosion and sediment yield in a watershed. *Water*, *12*, 2247. https://doi.org/10.3390/w12082247

Chen, X. L., Zhao, H. M., Li, P. X., and Yin, Z.-Y. (2006). Remote sensing image-based analysis of the relationship between urban heat island and land use/cover changes. *Remote Sensing of Environment*, *104*(2), 133–146. https://doi.org/10.1016/j.rse.2005.11.016

Ganasri, B. P. and Ramesh, H. (2016). Assessment of soil erosion by RUSLE model using remote sensing and GIS—A case study of Nethravathi Basin. *Geoscience Frontiers*, *7*, 953–961. https://doi.org/10.1016/j.gsf.2015.10.007

Hazbavi, Z., Sadeghi, S. H., and Gholamalifard, M. (2019). Dynamic analysis of soil erosion-based watershed health. *GES*, *12*, 43–59. https://doi.org/10.24057/2071-9388-2018-58

Kayet, N., Pathak, K., Chakrabarty, A., and Sahoo, S. (2018). Evaluation of soil loss estimation using the RUSLE model and SCS-CN method in hillslope mining areas. *International Soil and Water Conservation Research*, *6*, 31–42. https://doi.org/10.1016/j.iswcr.2017.11.002

Madhukar, A., Hari, N., Srivalli, R., and Neelima, T. L. (2023). Spatial estimation of soil erosion using the RUSLE model: A case study of Sangareddy, Telangana State, India. *International Journal of Plant & Soil Science*, *35*, 490–498. https://doi.org/10.9734/ijpss/2023/v35i183314

Nam, K.-H., Lee, D.-H., Chung, S.-R., and Jeong, G.-C. (2014). Effect of rainfall intensity, soil slope, and geology on soil erosion. *Journal of Engineering Geology*, *24*, 69. https://doi.org/10.9720/kseg.2014.1.69

Niyonsenga, J. D., Mugabowindekwe, M., and Mupenzi, C. (2020). Spatial analysis of soil erosion sensitivity using the Revised Universal Soil Loss Equation model in Nyamasheke District, Western Province of Rwanda. *Transactions in GIS*, *25*, 735–750. https://doi.org/10.1111/tgis.12701

Parveen, R. and Kumar, U. (2012). Integrated approach of Universal Soil Loss Equation (USLE) and Geographical Information System (GIS) for soil loss risk assessment in Upper South Koel Basin, Jharkhand. *Journal of Geographic Information Systems*, *4*, 588–596. https://doi.org/10.4236/jgis.2012.46061

Renard, K. G., Foster, G. R., Weesies, G. A., McCool, D. K., and Yoder, D. C. (1997). *Predicting soil erosion by water: A guide to conservation planning with the Revised Universal Soil Loss Equation (RUSLE)* (Issue Agricultural Handbook No. 703). USDA, Agricultural Research Service. NTIS PB97153704.

Sharma, N., Kaushal, A., Yousuf, A., Kaur, S., Sharda, R., Singh, S. P., Gupta, O., and Sood, A. (2023). Prioritization of erosion susceptible watersheds using morphometric analysis and PCA approach: A case study of the lower Sutlej River basin of Indian Punjab. *Watershed Ecology and Environment*, *5*, 209–224. https://doi.org/10.1016/j.wsee.2023.08.002

Sharma, P. D., Goel, A. K., and Minhas, R. S. (1991). Water and sediment yield into the Sutlej River from the high Himalaya. *Mountain Research and Development*, *11*, 87. https://doi.org/10.2307/3673569

Shen, H., Zheng, F., Wen, L., Han, Y., and Hu, W. (2016). Impacts of rainfall intensity and slope gradient on rill erosion processes at loessial hillslope. *Soil and Tillage Research*, *155*, 429–436. https://doi.org/10.1016/j.still.2015.09.011

Wischmeier, W. and Smith, D. (1978). *Predicting rainfall erosion losses—A guide to conservation planning* (Issue Agricultural Handbook No. 537). U.S. Department of Agriculture.

Wischmeier, W. H. (1965). *Predicting rainfall-erosion losses from cropland east of the rocky mountain: guide for selection of practices for soil and water conservation* (Issue Agricultural Handbook No. 282). U.S. Department of Agriculture.

12 AI Integration in Agriculture

Tools, Software, and Frameworks for Sustainable Farming

Atiya Khan, C H Patil, Amol D. Vibhute, and Shankar Mali

12.1 INTRODUCTION

Computer vision provides an essential framework for more intelligent and practical farming by enabling agricultural systems to perceive, comprehend, and react to a variety of field circumstances through complex algorithms. This technology enables machines to evaluate and interpret visual input from their surroundings, providing farmers with automated systems that are adaptive and feature specialized problem-solving capabilities. Plant diseases lower the number and quality of crops produced as agriculture tries to feed the world's rapidly expanding population. Post-harvest infections can cause significant issues, leading to economic losses in agriculture. An essential part of the country's economy is crop production. These days, it is widely accepted that agricultural production encompasses not only the initial stages of production but also advertising, analysis, distribution, and assistance following the purchase.

Crop cultivation and other core industries are being prioritized in areas with low real income per capita. It is evident that higher crop productivity and output typically make a significant contribution to a nation's total economic growth (Eli-Chukwu, 2019). The Internet of Things (IoT), a practical data analysis and storage technology with several applications in agriculture, soil, and weed management, is one area where artificial intelligence (AI) has been shown to serve an important role. They also need to be able to process and interpret large amounts of data in a timely and accurate manner (Zha, 2020; Khan et al., 2025). Future agricultural challenges will be increasingly addressed by machine learning (ML) and intelligent systems based on large datasets, which will be extensively utilized in every aspect of agricultural management and production. Combining computer vision technology with AI technologies will enhance agricultural automation systems' overall performance, coordination, robustness, and economic performance. The efficiency and caliber of

DOI: 10.1201/9781003616207-14

agricultural output will increase in the future as computer vision technology is applied and developed, and it will give farmers evidence-based practical recommendations for actions and decision support (Tian et al., 2019).

The development of plant phenotyping approaches that can be effectively applied to precision agriculture (PA) has been made possible by advancements in ML algorithms and modern data collection methods that utilize a range of cameras, sensors, and Drones (Unmanned Aerial Vehicles). There are three categories of plant phenotyping techniques: ground-based, aerial, and satellite-based, depending on how the data is collected. To quantify practical characteristics, including plant biomass, crop growth stage, plant height, and crop yield, high-precision sensors are installed on mobile vehicles or integrated into portable devices in ground-based phenotyping (Balasubramanian et al., 2020). An expert and farmer network can be used to help the AI model learn from a large amount of pertinent data and images. A farmer can then use the output to augment their expertise and knowledge, helping them make better decisions, such as which pesticide formula to use and how much to apply at what times. There is insufficient, inconsistent, and unreliable data on agricultural water management, all of which are necessary for effective AI solutions (Javaid et al., 2022).

Weather conditions and other meteorological events that could affect crops and farming operations can be predicted using deep learning (DL). By analyzing data from satellites, weather stations, and other sources, DL algorithms can be proficient in producing accurate and consistent weather forecasts. Making selections about whether to plant, water, and harvest crops can be assisted by this data for farmers and other stakeholders in agriculture (Adli et al., 2023). Images from a camera, along with additional sensors, for instance, can be used to train algorithms that recognize pests or diseases in crops. This information helps problems to be recognized early and remedial measures taken before they cause significant harm to the crops. With DL algorithms, marketplace conditions and consumer preferences can be forecasted. With this information, farmers could choose more wisely what crops to produce, when to sow them, and the best way to sell and market them (Lischeid et al., 2021). When it comes to object detection and classification, the Convolutional Neural Network (CNN) model operates most effectively. Although CNNs have numerous advantages, they also have disadvantages, such as the necessity for large datasets and an extensive training process. Images become more complex for model training since deep neural network models require low-level features from the images. These difficulties related to the dataset can be overcome via transfer learning strategies (Andrew et al., 2022).

The objective of this study is to examine High Performance Computing (HPC) and intelligent vision technology to speed up data processing and data analysis for predictive analytics in agriculture. The study of remote sensing imagery is necessary for a better understanding of crop health, changes in land cover, water use, and the impact of climate change on farming ecosystems. It also explores the advanced AI and DL techniques for smart agriculture to enable decision-support systems that can identify high-potential fields for sustainable farming.

The remaining section of the book chapter is structured as follows. The use of AI and DL in agriculture is covered in Section 12.3, while Section 12.2 defines the relevant subjects about HPC and innovative vision for smart agriculture. In Section 12.4, the challenges that can be faced with DL and HPC are presented. In Section 12.5, we present our conclusions and outline future research objectives.

12.2 USE OF INTELLIGENT VISION AND HPC FOR SMART AGRICULTURE

Intelligent vision and HPC are two innovative technologies with a higher rate of growth and advancement that have changed agriculture. By using these technologies in agriculture, farmers and researchers can improve their decision-making abilities and increase production and precision. Precision farming could provide comprehensive information to assist farmers in gathering data on weather, soil conditions, and crop health by utilizing imaginative vision, high-performance computing, sensor data processing, drones, and other monitoring tools.

- **Predictive analytics:** Farmers can create models that predict future climatic and agro-economic scenarios with advanced computer vision technologies and HPC systems by analyzing previous weather records of agricultural datasets. Also, these prediction tools support better forecasting for launch dates and harvest dates to make more precise decisions.
- **Enhanced resource management:** Farmers can use intelligent vision and HPC to analyze metrics such as soil moisture levels and nitrogen content. This enables optimized resource allocation, such as water and fertilizer. The potential financial savings, in addition to the lowered environmental impact, could be substantial.
- **Increased accuracy:** Surveillance, planting, and harvesting can all be automated using intelligent vision and HPC. This will not only increase the precision/recall metrics to improve accuracy and lower error rates but also reduce time and labor costs.
- **Strategic planning:** By employing HPC along with intelligent vision technologies, farmers can analyze vast volumes of data, enabling them to make more efficient decisions on when to sow, water, and reap crops. This is likely to result in higher crop yields, reduced waste, enhanced revenues, and improved profits. 5. Applications of AI and DL in agriculture.

12.3 USES OF AI AND DL IN AGRICULTURE

Scalable and robust frameworks, like fast computers powered by DL and AI, can handle and analyze vast volumes of agricultural data. AI is the capability of robots to do actions like pattern recognition and decision-making that often require human intelligence. Artificial neural networks (ANNs) are trained to find patterns in data in DL. In PA, it can be used to handle the high-dimensional datasets that include spatial and temporal variations produced by sensors and other equipment. It can also be used to simulate and model complex processes such as soil dynamics, weather patterns, and agricultural growth.

12.3.1 Crop Disease Detection

ML and DL are two new areas of technological innovation in agriculture that increase crop health and productivity by finding and controlling diseases. CNNs are a kind of deep neural network system that uses various filter layers. Large data sets of healthy and unhealthy plants can be used to train CNNs to differentiate between healthy and diseased plants. By providing customized treatments based on the disease type, these technologies can also contribute to a decrease in the use of hazardous chemicals in agriculture.

Humidity, light intensity, temperature, and soil moisture sensors are the most well-known meteorological sensors utilized for disease diagnosis. AI algorithms can develop multidimensional data sources and provide early predictions about agricultural diseases, giving better results than traditional ML algorithms (Ouhami et al., 2021).

Indeed, in the realm of agriculture, neural networks and deep neural network models have demonstrated a notable ability to track crop growth healthily and acquire valuable insights. The tomato leaf database (Harakannanavar et al., 2022) considers plants affected by various illnesses. To research leaf disease identification, images of leaves of tomatoes with six distinct diseases are collected. The database's disease leaf image examples are shown in Figure 12.1.

To find the greatest classification and recognition accuracy for multi-class plant ailments, experiments were conducted using the Plant Village dataset and pre-trained CNN models, like DenseNet-121, Inception V4, ResNet-50, and VGG-16

FIGURE 12.1 An illustration of diseased tomato leaves.

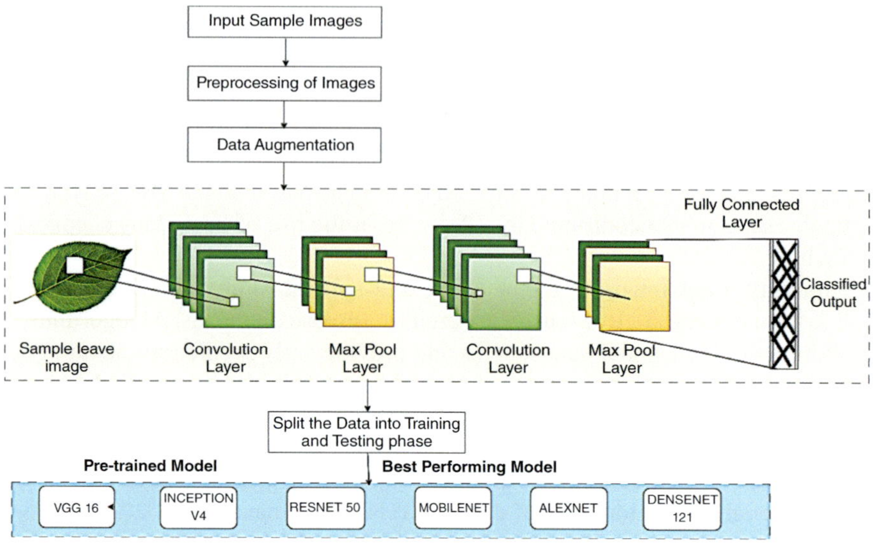

FIGURE 12.2 Overall process of CNN model for leaf disease detection.

(Andrew et al., 2022). The complete flow for predicting the type of leaf disease with the best-performing DL model is shown in Figure 12.2.

12.3.2 Crop Detection and Segmentation

Crop detection and segmentation refer to the process of identifying and separating crop areas from non-crop areas in an image. Utilizing this technology will enable farmers to identify crop stress, manage crop health, and maximize resource use. Crop identification and segmentation frequently utilize computer vision methods, such as object detection and image segmentation, which are designed to recognize and separate different elements in an image. By accurately identifying and separating crop areas from non-crop areas, farmers can apply resources such as fertilizer, pesticides, and water more efficiently and effectively. Monitoring crop condition and stress allows farmers to take preventative measures against crop damage and optimize yield.

Crop Deep (Zheng et al., 2019) is one of the rare rich datasets available for species detection and classification, with 31,147 images and nearly 49,000 labeled instances categorized in 31 different classes, to support future research in difficult real-world agricultural situations. Unlike current vision datasets, images were collected in a variety of circumstances using a variety of cameras and equipment in greenhouses (Zheng et al., 2019). Figure 12.3 shows a grid of annotated small images, most likely representing different crops or vegetables. Every little image in the grid seems to represent an example from a dataset, and some of them contain annotations or text overlays. However, the resolution of the image and the tiny dimensions of the individual panels make the text largely unreadable.

FIGURE 12.3 Some annotations of the results for different crop classes.

12.3.3 Seed Quality Investigation

The goal of the commercialized seed industry is to provide farmers with the correct quantity and quality of seeds at the appropriate time. In addition to being time-consuming and labor-intensive, separating low-quality seeds from high-quality ones requires specialized equipment, infrastructure, and resources. AI systems based on DL have significantly boosted the performance of computer vision, leading to a broader application in seed variety identification (Dhanya et al., 2022). A systematic method for using computer vision to evaluate seed quality is shown in Figure 12.4,

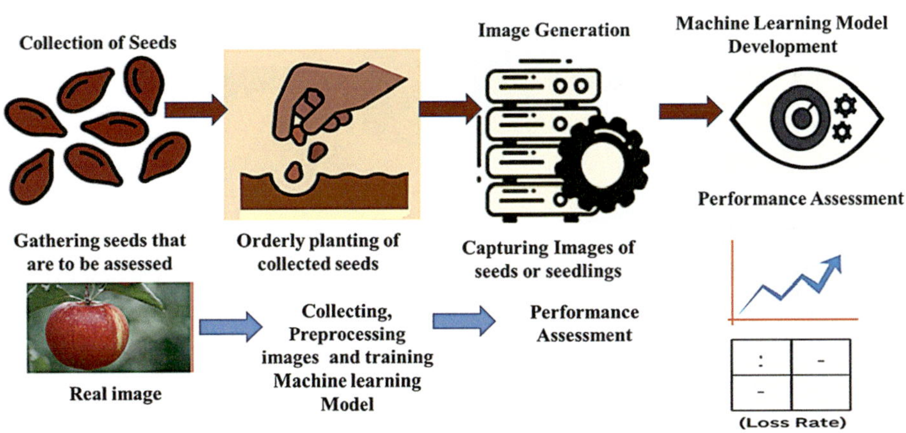

FIGURE 12.4 Assessment of seed quality based on computer vision.

which proceeds from handling seeds physically to creating data, building models, and assessing performance at the end. The addition of "Discriminator" and "Real Sample" (apple photos) implies that ML—possibly generative adversarial networks, or GANs—or comparable discriminative models—will be used for the creation and evaluation of image data.

12.3.4 Soil Analysis

Soil analysis using computer vision involves using digital image processing techniques to analyze the physical properties of soil samples. Some of the key parameters that can be analyzed using computer vision include soil color, texture, structure, and moisture content. By analyzing these parameters, anyone can gain an understanding of the fertility, nutrient content, and physical properties of the soil. With the use of this data, irrigation and fertilization schedules can be optimized for higher crop yields.

The authors (Sheeba et al., 2022) in their work have classified soil based on several attributes, such as the soil response parameter and fertility indices for potassium, carbon, phosphorus, boron, and accessible boron. For cross-validation purposes, a total of ten steps is used, with 10% of the data used for validation and the remaining for training purposes. The data is used to train an extreme learning method (ELM), a quick classification technique, to recognize the micronutrients found in the soil. Figure 12.5 illustrates how basic soil parameters are analyzed to determine pH levels and soil micronutrient fertility using an ML pipeline. After that, this data is separated

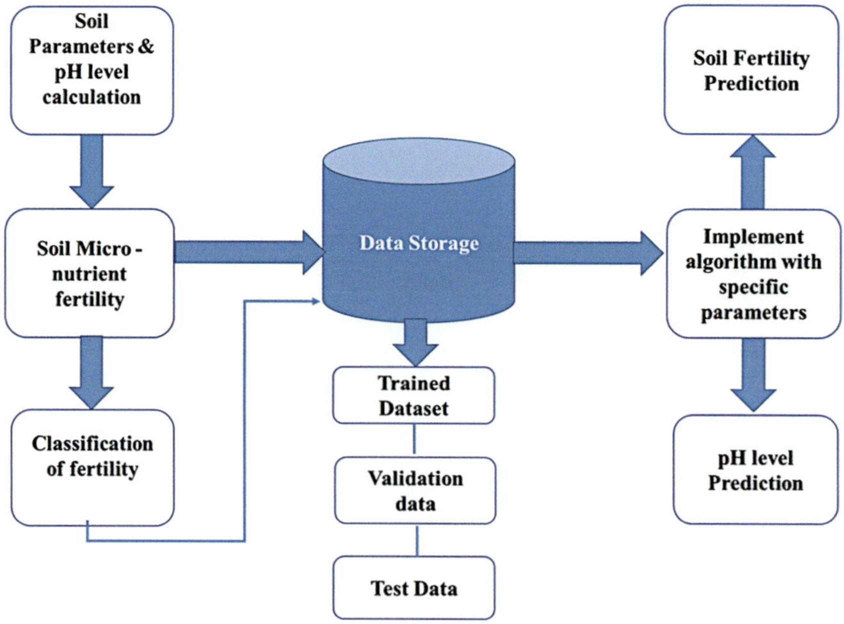

FIGURE 12.5 Methods for estimating soil fertility and pH content.

into validation, test, and training sets and stored. Both the fertility of the soil and pH level predictions are then made using an ELM algorithm on these datasets.

To classify the pH and level of soil nutrients, a unique ISNpHC-WVE method was developed. Additionally, three DL models were applied for the predictive analysis: gated recurrent unit, the deep belief network, and bidirectional long short-term memory (Escorcia-Gutiérrez et al., 2022).

12.3.5 Crop Management and Monitoring

Utilizing high-performance computation and intelligent vision, integrated with modern technology, is used in farming practices to increase agricultural yield, increase crop production, and reduce crop damage from pests, diseases, and environmental factors. Image analysis can be used to identify weeds, track crop growth, and spot symptoms of illness or stress in plants. In contrast, high-performance computing leverages parallel processing and supercomputers to swiftly handle massive volumes of data. Due to the mobility and ability to fly at low altitudes, uncrewed aerial vehicles (UAVs) enable the capture of images with excellent spatial and temporal resolution from satellite imagery or images obtained from human-crewed aircraft. Weeds, pests, diseases, nutrients, hydric stress, and other issues are identified and managed in PA. Selecting the best phenotypic features in seeds, such as height and biomass results, is known as crop phenotyping (Cuaran & Leon, 2021). As illustrated in Figure 12.6, it has been discovered that cereal crops, including wheat, maize, and barley, have been observed to grow and monitored using remote sensing techniques. Drone surveillance might be less expensive than other approaches because it takes less time and labor than traditional techniques.

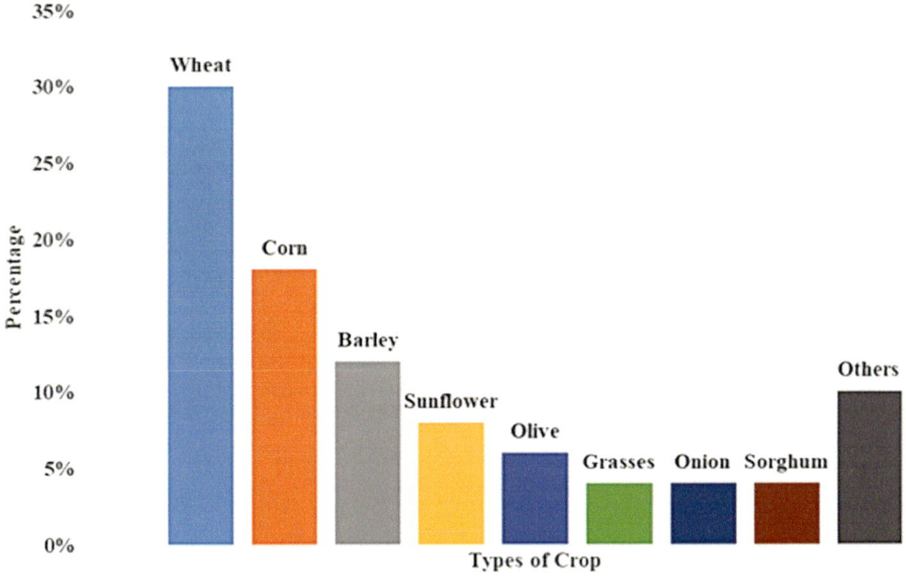

FIGURE 12.6 Crops monitored using remote sensing.

12.3.6 Screening and Analyzing Fruits and Vegetables for Freshness and Quality

Computer vision is utilized for assessing the nutritional content of vegetables and fruits through image analysis techniques. Fruits and vegetables can have their overall quality, flaws, and ripening assessed using images. Computer vision systems regularly use ML algorithms for recognizing and categorizing a variety of anomalies, including wounds, discolorations, and bruises, as shown in Figure 12.7. Computer vision also has the potential to assess the maturity of fruits and vegetables, as well as identify defects. Images of vegetables and fruits can be described using a range of attributes, such as size, form, texture, and color, using image processing techniques. These traits are employed to build the set of training data, and a classification technique is applied to obtain the knowledge base that determines the outcome of an ambiguous case. A wide range of techniques, including DL/CNN, SVM, KNN, and ANN, were created for classification while utilizing computer vision systems to assess the quality of food (Bhargava & Bansal, 2018). To achieve high-precision efficiency using cross-validation metrics for dimensional reduction, image preprocessing methods, such as normalization and data augmentation, are employed, along with Principal Component Analysis with DL (Tata et al., 2022). Crop sorting and grading are important steps in the agricultural supply chain. It involves the process of separating harvested crops into different grades based on quality, size, and other criteria. This helps ensure that

FIGURE 12.7 Test result for quality assurance of fruits.

only high-quality crops make it to market and can fetch the highest prices. Figure 12.7 illustrates how well a quality evaluation approach works. The fact that the model's prediction (Pred:) and actual classification (Real:) agree in every example that is shown shows that the model is doing a good job of recognizing the type of fruit and its quality (fresh or spoiled) in these particular test scenarios. "s_" and "f_" prefixes efficiently classify the quality of the displayed produce.

12.3.7 Livestock Monitoring

CNN architectures have applications in poultry, location estimation, and anomaly detection, but their use in grading applications requires greater attention to address knowledge gaps. Numerous uses of AI and HP can enhance fish, poultry, and cow husbandry. Computer vision can be used to monitor the actions and well-being of cattle using cameras and sensors. This helps farmers identify and treat sick animals faster. Multidisciplinary collaboration to produce innovative ideas and enhance management techniques for the livestock and poultry industries is encouraged (Li et al., 2021).

Here are some examples:

12.3.7.1 Livestock Health Monitoring

Animal health can be monitored using AI and HPC to detect diseases and prevent outbreaks. Sensors and cameras can be placed in barns and other facilities to collect data on animal behavior, such as movement patterns, feeding habits, and temperature. Then, in order to identify correlations and trends that are statistically significant and identify variations, ML algorithms can analyze data in real time. HPC can accelerate this process, allowing farmers to take immediate action when necessary (Farooq et al., 2022). Figure 12.8 shows that the agricultural community can examine

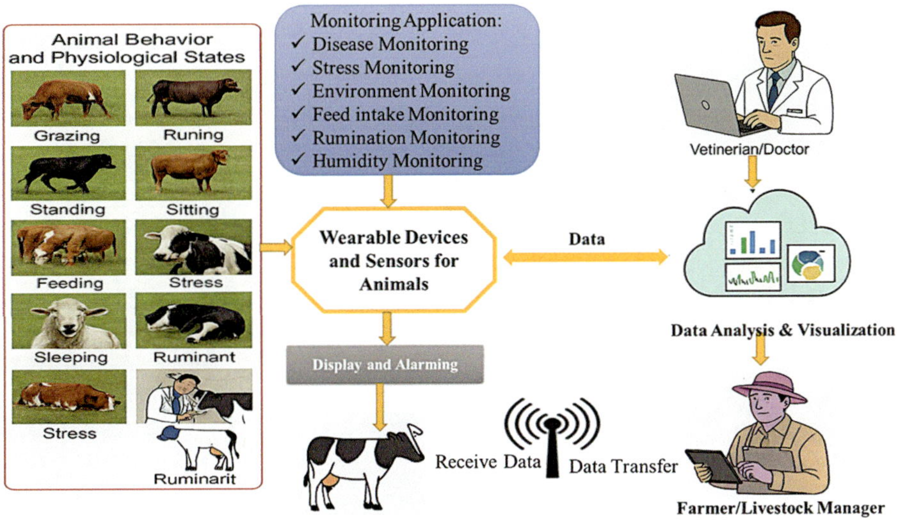

FIGURE 12.8 A perfect wearable sensor and device solution for livestock monitoring.

all aspects of farm conditions without being physically present by utilizing IoT technologies. The majority of current research on IoT-enabled devices' use in intelligent livestock farming focuses on tracking, monitoring, and predicting variables related to cattle health in order to track and grow animals.

Planned architecture enables the integration of data extraction from Internet of Things devices, storage, feature design, preprocessing, modeling, and visualization into a comprehensive solution. IoT device data streaming to live interactive reports and pre-built ML models is also included. The suggested intelligent monitoring system allows for sophisticated dynamic dashboard visualization of the ML modeling results and the gathered data. Users may monitor various farm characteristics on the dashboard and see trends, patterns, and outliers in the data (Dineva & Atanasova, 2022). A conceptual overview of the combination of cloud computing, data visualization, and Internet of Things (IoT) technologies for intelligent livestock monitoring is shown in Figure 12.9. An interactive data visualization dashboard is then used to deliver these insights to the farmers, empowering them to make well-informed decisions about farm operations, animal management, and health.

12.3.7.2 Poultry Farming

AI can be used in poultry farming to optimize the production process. ML algorithms can be used to predict bird growth rates, identify the best time to harvest, and optimize feed formulations. The creation of prediction models using DL and high-performance computing (HPC) could be applied to edge AI devices with cameras installed in chicken farms. These clever sensors enable tasks like weighing, identifying deceased chickens, counting birds, and even identifying uneven growth. The best

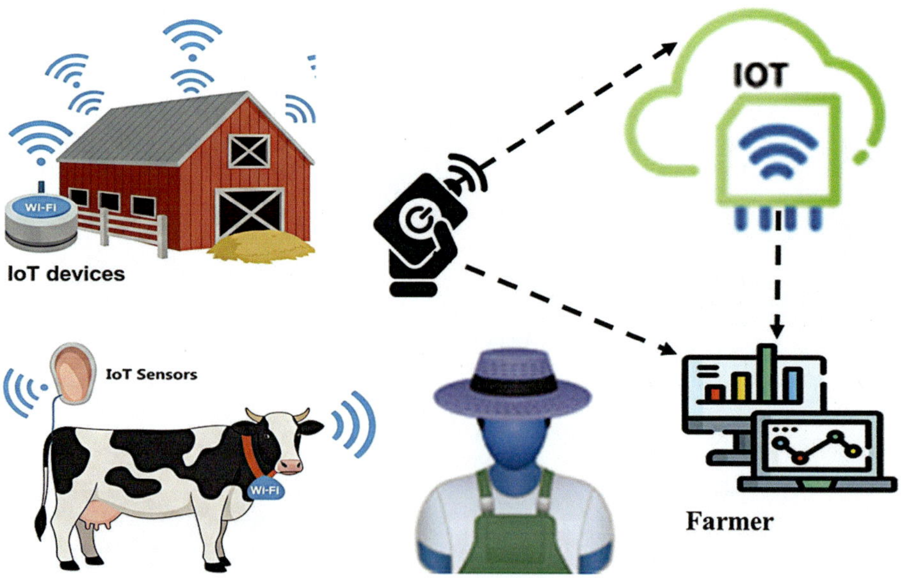

FIGURE 12.9 Smart livestock monitoring concept.

framework was identified using AutoML for segmenting and recognizing chickens in the provided dataset. Faster R-CNN architecture was the primary focus of the inquiry (Cakic et al., 2023). For tracking, counting, or evaluating the behavior or health of individual birds in a large flock, Figure 12.10 shows how computer vision—more especially, object detection—can be successfully applied to livestock monitoring. The image has colored bounding boxes surrounding individual chickens superimposed on it. Each box has a label that reads "Chicken" and a percentage that represents the object identification model's confidence score.

12.3.7.3 Fish Farming

To monitor fish behavior and enhance fish health in fish farming, HPC and AI can be applied. Temperature, water quality, and fish behavior can all be measured in fish tanks with the use of sensors. Then, through analyzing this data, AI algorithms can spot trends and forecast fish behavior. HPC can accelerate this process, enabling farmers to act promptly in the case of an issue.

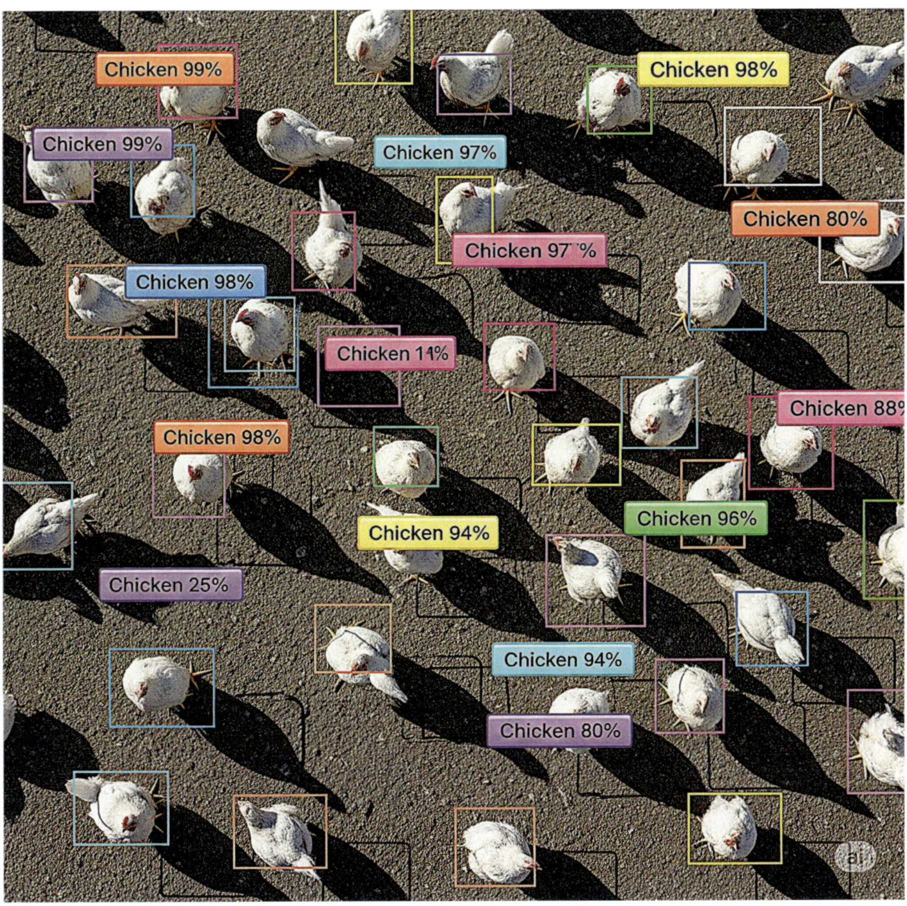

FIGURE 12.10 Object identification and confirmation at the user's site with the AI device.

12.3.7.4 For the Mutation of the Feed

AI and HPC can be used to enhance the composition of feed for animals, poultry, and fish. ML methods can analyze data on fish and animal genetics, feed mix, and nutritional requirements. This could result in improved feed compositions that adapt to the unique needs of each species of fish or animal.

12.3.7.5 Predictive Management

Predictive maintenance with AI and HPC can help the tractor, harvesting equipment, irrigation systems, and other farm machinery. To collect performance information and other factors, we can use sensor-equipped devices. The output can be increased and downtime reduced by using ML algorithms to examine this data and predict when service is required.

12.3.8 Yield Estimation

Agricultural production forecasting involves projecting the amount of crop that will be produced over a specific growing season. There are several ways to forecast agricultural yield, including statistical techniques, ML, and remote sensing technologies. A variety of data sources are used in these procedures, such as satellite imagery, crop parameters, soil data, and historical weather and climate data. By using statistical methods to analyze previous weather and harvest data, one can predict expected agricultural yields by identifying patterns and connections. The monitoring of crops with human aid becomes extremely difficult due to various elements such as terrain diversity, climate fluctuations, geographic locations, resilience to natural catastrophes, and resistance to diseases. During bad weather, traditional methods of yield estimation and crop classification are ineffective.

DL is regarded as very promising and has lately been applied to the challenge of predicting agricultural productivity. Due to DL algorithms' higher performance in other problem domains, we anticipate that more studies will be carried out based on DL techniques in crop production prediction in the near future (Van Klompenburg et al., 2020; Asole et al., 2024). To find a correct forecast, the author presented an improved optimizer function (IOF), which was then applied with the Long Short-Term Memory (LSTM) model. The results were expressed in relative mean square error (RMSE) after a comparison was made between the actual yield and the projected crop production. The suggested IOFLSTM can forecast crop yield better than CNN, Recurrent NN (RNN), and LSTM, according to the reduced RMSE for the model (Bhimavarapu et al., 2023).

12.3.9 Autonomous Farming

Computer vision can be used to enable autonomous farming by providing machines with the ability to perceive and interpret their environment. This may contribute to lower labor expenses and more productive farming practices. In various PA applications, blockchain is quickly emerging as a crucial tool. In light of the necessity for innovative peer-to-peer systems for the monitoring, security, and analysis of agricultural data, researchers are considering developing blockchain-based IoT systems for PA (Durai & Shamili, 2022). IoT is an up-and-coming technology since embedded technologies and sensors will continue to be increasingly widely available and affordable. Because of the advancements in AI and complex algorithms, intelligent applications can make more intelligent choices (Subeesh & Mehta, 2021). Figure 12.11 provides a clear and concise overview of how AI and IoT are promoting automation throughout several areas of modern agriculture, from fieldwork and managing livestock to post-harvest preservation and controlled environment farming. The image shows four significant fields of agricultural automation made possible by the blend of AI and IoT. The image provides an informative and precise summary of how AI and IoT are automating various aspects of contemporary agriculture, including post-harvest storage, controlled environment farming, field operations, and livestock management.

12.3.10 Pesticide Spraying with Drones

Pesticide spraying with drones is a modern agricultural practice that involves using UAVs or drones to spray pesticides over crops. This technology offers a safer, more efficient, and precise method of pesticide application, as compared to traditional ground-based methods.

Airborne remote sensing is one of the most important advancements in smart farming and PA. Drone surveillance technologies (Hoummaidi et al., 2021) have made it possible for farmers to view aerial images of their harvested crops. This provides information on various soil types, pests, fungal diseases, and the water system. Drone crop images capture information in the visible and infrared spectral bands. Drone-mounted nozzles have improved chemical potency, intensified, and simplified the spraying process in the field (Hafeez et al., 2022).

The author's work (Talaviya et al., 2020) evaluates the numerous applications of AI in farming, including spraying, weeding, and irrigation, utilizing sensors and other instruments integrated into drones and robotic vehicles. This is a primary objective. These advances minimize the use of pesticides, excessive water, promote

FIGURE 12.11 Agriculture automation applications using AI-IoT.

production, enhance product quality, and preserve soil fertility. They also make efficient use of labor.

12.3.11 Security Monitoring for Remote Farms

Security monitoring for remote farms using ML can be a highly effective way to enhance the security of these sites. ML analyzes data from several sources, including cameras, sensors, and other monitoring devices, to help detect potential dangers and anomalies. The practice of machine vision has been amplified recently in response to the growing need for efficient and precise methods to monitor grain production (Patrício and Rieder, 2018).

Additionally, an interactive animal health checking system based on the IoT was projected by the author (Chaudhry et al., 2020). The system under consideration would consist of a multi-sensor board that has been specifically built to monitor several physiological data, such as heart rate, skin temperature, and observations of ambient temperature and humidity.

Security monitoring for remote farms using ML can be a highly effective way to enhance the security of these sites. ML can help identify potential threats and anomalies by analyzing data from various sources such as sensors, cameras, and other monitoring devices. Security monitoring for remote farms using ML can be a highly effective way to enhance the security of these sites. ML can help identify potential threats and anomalies by analyzing data from various sources such as sensors, cameras, and other monitoring devices. Security monitoring for remote farms using ML can be a highly effective way to enhance the security of these sites. ML can help identify potential threats and anomalies by analyzing data from various sources such as sensors, cameras, and other monitoring devices. Security monitoring for remote farms using ML can be a highly effective way to enhance the security of these sites. ML can help identify potential threats and anomalies by analyzing data from various sources such as sensors, cameras, and other monitoring devices. Security monitoring for remote farms using ML can be a highly effective way to enhance the security of these sites. ML can help identify potential threats and anomalies by analyzing data from various sources such as sensors, cameras, and other monitoring devices.

12.3.12 Phenotyping with Computer Vision

Computer vision phenotyping, as used in farming, is the process of assessing and identifying the physical traits of plants using ML and computer vision methods. The technique of obtaining and evaluating complex plant traits, such as yield, yield tolerance, morphological structure, biochemical properties, disease resistance, and stress tolerance from abiotic sources, is known as phenotyping. Both ML and remote sensing have made significant contributions to high-throughput phenotyping. Depending on the desired feature, a variety of spectral and spatial resolutions can be achieved through the simultaneous use of numerous sensors, such as chlorophyll fluorescence, light detection, high-resolution RGB, and ranging (LiDAR), multispectral, and hyperspectral. Plant phenotyping, particularly in genetic and breeding research, has significant use in PA (C. Zheng et al., 2021).

12.3.13 Precision Agriculture

PA is a farming practice that can assist farmers in making decisions about crop management, such as how to irrigate, fertilize, and manage pests. PA, along with remote sensing, has been promoted by the introduction of UAV-based remote sensing systems. The prospective application of remote sensing in precision farming is viewed as being built on UAV-based IoT technology (Tsouros et al., 2019). Precision farming has a significant impact on the result and utilizes minimal input efficiently and productively. Weather forecasting, advanced analytics, and IoT are some of the significant advancements. Drones are utilized to advance accurate measurement, continuous gathering of data, and effective crop growth and development, in addition to agricultural productivity, precision, and the ability to overcome hurdles (Refaai et al., 2022; Khan et al., 2022).

12.3.14 Object Tracking in Agriculture

Recently, proposals for robotic platforms have been made that more effectively combine DL, robotics, and vision systems. These robots can repeatedly appraise the state of the field in addition to carrying out a specified activity or intervention. As an example, we introduced PATHoBot, which combines robot (platform) data and DL innovations to provide a more reliable technique for monitoring crops. The masks discovered by a DL model using wheel odometry and stereo data were projected, improving a tracking algorithm. However, only one robot (PATHoBot) and one crop (sweet pepper) were used to test the proposed robotic vision algorithm (Halstead et al., 2021). The author (Pratama et al., 2020) has proposed a DL technique that automatically detects blossom and seedpod regions from pictures of real soybean fields taken in an uncontrolled environment. RetinaNet, Cascade R-CNN, and Faster R-CNN are a few of the object recognition methods that are compared. The effectiveness of these techniques on flowers and seedpods across a variety of parameters is assessed using ablation studies (Pratama et al., 2020).

12.4 CHALLENGES

HPC, ML, DL, and Intelligent Vision have the potential to enhance agriculture's conservation, output, and effectiveness significantly. However, in order to take full benefit of these, specific problems must also be addressed.

i. **Annotation of the dataset:** Building strong CNN models requires the annotation of datasets. Since CNNs must complete a variety of computer vision tasks, instance-level and pixel-level annotations are needed for the pertinent datasets (Wang & Su, 2022; Khan et al., 2025).
ii. **Data Management:** Data from soil sensors, satellite imaging, climate, and crop production are just a few of the massive amounts of data produced by farming. It might be difficult and complex to manage and process this data. It is not easy to effectively and reliably store, integrate, and analyze this data.

iii. **Scalability:** The large-scale data processing and analysis required in agriculture can be carried out using high-performance computing systems. The problem arises in creating scalable algorithms that manage large, diverse datasets that are high in volume and make effective use of the computational resources at hand.
iv. **Data Quality:** The accuracy as well as reliability of models based on ML can be highly affected by the excellence and consistency of the data. Data used in agricultural analysis are frequently erroneous, noisy, and inadequate.
v. **Interoperability:** The agricultural data originates from a diverse set of sources and formats, making it difficult to combine and evaluate them. Establishing customs and norms for data interoperability is necessary to provide easier accessibility and analysis.
vi. **Training Data Availability:** To attain high accuracy, ML models want a lot of superior training data. Large-scale training data collection and labeling in agriculture, however, can be costly and time-consuming. Creating techniques to decrease the quantity of training data required while preserving model accuracy is the problematic part.
vii. **Limited Connectivity:** In many rural areas, internet connectivity is limited, which can be a significant barrier to accessing cloud-based HPC and ML resources. The challenge is to develop solutions that can work effectively in low-connectivity environments.
viii. **Agricultural robot** research has advanced significantly, yet there are currently no commercially produced robots that can operate in challenging agricultural environments. The main reason was that there were no algorithms that could handle the unpredictable and unmanaged practical agricultural situation. Other factors, such as the changing seasons in agriculture, additionally set the real world beyond lab settings. In order to find creative solutions that can accelerate the use of HPC and ML in agriculture, researchers, business leaders, and legislators must work together to address these issues (Zha, 2020).

12.5 CONCLUSION

AI in agriculture is one of the most appropriate and effective ways to decrease food shortages and meet the demands of a growing population. AI has been developed and refined for years by professionals from around the world. The use of AI in the agricultural sector is summarized in this study. Grain crops can be efficiently classified and characterized using high-throughput phenotyping with computer vision combined with a deep CNN. This study provides a summary of AI and DL applications in agricultural domains. In addition to quickly gathering data on plants cultivated and separating crops from weeds, these methods have also been used to identify grain varieties and detect pests and plant diseases.

Agricultural efficiency, long-term sustainability, and productivity can all be enhanced using HPC and ML. Data quality, compatibility, scalability, accessibility of training data, and limited connectivity are among the issues that must be resolved. Integrating AI with other technologies, such as generative AI, is necessary to develop innovative ideas tailored explicitly for agriculture, which can boost output, reduce expenses, and enhance yield. The advantages of various technologies can be used to create creative solutions that address complex problems in agriculture and contribute to enhancing food productivity to meet increased food demand. Real-time expert systems may be used in the future as a result of developing flexible AI-integrated models suited for various agricultural environment conditions that can be beneficial to small and medium-sized agricultural farms for increasing profits. Moreover, automation in agriculture can demonstrate improved resource utilization to make it more efficient and scalable by combining these techniques with cloud-based computing, IoT, and DL within scalable frameworks.

REFERENCES

Adli, H. K., Remli, M. A., Wong, K. N. S. W. S., Ismail, N. A., González-Briones, A., Corchado, J. M., & Mohamad, M. S. (2023). Recent advancements and challenges of AIoT application in smart agriculture: A review. *Sensors*, 23(7), 3752. https://doi.org/10.3390/s23073752

Andrew, J., Eunice, J., Popescu, D. E., Chowdary, M. K., & Hemanth, J. (2022). Deep learning-based leaf disease detection in crops using images for agricultural applications. *Agronomy*, 12(10), 2395. https://doi.org/10.3390/agronomy12102395

Asole, A., Mundhada, D., Chourasia, I., Banode, S., & Khan, A. (2024). Monitoring urban growth and infrastructure evolution with a U-Net deep learning model. *2024 First International Conference for Women in Computing (InCoWoCo)*, Pune, 1–5. https://doi.org/10.1109/incowoco64194.2024.10863148

Balasubramanian, V. N., Guo, W., Chandra, A. L., & Desai, S. V. (2020). Computer vision with deep learning for plant phenotyping in agriculture: A survey. *Advanced Computing Communications*. https://doi.org/10.48550/arXiv.2006.11391

Bhargava, A., & Bansal, A. (2018). Fruits and vegetables quality evaluation using computer vision: A review. *Journal of King Saud University - Computer and Information Sciences*, 33(3), 243–257. https://doi.org/10.1016/j.jksuci.2018.06.002

Bhimavarapu, U., Battineni, G., & Chintalapudi, N. (2023). Improved optimization algorithm in LSTM to predict crop yield. *Computers*, 12(1), 10. https://doi.org/10.3390/computers12010010

Cakic, S., Popovic, T., Krco, S., Nedic, D., Babic, D., & Jovovic, I. (2023). Developing Edge AI computer vision for smart poultry farms using deep learning and HPC. *Sensors*, 23(6), 3002. https://doi.org/10.3390/s23063002

Chaudhry, A. A., Mumtaz, R., Zaidi, S. M. H., Tahir, M. A., & School, S. H. M. (2020). Internet of things (IoT) and machine learning (ML) enabled livestock monitoring. *HONET 2020 - IEEE 17th Int. Conf. Smart Communities Improv. Qual. Life Using ICT, IoT, AI*, 151–155. https://doi.org/10.1109/honet50430.2020.9322666

Cuaran, J., & Leon, J. (2021). Crop monitoring using unmanned aerial vehicles: A review. *Agricultural Reviews*. https://doi.org/10.18805/ag.r-180

Dhanya, V., Subeesh, A., Kushwaha, N., Vishwakarma, D. K., Kumar, T. N., Ritika, G., & Singh, A. (2022). Deep learning based computer vision approaches for smart agricultural applications. *Artificial Intelligence in Agriculture*, 6, 211–229. https://doi.org/10.1016/j.aiia.2022.09.007

Dineva, K., & Atanasova, T. (2022). Cloud data-driven intelligent monitoring system for interactive smart farming. *Sensors*, 22(17), 6566. https://doi.org/10.3390/s22176566

Durai, S. K. S., & Shamili, M. D. (2022). Smart farming using machine learning and deep learning techniques. *Decision Analytics Journal*, 3, 100041. https://doi.org/10.1016/j.dajour.2022.100041

Eli-Chukwu, N. C. (2019). Applications of artificial intelligence in agriculture: A review. *Engineering Technology & Applied Science Research*, 9(4), 4377–4383. https://doi.org/10.48084/etasr.2756

Escorcia-Gutiérrez, J., Gamarra, M., Soto-Díaz, R., Pérez, M., Madera, N., & Mansour, R. F. (2022). Intelligent agricultural modelling of soil nutrients and pH classification using ensemble deep learning techniques. *Agriculture*, 12(7), 977. https://doi.org/10.3390/agriculture12070977

Farooq, M. S., Sohail, O. O., Abid, A., & Rasheed, S. (2022). A survey on the role of IoT in agriculture for the implementation of a smart livestock environment. *IEEE Access*, 10, 9483–9505. https://doi.org/10.1109/access.2022.3142848

Hafeez, A., Husain, M. A., Singh, S., Chauhan, A., Khan, M. T., Kumar, N., Chauhan, A., & Soni, S. (2022). Implementation of drone technology for farm monitoring & pesticide spraying: A review. *Information Processing in Agriculture*, 10(2), 192–203. https://doi.org/10.1016/j.inpa.2022.02.002

Halstead, M., Ahmadi, A., Smitt, C., Schmittmann, O., & McCool, C. (2021). Crop agnostic monitoring driven by deep learning. *Frontiers in Plant Science*, 12. https://doi.org/10.3389/fpls.2021.786702

Harakannanavar, S. S., Rudagi, J. M., Puranikmath, V. I., Siddiqua, A., & Pramodhini, R. (2022). Plant leaf disease detection using computer vision and machine learning algorithms. *Global Transitions Proceedings*, 3(1), 305–310. https://doi.org/10.1016/j.gltp.2022.03.016

Hoummaidi, L. E., Larabi, A., & Alam, K. (2021). Using unmanned aerial systems and deep learning for agriculture mapping in Dubai. *Heliyon*, 7(10), e08154. https://doi.org/10.1016/j.heliyon.2021.e08154

Javaid, M., Haleem, A., Khan, I. H., & Suman, R. (2022). Understanding the potential applications of artificial intelligence in the agriculture sector. *Advanced Agrochem*, 2(1), 15–30. https://doi.org/10.1016/j.aac.2022.10.001

Khan, A., Patil, C. H., Vibhute, A. D., & Mali, S. (2025). Machine learning methods for crop yield prediction across diverse agricultural environments. In: Kulkarni, A.J., Hajiaghaei-Keshteli, M. (eds) *Advances in Intelligent Systems for Sustainable Agriculture. ICISA 2023*. Lecture Notes in Networks and Systems, Vol. 1201, 107–126. https://doi.org/10.1007/978-981-97-9839-1_7

Khan, A., Vibhute, A. D., Mali, S., & Patil, C. (2022). A systematic review on hyperspectral imaging technology with a machine learning and deep learning methodology for agricultural applications. *Ecological Informatics*, 69, 101678. https://doi.org/10.1016/j.ecoinf.2022.101678

Li, G., Huang, Y., Chen, Z., Chesser, G. D., Purswell, J. L., Linhoss, J., & Zhao, Y. (2021). Practices and applications of convolutional neural network-based computer vision systems in animal farming: A review. *Sensors*, 21(4), 1492. https://doi.org/10.3390/s21041492

Lischeid, G., Webber, H., Sommer, M., Nendel, C., & Ewert, F. (2021). Machine learning in crop yield modelling: A powerful tool, but no surrogate for science. *Agricultural and Forest Meteorology*, 312, 108698. https://doi.org/10.1016/j.agrformet.2021.108698

Ouhami, M., Hafiane, A., Es-Saady, Y., Hajji, M. E., & Canals, R. (2021). Computer vision, IoT, and data fusion for crop disease detection using machine learning: a survey and ongoing research. *Remote Sensing*, 13(13), 2486. https://doi.org/10.3390/rs13132486

Patrício, D. I., & Rieder, R. (2018). Computer vision and artificial intelligence in precision agriculture for grain crops: A systematic review. *Computers and Electronics in Agriculture*, 153, 69–81. https://doi.org/10.1016/j.compag.2018.08.001

Pratama, M. T., Kim, S., Ozawa, S., Ohkawa, T., Chona, Y., Tsuji, H., & Murakami, N. (2020). Deep learning-based object detection for crop monitoring in soybean fields. *2022 International Joint Conference on Neural Networks (IJCNN)*. https://doi.org/10.1109/ijcnn48605.2020.9207400

Refaai, M. R. A., Dattu, V. S., Gireesh, N., Dixit, E., Sandeep, C., & Christopher, D. (2022). Application of IoT-Based drones in precision agriculture for pest control. *Advances in Materials Science and Engineering*, 2022, 1–12. https://doi.org/10.1155/2022/1160258

Sheeba, T. B., Anand, L. D. V., Manohar, G., Selvan, S., Wilfred, C. B., Muthukumar, K., Padmavathy, S., Kumar, P. R., & Asfaw, B. T. (2022). Machine Learning algorithm for soil analysis and classification of micronutrients in IoT-enabled automated farms. *Journal of Nanomaterials*, 2022(1). https://doi.org/10.1155/2022/5343965

Subeesh, A., & Mehta, C. (2021). Automation and digitization of agriculture using artificial intelligence and the internet of things. *Artificial Intelligence in Agriculture*, 5, 278–291. https://doi.org/10.1016/j.aiia.2021.11.004

Talaviya, T., Shah, D., Patel, N., Yagnik, H., & Shah, M. (2020). Implementation of artificial intelligence in agriculture for optimisation of irrigation and application of pesticides and herbicides. *Artificial Intelligence in Agriculture*, 4, 58–73. https://doi.org/10.1016/j.aiia.2020.04.002

Tata, J. S., Kalidindi, N. K. V., Katherapaka, H., Julakal, S. K., & Banothu, M. (2022). Real-time quality assurance of fruits and vegetables with artificial intelligence. *Journal of Physics Conference Series*, 2325(1), 012055. https://doi.org/10.1088/1742-6596/2325/1/012055

Tian, H., Wang, T., Liu, Y., Qiao, X., & Li, Y. (2019). Computer vision technology in agricultural automation — A review. *Information Processing in Agriculture*, 7(1), 1–19. https://doi.org/10.1016/j.inpa.2019.09.006

Tsouros, D. C., Bibi, S., & Sarigiannidis, P. G. (2019). A review of UAV-based applications for precision agriculture. *Information*, 10(11), 349. https://doi.org/10.3390/info10110349

Van Klompenburg, T., Kassahun, A., & Catal, C. (2020). Crop yield prediction using machine learning: A systematic literature review. *Computers and Electronics in Agriculture*, 177, 105709. https://doi.org/10.1016/j.compag.2020.105709

Wang, Y., & Su, W. (2022). Convolutional neural networks in computer vision for grain crop phenotyping: A review. *Agronomy*, 12(11), 2659. https://doi.org/10.3390/agronomy12112659

Zha, J. (2020). Artificial intelligence in agriculture. *Journal of Physics Conference Series*, 1693(1), 012058. https://doi.org/10.1088/1742-6596/1693/1/012058

Zheng, C., Abd-Elrahman, A., & Whitaker, V. (2021). Remote sensing and machine learning in crop phenotyping and management, with an emphasis on applications in strawberry farming. *Remote Sensing*, 13(3), 531. https://doi.org/10.3390/rs13030531

Zheng, Y., Kong, J., Jin, X., Wang, X., Su, T., & Zuo, M. (2019). CROpDeEp: The crop vision dataset for deep-learning-based classification and detection in precision agriculture. *Sensors*, 19(5), 1058. https://doi.org/10.3390/s19051058

13 Agriculture Using Generative AI for Crop Management

Mahesh R and Sakshi Rawat

13.1 INTRODUCTION

The global agricultural sector is currently shifting from conventional practices to technological innovations, gaining more and more efficiency, sustainability, and adaptability. The increasing demand for smart farming is mainly due to climate change, soil depletion, growing population, and non-renewable resources. This leads to a shift in the next phase of agriculture, Agriculture 5.0. This paradigm mainly uses real-time sensing, AI-powered decision support systems, and intelligent automation to create a fully automatic AI-driven farming ecosystem. To optimise the agricultural process from cultivation to post-production. The applications of Generative Artificial Intelligence (GenAI) include crop yield forecasting, precise nutrient application, soil health assessment, detection of the faults/disorders that impact post-harvest life of most horticulture crops, and disease and pest identification and management, which have proved to be novel in farming. GenAI uses models like Generative Adversarial Networks (GANs) and Variational Autoencoders (VAEs) to analyse massive datasets from satellite imagery, drones, and IoT sensors. They result in in-depth insights for smart farming, which leads to increased productivity and efficient resource management.

GenAI is an integral part of Agriculture 5.0 as it moves the field from predictive analytics to automatic systems for decision-making. GANs and VAEs, popular models of GenAI, are utilised in evaluating crop health, forecasting yields, and optimising resources using data from satellites, drones, Unmanned Aerial Vehicles (UAVs), and IoT sensors with high precision (Sengar et al., 2024; Omia et al., 2023). The growing need for potent AI models is forecasted to reach $4.7 billion by 2028 in the agriculture market (MarketsandMarkets, 2023). For the AI models to perform successfully, high-quality, precise data is essential for accurate analysis. In data-scarce scenarios, synthetic datasets generated by generative AI offer a practical solution. Transitioning from reactive to proactive farming, enhancement of low-resolution images, simulation of complex agro-ecological systems, and bridging of data gaps are crucial (Bai et al., 2024; Khanna et al., 2023).

To enhance pattern recognition and prediction accuracy, recent studies have begun exploring the integration of sophisticated AI. Integration of deep learning has shown

 DOI: 10.1201/9781003616207-15

the revolutionary capabilities of GenAI; for instance, GANs have achieved 92–97% precision in crop disease detection (Olaniyi et al., 2022), VAEs lower yield prediction errors by 6.3–18.7% (Cao et al., 2022; Liu and Song, 2025), and diffusion models like DiffusionSat perform better in temporal forecasting than conventional techniques (Khanna et al., 2023). Transformer architectures, another important element of GenAI, have reformed sequential data processing, with Vision Transformers (ViTs) showing 15.8% improvement in soybean yield forecasting over long short-term memories (LSTMs) (Zhao et al., 2023). Case studies across US soybean farms and Indian organic cotton fields validate GenAI's scalability, with CGANa-CBR frameworks fulfilling harvest-period data gaps (Bai et al., 2024) and integrated ANN-GAN systems achieving 97% real-time detection accuracy (Suganthi et al., 2025).

Although GenAI holds substantial potential in agricultural remote sensing, its complete capabilities are still mostly unexploited. Comprehensive evaluations of how GenAI models can be successfully integrated with remote sensing procedures to solve problems like scarce data, fluctuating image quality, and domain-specific flexibility are lacking in current research. Furthermore, rather than developing complete, scalable, and autonomous systems that integrate many imaging techniques and sensor data streams, present research usually focuses on single applications (such as forecasting agricultural yields or identifying crop diseases). There is an urgent need to address the aforementioned gaps by modifying existing methods, thoroughly assessing GenAI models in agricultural settings, and suggesting frameworks for real-time and effective decision support systems. With a notable number of benefits, GenAI also has several serious problems that limit its widespread application. These include concerns with generalisation, temporal-spatial disparities, data restrictions (Zhang et al., 2022), computational limits (Wu et al., 2021), and ethical considerations (FAO, 2022). This chapter primarily explores how precision agriculture can enhance crop management by integrating GenAI and remote sensing technologies. It aims to (a) introduce core GenAI models and their agricultural applications, (b) examine remote sensing technologies and their data types, (c) investigate the combination of GenAI with remote sensing, (d) present real-world case studies, and (e) discuss key challenges and future research directions.

13.2 GENERATIVE ARTIFICIAL INTELLIGENCE

GenAI is a branch of AI that generates new data (text, videos, audio) based on patterns found in current data. After training, these models can produce data that feels both realistic and relevant to the context (Sengar et al., 2024). GenAI models mimic data structures to generate new, coherent, and contextually appropriate outputs, whereas conventional AI focuses on prediction and classification (Goodfellow et al., 2014). Most widely used GenAI models are GANs (Goodfellow et al., 2014), Transformer-Based Models (TRMs) (Vaswani et al., 2017), VAEs (Kingma & Welling, 2013), and autoregressive models (Gregor et al., 2014). In Table 13.1, a brief account of the foundational models is discussed.

The practicality of GenAI has swept away several other industries. Models like GPT-3 are used to produce text that reads naturally and makes sense in almost any

TABLE 13.1
Summary of All the Foundational Models Discussed

Model	Core Architecture	Strengths	Limitations	Agricultural Applications in Remote Sensing	References
Generative Adversarial Networks	Comprises a generator and a discriminator in a minimax game framework. The generator produces synthetic data, while the discriminator tests its authenticity.	- Capable of generating high-fidelity and realistic images - Effective for data augmentation, enhancing model performance with limited datasets.	- Training can be unstable and sensitive to hyperparameters. - Susceptible to mode collapse, leading the generator to create a restricted range of outputs.	- **Crop Classification:** Enhanced classification of crops like corn and soybeans using time-series satellite imagery by integrating GAN with CNN and LSTM architectures (Li et al., 2021). - **Image Augmentation:** Augmenting agricultural datasets to improve model robustness in tasks like plant disease detection and fruit quality assessment (Olaniyi et al., 2022). - **Pan-Sharpening:** Improvement in the spatial resolution of satellite images for better crop monitoring (Liu et al., 2018).	Li et al. (2021); Olaniyi et al. (2022); Liu et al. (2018); Zhang et al. (2022); Radford et al., 2015

Variational Autoencoders	Encoder-decoder architecture that learns latent representations of data by optimising a variational lower bound.	- Provides a probabilistic framework for data generation. - Effective in learning meaningful latent spaces for complex data distributions.	- Generated outputs may be blurrier compared to GANs. - Less effective in capturing sharp details in images.	- **Yield Prediction:** Utilising VAE-based multiple instance regression models to predict large-scale corn yields from remote sensing data, addressing issues like mixed pixels (Cao et al., 2022).	Cao et al. (2022); Kingma et al. (2016); Liu and Song (2025)
Autoregressive Models (e.g., PixelCNN, GPT)	Models that generate data sequentially, predicting each element based on previous ones.	- High accuracy in modelling sequential data. - Effective in capturing complex dependencies in data sequences.	- Computationally intensive due to the sequential nature.Slower generation times compared to parallel models.	- **Crop Monitoring:** Potential in modelling temporal sequences of crop growth and health indicators from satellite data (van den Oord et al., 2016).	van den Oord et al. (2016)
Transformers/ Diffusion Models	Transformers utilise self-attention mechanisms to capture global dependencies; diffusion models generate data by reversing a gradual noising process.	- Excellent at capturing long-range dependencies in data. - Capable of generating high-quality and diverse outputs.	- Require large datasets and significant computational resources. - Training can be complex and time-consuming.	- **Satellite Image Generation:** Diffusion models like DiffusionSat are tailored for generating realistic satellite imagery, aiding in tasks such as environmental monitoring and crop yield prediction (Khanna et al., 2023). - **Remote Sensing Analysis:** Transformers have been applied in multiple remote sensing tasks, including land cover classification and change detection, which are pertinent to agricultural monitoring (Aleissaee et al., 2022).	Khanna et al. (2023). Aleissaee et al. (2022); Dosovitskiy et al. (2020)

situation in the field of natural language processing. In computer vision, GenAI can enhance images or produce realistic images. Recently, it has been used to produce human-like audio and videos. GenAI with remote sensing boosts precision farming by assessing crop vitality, forecasting production levels, and identifying diseases or pests early, allowing for quick actions. A real-world example is the collaboration of the GOTS with the European Space Agency and the AI company Marple. By utilising satellite data, they used AI models to find organic cotton fields in India, to make certification precise while reducing fraud (Vogue Business, 2023). GANs have also gained attention in mapping cities and image classification (Zhang et al., 2022). These applications demonstrate the vast potential of GenAI across various industries, including agriculture. The following section takes a closer look at the architectures behind some of the most important generative AI models.

13.2.1 Generative Adversarial Networks

Goodfellow et al. (2014) first introduced GANs. Since their innovation, GANs have become extensively used in GenAI (Figure 13.1). It is composed of two neural networks, the discriminator and the generator. Both networks compete with each other, where the generator creates false data and the discriminator verifies its validity. Through training, the generator improves itself. Eventually, the discriminator cannot distinguish between actual and fake data. GANs do not need large labelled datasets for training, unlike traditional deep learning models. They are architecturally distinct from traditional deep neural networks (Creswell et al., 2018).

The following are the variants of GANs:

1. Conditional GANs (cGANs): These models use data like class labels or metadata to guide the data generation process. This enables more precise and regulated results (Mirza & Osindero, 2014).
2. Deep Convolutional GANs (DCGANs): DCGANs improve the overall quality of generated image data by using convolutional neural networks (CNNs) (Radford et al., 2015).

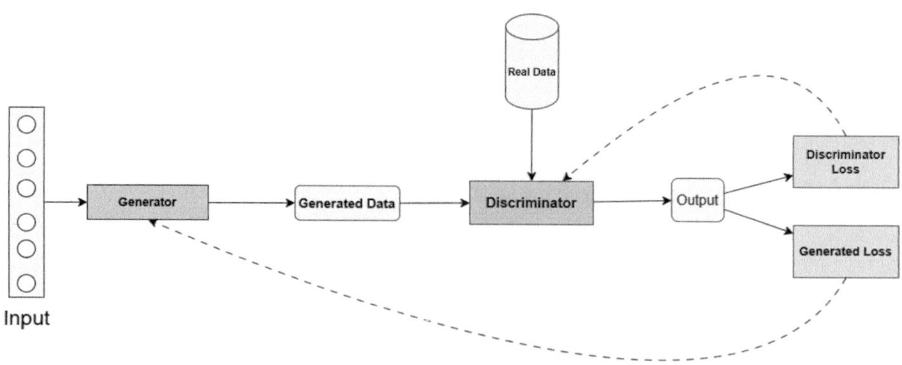

FIGURE 13.1 GAN architecture illustrating the generator and discriminator.

3. Wasserstein GANs (WGANs): These models use a Wasserstein distance-based loss function to address mode collapse and improve training stability (Arjovsky et al., 2017).
4. Self-Attention GANs (SAGANs): Self-attention GANs will be incorporated to capture long-range spatial correlations in data (Zhang et al., 2019).

13.2.2 Transformer-Based Models

In the research of Vaswani et al. (2017), the transformer architecture utilised a self-attention mechanism that successfully captured the long-range dependencies in the data previously difficult to model using traditional sequential architectures such as recurrent neural networks (RNNs) and LSTMs. The encoder and decoder transformer architecture is illustrated in Figure 13.2. The transformers are successfully used in OpenAI's ChatGPT and BERT for generative natural language processing tasks (Devlin et al., 2019). For the vision, the Vision Transformers (ViTs) have been widely adopted for a generative task (Dosovitskiy et al., 2020). The ViTGAN, a combination of ViT and GANs, is used to improve image synthesis (Lee et al., 2021). Similarly, DiffiT combines ViTs with diffusion-based generative learning to generate high-quality images (Zhou et al., 2023). These models highlight the versatility of transformers in generative applications.

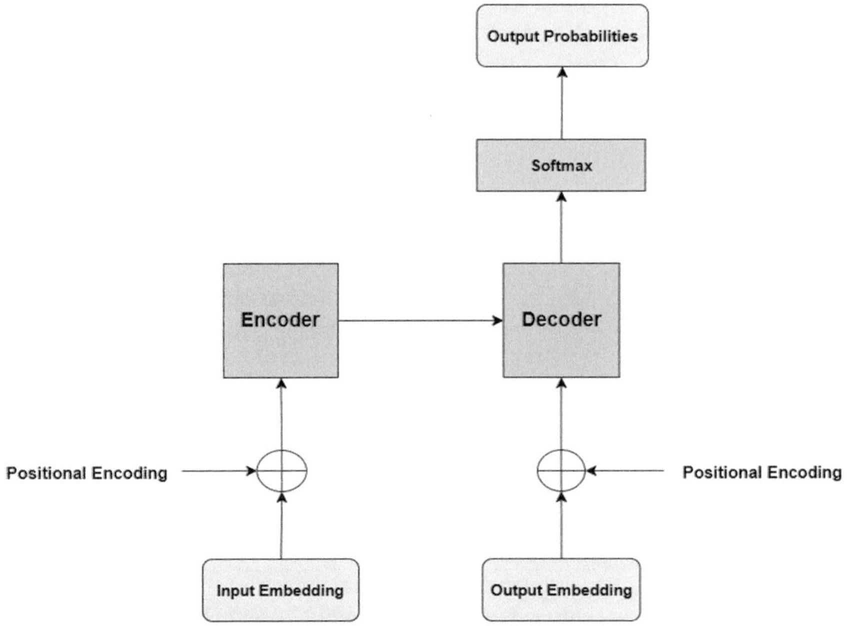

FIGURE 13.2 The transformer encoder-decoder model.

Adapted from Vaswani et al., 2017.

13.2.3 Variational Autoencoders

The VAEs with the input data in the form of images or texts are transformed into a latent space, and it is mainly a compressed representation of the data (Kingma & Welling, 2013). The latent space is typically Gaussian, which enables VAEs to create similar data samples from a distribution. In these processes, the distribution is learnt by sampling points and recreating the original data. The VAEs are commonly used for tasks like anomaly detection, text generation, and image synthesis. They are helpful for job production and reconstruction because of their capacity to develop meaningful latent representations.

13.2.4 Autoregressive Models

The AR models are used to create the data sequentially, and each new data point is generated by utilising their conditional probabilities depending on what has already been generated. Van den Oord et al. (2016) developed PixelRNN and Pixel CNN, where each pixel value in the generated image depends on the pixels created before it. Gregor et al. (2014) presented the Deep Autoregressive Networks (DARN), which combines a deep generative autoencoder with an autoregressive model. The model can handle randomness with flexible hidden layers. This enables it to generate high-quality, realistic samples for well-known datasets like Modified National Institute of Standards and Technology (MNIST) and other collections from the University of California, Irvine (UCI) repository. This next section explores how remote sensing data forms the backbone of agricultural monitoring and what the various types of remote sensing data are currently utilised for.

13.3 REMOTE SENSING IN AGRICULTURE

Remote sensing serves a pivotal role in modern agriculture. It provides continuous large-scale data for crop vigour, soil properties, and environmental conditions. Remote sensing blended with GenAI can help farmers and agronomists to make smarter decisions on irrigation, fertilisation, and disease management. As data lies at the core of remote sensing, it is important to understand different types of remote-sensing inputs (satellites, UAVs, IoT sensors).

13.3.1 Types of Remote Sensing Data

1. Satellite Imagery: Satellites offer a broad spatial coverage and frequent revisit cycles, which enable farmers and researchers to monitor crops on a large scale. They gather data in different spectral ranges, which helps in tracking vegetation health, checking soil moisture, and detecting changes in land use. Satellites such as Landsat and Sentinel provide open-source data, whereas Plantescope is a commercial satellite.
2. Unmanned Aerial Vehicles (UAVs): Drones are the most commonly used UAVs in agriculture to get a close-up view of fields. They are fitted with

thermal, multispectral, or hyperspectral sensors that capture high-resolution images for a field-specific scale. These UAVs help identify crop diseases, disorders, and nutrient deficiencies, and monitor plant health. These are also used to check the irrigation systems' work efficiency.

3. Internet of Things (IoT)-Based Sensors: Through IoT, real-time data is collected through sensors embedded in weather stations and fields. Constant monitoring of conditions like soil pH, soil EC, soil moisture, soil temperature, atmospheric temperature, relative humidity, light intensity, light quality, and pest activity is carried out. This helps farmers make precise decisions about watering crops, targeting fertiliser, and catching infestations early. According to this recent research, the approach is transforming traditional farming into a more efficient practice (Rajak et al., 2023).

13.3.2 Spectral Imaging: Multispectral and Hyperspectral

The spectral imaging technology captures real-time data at different wavelengths of light, which are beyond the human eye's visible spectrum. By analysing this hidden light, farmers can detect early signs of crop stress, monitor plant health, and even identify subtle changes in a plant's biological makeup.

a. Multispectral Imaging: The multispectral imaging technology can capture multiple spectral bands, which typically include visible and near-infrared wavelengths. By analysing spectral reflectance, this provides valuable insights into crop vigour, stress levels, and growth dynamics. Calculating NDVI using multispectral imaging indicates plant health (EOS Data Analytics, 2023).
b. Hyperspectral Imaging: The sensors with hyperspectral imaging record numerous narrow, adjoining spectral bands and provide highly detailed spectral information. The elevated spectral resolution can discern distinct subtle variations in plant physiology, rendering it helpful in pinpointing nutrient deficiencies, disease indicators, and crop stress (Ram et al., 2024).

13.4 INTEGRATION OF GENAI WITH REMOTE SENSING FOR CROP MANAGEMENT

Remote sensing and generative AI are converging to create unique capabilities for precision agriculture. By fusing multi-platform imagery (satellite, UAV, SAR, LiDAR) and real-time sensor streams, AI models can generate high-resolution, cloud-free datasets, synthesise realistic imagery where data are missing, and produce synthetic training examples that fill gaps in labelled data. These innovations underpin five key applications: (1) AI-driven data fusion across sensors, (2) GAN-based image synthesis and super-resolution, (3) synthetic data generation for model training, (4) GenAI-enhanced land cover classification, and (5) VAE- and Transformer-based time-series forecasting of crop yield and growth.

13.4.1 AI-Driven Data Fusion from Different Sources

Effective combining of remote sensing and GenAI models is transforming precision agriculture by enhancing data quality and predictive capabilities. Satellites (e.g., Sentinel-2, Landsat) offer broad spatial coverage and frequent revisits, while UAVs provide ultra-high spatial resolution at localised scales (Omia et al., 2023; Alsadik et al., 2023). However, differences in spectral resolutions, timing, and image quality complicate direct integration. AI-driven fusion techniques address these challenges by learning optimal combinations of multi-source inputs. Deep neural networks, such as convolutional encoders or attention-based models, dynamically weigh sensor contributions to achieve more accurate, context-aware fusion (Mena et al., 2024). Multi-view gated fusion networks, for example, have improved crop-yield predictions by merging satellite reflectance and drone-derived vegetation indices (Mena et al., 2024). AI models also increase fusion of optical and Synthetic Aperture Radar (SAR) imagery, mitigating challenges like cloud interference and data gaps (Mirzaei et al., 2023). Coherent, enriched datasets created through such fusion enable more effective application of GenAI methods, including GANs, VAEs, and transformers, downstream tasks like image super-resolution, synthetic data generation, and forecasting crop yield. The coherent and enriched multi-source dataset in the next step involves harnessing GenAI techniques such as GANs to synthesise high-quality images and also enhance spatial resolution and fill observational gaps critical for crop monitoring and management.

13.4.2 Image Synthesis and Super-Resolution Using GANs

The GANs have emerged as a powerful, transformative tool in remote sensing, particularly for reconstructing missing imagery data and enhancing the quality of satellite and drone imagery in agriculture. In many farming regions, persistent cloud cover, the slow return time of satellites, and the coarse detail of some imagery mean that important parts of a crop's growth cycle are not fully observed. During these missing periods, GANs can be used to fill in the picture by producing high-resolution synthetic images from whatever information is available at the time (Goodfellow et al., 2014; Suganthi et al., 2025). In practice, these extra images become part of a much larger training dataset for models such as CNNs and the newer transformer architectures. Having more and more detailed data allows these models to make better estimates of yield, pick up early signs of crop stress, and provide a clearer picture of soil conditions. From my experience, combining the synthetic imagery with actual field or satellite data tends to make the models more reliable and better able to cope with changing conditions. In situations where field data is limited or hard to collect, GenAI can even create complete, stand-alone datasets. When the output from GANs is linked with IoT sensor readings and ongoing satellite feeds, it becomes possible to keep track of things like soil moisture patterns, nutrient distribution, outbreaks of disease, or pest build-up almost as they happen (Suganthi et al., 2025). Spotting these problems early gives farmers a chance to respond quickly, whether that means adjusting irrigation in a dry spell or applying pest control only where it is needed so crops have a better chance of coping with stress and reaching their potential yield.

13.4.3 Synthetic Data Generation for Model Training

In applying deep learning to agricultural remote sensing, one persistent difficulty is the limited supply of annotated datasets that span different crop types, seasons, and weather conditions. In many cases, collecting such data directly in the field is not only expensive but also slowed by infrequent satellite passes and unfavourable weather. Over the past few years, GenAI methods, particularly VAEs, GANs, and diffusion models, have emerged as a practical way to bridge these gaps (Goodfellow et al., 2014; Kingma & Welling, 2013; Rombach et al., 2022). By producing synthetic datasets, these methods reduce reliance on costly and often incomplete real-world observations. Different GenAI techniques bring different strengths. Conditional GANs, for instance, can be trained to produce imagery that reflects particular agricultural conditions, such as prolonged drought, nutrient deficiencies, or pest damage. In the poorly presented classes, this capability helps train a dataset (Zhang et al., 2022). The problem was addressed differently by VAE's approach by learning a compact latent representation of multispectral imagery and then sampling from this learnt space to create new examples (Cao et al., 2022). The contrasted images resulting from the diffusion models showed realism and diversity in synthetic satellite imagery, which is difficult to achieve with other methods (Khanna et al., 2023). The produced synthetic datasets are combined with actual observations, resulting in an outcome with a more adaptable and robust monitoring system. These systems are equipped to manage the practical challenges in detecting subtle shifts in land cover, precise mapping, or fine-tuning the allocation of resources, that is, water and fertilisers.

13.4.4 GenAI-Enhanced Land Cover Classification Applications

For agricultural monitoring and effective resource management, the land cover classification is the core requirement. However, models that rely heavily on manual feature extraction often struggle when applied to landscapes with highly variable land-use patterns or to areas undergoing seasonal phenological changes. Over the past few years, developments in GenAI, particularly transformer-based architectures and GANs, have helped to overcome some of these limitations. These advances have improved classification accuracy by enhancing data quality, extracting more meaningful features, and sharpening boundary delineation. Several recent studies illustrate this trend. While classification remains a significant focus, other applications, such as predicting crop growth or estimating yields, require modelling how conditions change over time. This temporal dimension is handled effectively by VAEs and Transformer-based models, both of which are well-suited to decoding complex patterns in agricultural time-series datasets derived from remote sensing. In practice, this means the same GenAI tools that improve classification can also form the backbone of more predictive, forward-looking agricultural monitoring systems. The combined Sentinel-2 time-series data with GAN-generated residuals, achieving a 94% classification accuracy even in fragmented agricultural regions (Li et al., 2022). The boosted IoU scores are achieved by refining a U-Net model with GAN components, utilising the generalised vegetative index (Sheng et al., 2020) as input. More recently, the hybrid approaches such as TransGAN have paired adversarial learning with self-attention mechanisms, delivering fine-grained local segmentation while simultaneously capturing broader spatial patterns. These methods tend to perform best when supported by high-resolution data from drones or advanced satellite platforms.

13.4.5 VAEs and Transformers for Time-Series Forecasting of Crop Yield and Growth

The direction of precision agriculture is moving towards deep generative models and attention-based architectures, driven by rapid advances in remote sensing (Figure 13.3). Traditional methods such as LSTMs and RNNs (Recurrent Neural Network) remain in use but often struggle with long-term dependencies and uncertainty estimation. To overcome these limitations, researchers are turning to modern approaches like VAEs and transformer-based models. Both have proven effective for agricultural time-series analysis, whether for yield prediction, plant growth monitoring, or tracking productivity shifts. VAEs are valuable for uncovering hidden data structures and producing realistic variations (Kingma & Welling, 2013). A conditional VAE to improve crop forecasts, predicting harvest yields by capturing temporal satellite data (Liu and Song, 2025). Similarly, the VAE-based multiple instance regression (VAEMIR) model outperformed traditional regression in country-level corn yield prediction developed by Cao et al. (2022). The transformers, with their self-attention mechanism, capture long-range dependencies more effectively, enabling more accurate sequence modelling. The ViTs surpassed LSTMs in early soybean yield prediction using UAV imagery (Vaswani et al., 2017). Diffusion-based Transformer models in DiffusionSat forecast multi-spectral crop growth patterns up to 30 days ahead

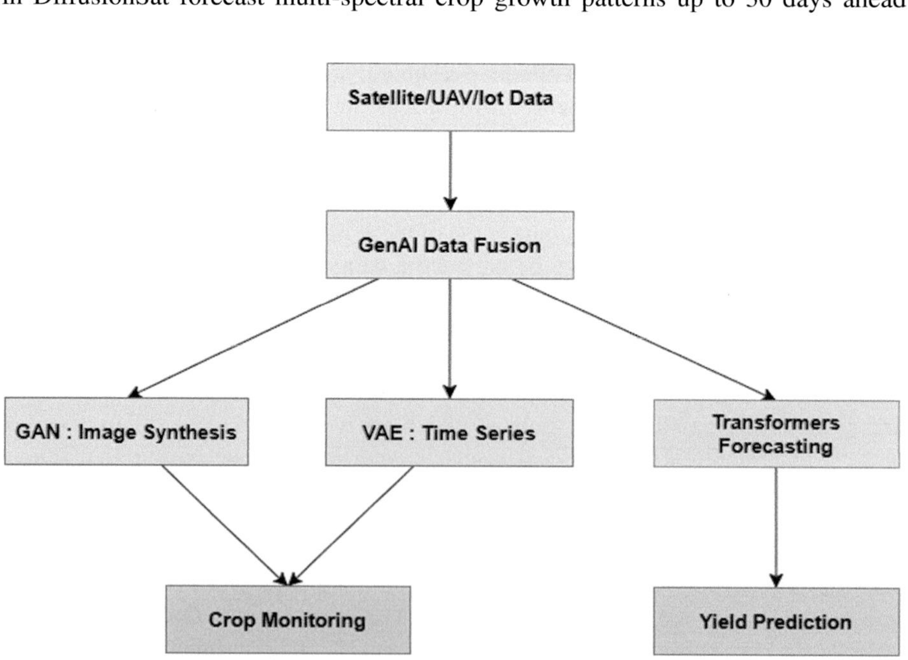

FIGURE 13.3 Integrated GenAI with a remote sensing framework.

(Khanna et al., 2023). Additionally, Kamangir et al. (2024) proposed CMAViT, integrating remote sensing, weather, and management data to predict vineyard yields at the pixel level, achieving $R^2 = 0.84$ and MAPE = 8.22%. VAEs and transformers are modern GenAI models that allow more accurate, uncertainty-aware forecasting, empowering proactive and data-driven crop management.

The fusion of generative AI techniques and remote sensing technologies is revolutionising precision agriculture, offering innovative solutions to traditional challenges. As shown in Figure 13.3, this integrated framework leverages AI-driven data fusion, high-fidelity image synthesis using GANs, and synthetic dataset generation using VAEs and diffusion models to overcome limitations like data scarcity, low resolution, and missing observations. Additionally, the application of advanced architectures like transformers for time-series forecasting enables more precise, proactive decision-making under varying farming conditions. Real-world case studies across different crops and regions demonstrate that these GenAI approaches increase predictive performance and support resource optimisation and sustainability by enabling early crop stress detection, diseases, and yield fluctuations.

13.5 CASE STUDIES

13.5.1 Yield Prediction Using GANs with Deep Learning and Remote Sensing in Precision Agriculture

Pre-training on various remote sensing datasets, including fMoW, Satlas, and SpaceNet (Christie et al., 2018; Bastani et al., 2022), helps in the generalisability of the model. Beyond the single-image synthesis, DiffusionSat employs a 3D ControlNet-inspired architecture (Blattmann et al., 2023) for tasks like multi-spectral super-resolution, temporal forecasting, and inpainting. For instance, it can upsample low-resolution multispectral data to high-resolution RGB or predict future imagery, critical for agriculture and disaster management. Experiments showed the significant improvements over baselines. Its performance highlights include achieving a Fréchet Inception Distance (FID) of 15.80 on the fMoW dataset, surpassing fine-tuned Stable Diffusion variants (Khanna et al., 2023). Outperformed Deep Back-Projection Network (DBPN) and Pix2Pix in perceptual quality (SSIM, LPIPS). Excelled in temporal prediction over Spatial–Temporal Super-Resolution (STSR) and Masked Conditional Video Diffusion (MCVD), better capturing seasonal dynamics. DiffusionSat captured seasonal dynamics likely more effectively than STSR (He et al., 2021) and MCVD (Voleti et al., 2022), offering higher predictive accuracy. By integrating metadata conditioning, latent diffusion, and 3D temporal learning, DiffusionSat advances GenAI systems for remote sensing, enabling high-fidelity satellite imagery for applications like crop monitoring and emergency disaster response (Khanna et al., 2023). Suganthi et al. (2025) developed an innovative farming system that generates its data to tackle data limitations. Their research integrates geospatial technologies and deep learning for crop growth and yield prediction. In collaboration with the global positioning system and geographic information system, their systems use ANNs, GANs, and YOLO object detection models. Synthetic images are generated with GANs to overcome data scarcity and regional variations.

13.5.2 Gen AI Application for Synthetic Satellite Imagery

The main application of the GenAI application in agriculture is for yield estimation. The CGAN component uses a generator to produce harvest-season images from historical data, while a discriminator evaluates their authenticity to improve accuracy and robustness in yield estimation (Goodfellow et al., 2014). The Case-Based Reasoning (CBR) component adds another layer by retrieving past images that closely match current conditions, helping the generator learn from relevant historical patterns. Similar approaches have been used in meteorology, where GAN-based systems generate synthetic radar data for forecasting and nowcasting (Ravuri et al., 2021). Bai et al. (2024) proposed a framework that blends CGANs with CBR to tackle the common problem of missing remote sensing data during the harvest season. Such gaps often occur due to delayed satellite passes or extended cloud cover, which can leave critical periods unobserved. When actual harvest-season images were missing, the CGAN produced synthetic replacements, which were then combined with real data to train a CNN for county-level yield forecasts. The results were striking on average; forecast relative mean square error (RMSE) fell by 6.3% compared with models trained only on real data. These findings show that advanced GenAI techniques can effectively bridge data gaps, improving prediction accuracy and reliability even when minimal field observations are available.

13.5.3 DiffusionSat Generative Model for Satellite Imagery Challenges

DiffusionSat is the first specialised generative model used to understand remote sensing imagery. It can handle complexities related to satellite imagery like multi-spectral bands, irregular time intervals, and missing text annotations (Khanna et al., 2023). These intricacies are often overlooked in traditional diffusion models like Stable Diffusion, which is trained on natural images. The Latent Diffusion models (LDMs) first compress the satellite images in a compact latent space (Vahdat et al., 2021; Rombach et al., 2022). Diffusion is applied after encoding images to handle a large volume of data. In cases where text descriptions are missing in the dataset, the model uses numerical metadata directly. This is done through sinusoidal projections on learnt embeddings, where the model assimilates contextual information (Radford et al., 2021).

13.6 CONCLUSION

The integration of GenAI with remote sensing is transforming precision agriculture by significantly improving the automation efficiency, expanding the scalability of monitoring systems across large agricultural areas and enhancing the accuracy of predictive insights such as yield estimation, crop production management The advanced generative models like generative transformers, adversarial networks, diffusion models and VAEs enable the synthesis of missing satellite or drone imagery, enhance image resolution and generate diverse and high-quality datasets for training robust agricultural AI systems. These modern technologies tackle critical challenges such as cloud cover obstructions, limited labelled data, and irregular temporal gaps, while improving crop yield predictions and foundational models like DiffusionSat-highlight's ability to enhance both data quality and decision-making. The practical implementations include GAN-based precision irrigation, targeted fertilisation, pest and disease monitoring, and climate-resilient strategies for crop improvement. GenAI-powered remote sensing enables smarter, more sustainable farming. Across diverse agro-ecological regions, these systems are proving essential for maximising productivity while minimising the environmental impact.

REFERENCES

Aleissaee, A. A., Kumar, A., Anwer, R. M., Khan, S., Cholakkal, H., Xia, G.-S., & Khan, F. S. (2022). Transformers in remote sensing: A survey. *arXiv*. https://arxiv.org/abs/2209.01206

Alsadik, B., Dadrass Javan, F., & Nex, F. C. (2023). UAV remote sensing for smart agriculture. *GIM International*, *36*(1), 14–17. https://www.gim-international.com/content/article/uav-remote-sensing-for-smart-agriculture

Arjovsky, M., Chintala, S., & Bottou, L. (2017). Wasserstein GAN. *arXiv*. https://arxiv.org/abs/1701.07875

Bai, Y. B., Chen, D., & Zhang, Z. Y. (2024). A conditional generative adversarial networks-augmented case-based reasoning framework for crop yield predictions with time-series remote sensing data. In *Proceedings of the 32nd International Conference on Case-Based Reasoning (ICCBR)*, Merida, Mexico (Vol. 3708, pp. 192–205).

Bastani, F., Wolters, P., Gupta, R., Ferdinando, J., & Kembhavi, A. (2022). Satlas: A large-scale, multi-task dataset for remote sensing image understanding. *arXiv*. https://arxiv.org/abs/2211.15660

Blattmann, A., Rombach, R., Ling, H., Dockhorn, T., Kim, S. W., Fidler, S., & Kreis, K. (2023). Align your latents: High-resolution video synthesis with latent diffusion models. *arXiv*. https://arxiv.org/abs/2304.08818

Cao, Z., Ma, Y., & Zhang, Z. (2022). Corn yield prediction based on remotely sensed variables using variational autoencoder and multiple instance regression. *arXiv*. https://arxiv.org/abs/2211.13286

Christie, G., Fendley, N., Wilson, J., & Mukherjee, R. (2018). Functional map of the world. In *Proceedings of the IEEE Conference on Computer Vision and Pattern Recognition* (pp. 2466–2475).

Creswell, A., White, T., Dumoulin, V., Arulkumaran, K., Sengupta, B., & Bharath, A. A. (2018). Generative adversarial networks: An overview. *IEEE Signal Processing Magazine*, *35*(1), 53–65. https://doi.org/10.1109/MSP.2017.2765202

Devlin, J., Chang, M. W., Lee, K., & Toutanova, K. (2019). BERT: Pre-training of deep bidirectional transformers for language understanding. In *Proceedings of the 2019 Conference of the North American Chapter of the Association for Computational Linguistics: Human Language Technologies* (pp. 4171–4186). https://doi.org/10.18653/v1/N19-1423

Dosovitskiy, A., Beyer, L., Kolesnikov, A., Weissenborn, D., Zhai, X., Unterthiner, T., … & Houlsby, N. (2020). An image is worth 16x16 words: Transformers for image recognition at scale. *arXiv*. https://arxiv.org/abs/2010.11929

EOS Data Analytics. (2023). *Multispectral vs. hyperspectral imaging: Differences and uses*. https://eos.com/blog/multispectral-vs-hyperspectral-imaging/

Food and Agriculture Organization of the United Nations (FAO). (2022). *Ethical and Legal Issues in Agricultural Data Sharing*. Rome. http://www.fao.org/3/cb8661en/cb8661en.pdf

Goodfellow, I. J., Pouget-Abadie, J., Mirza, M., Xu, B., Warde-Farley, D., Ozair, S., Courville, A., & Bengio, Y. (2014). Generative adversarial networks (Version 1). *arXiv*. https://doi.org/10.48550/ARXIV.1406.2661

Gregor, K., Danihelka, I., Mnih, A., Blundell, C., & Wierstra, D. (2014, June). Deep autoregressive networks. In *International Conference on Machine Learning* (pp. 1242–1250). PMLR. https://proceedings.mlr.press/v32/gregor14.html

He, Y., Wang, D., Lai, N., Zhang, W., Meng, C., Burke, M., Lobell, D., & Ermon, S. (2021). Spatial-temporal super-resolution of satellite imagery via conditional pixel synthesis. *Advances in Neural Information Processing Systems*, *34*, 27903–27915.

Kamangir, H., Sams, B. S., Dokoozlian, N., Sanchez, L., & Earles, J. M. (2024). CMAViT: Integrating climate, management, and remote sensing data for crop yield estimation with multimodal vision transformers. *arXiv*. https://arxiv.org/abs/2411.16989

Khanna, S., Liu, P., Zhou, L., Meng, C., Rombach, R., Burke, M., Lobell, D., & Ermon, S. (2023). DiffusionSat: A generative foundation model for satellite imagery. *arXiv*. https://arxiv.org/abs/2312.03606

Kingma, D. P., Salimans, T., Jozefowicz, R., Chen, X., Sutskever, I., & Welling, M. (2016). Improving variational inference with inverse autoregressive flow (Version 2). *arXiv*. https://doi.org/10.48550/ARXIV.1606.04934

Kingma, D. P., & Welling, M. (2013). Auto-encoding variational Bayes. *arXiv*. https://arxiv.org/abs/1312.6114

Lee, Y., Kim, J., & Kim, G. (2021). ViTGAN: Training GANs with vision transformers. *arXiv preprint arXiv:2107.04589*. https://arxiv.org/abs/2107.04589

Li, J., Shen, Y., & Yang, C. (2021). An adversarial generative network for crop classification from remote sensing time-series images. *Remote Sensing*, 13(1), 65. https://doi.org/10.3390/rs13010065

Li, X., Sun, C., Meng, H., Ma, X., Huang, G., & Xu, X. (2022). A novel efficient method for land cover classification in fragmented agricultural landscapes using Sentinel satellite imagery. *Remote Sensing*, 14(9), 2045. https://doi.org/10.3390/rs14092045

Liu, L., & Song, X. (2025). Accurate crop yield prediction via temporal convolutional network and variational autoencoder. *Turkish Journal of Agriculture and Forestry*, 49(3), 612–623. https://doi.org/10.55730/1300-011x.3290

Liu, Q., Zhou, H., Xu, Q., Liu, X., & Wang, Y. (2018). PSGAN: A generative adversarial network for remote sensing image pan-sharpening. *arXiv preprint arXiv:1805.03371*. https://arxiv.org/abs/1805.03371

MarketsandMarkets. (2023). Artificial Intelligence in Agriculture Market by Technology (Machine Learning, Computer Vision), Application (Precision Farming, Livestock Monitoring, Drone Analytics), Offering (Software, Hardware, AI-as-a-Service), and Geography - Global Forecast to 2028. https://www.marketsandmarkets.com/Market-Reports/ai-in-agriculture-market-159957009.html

Mena, F., Pathak, D., Najjar, H., & Tiwari, A. (2024). Adaptive fusion of multi-view remote sensing data for optimal sub-field crop yield prediction. *arXiv preprint arXiv:2401.11844*. https://arxiv.org/abs/2401.11844

Mirza, M., & Osindero, S. (2014). Conditional generative adversarial nets. *arXiv preprint arXiv:1411.1784*. https://arxiv.org/abs/1411.1784

Mirzaei, A., Bagheri, H., & Khosravi, I. (2023). Enhancing crop classification accuracy through synthetic SAR-optical data generation using deep learning. *ISPRS International Journal of Geo-Information*, 12(11), 450. https://doi.org/10.3390/ijgi12110450

Olaniyi, E., Chen, D., Lu, Y., & Huang, Y. (2022). Generative adversarial networks for image augmentation in agriculture: A systematic review. *Computers and Electronics in Agriculture*, 198, 107017. https://doi.org/10.1016/j.compag.2022.107017

Omia, E., Bae, H., Park, E., Kim, M. S., Baek, I., Kabenge, I., & Cho, B.-K. (2023). Remote sensing in field crop monitoring: A comprehensive review. *Remote Sensing*, 15(2), 354. https://doi.org/10.3390/rs15020354

Radford, A., Kim, J. W., Hallacy, C., Ramesh, A., Goh, G., Agarwal, S., Sastry, G., Askell, A., Mishkin, P., Clark, J., Krueger, G., & Sutskever, I. (2021). Learning transferable visual models from natural language supervision. In *Proceedings of the 38th International Conference on Machine Learning* (pp. 8748–8763). https://arxiv.org/abs/2103.00020

Radford, A., Metz, L., & Chintala, S. (2015). Unsupervised representation learning with deep convolutional generative adversarial networks. *arXiv preprint arXiv:1511.06434*. https://arxiv.org/abs/1511.06434

Rajak, P., Ganguly, A., Adhikary, S., & Bhattacharya, S. (2023). Internet of Things and smart sensors in agriculture: Scopes and challenges. *Journal of Agriculture and Food Research*, 14, 100776. https://doi.org/10.1016/j.jafr.2023.100776

Ram, B. G., Oduor, P., Igathinathane, C., Howatt, K., & Sun, X. (2024). A systematic review of hyperspectral imaging in precision agriculture: Analysis of its current state and prospects. *Computers and Electronics in Agriculture*, 222, 109037. https://doi.org/10.1016/j.compag.2024.109037

Ravuri, S., Mohamed, S., & Mohamed, A. (2021). Skilful precipitation nowcasting using deep generative models of radar. *Nature*, 597(7878), 672–677. https://doi.org/10.1038/s41586-021-03854-z

Rombach, R., Blattmann, A., Lorenz, D., Esser, P., & Ommer, B. (2022). High-resolution image synthesis with latent diffusion models. In *Proceedings of the IEEE/CVF Conference on Computer Vision and Pattern Recognition* (pp. 10684–10695). https://doi.org/10.48550/arXiv.2112.10752

Sengar, S. S., Hasan, A. B., Kumar, S., & Carroll, F. (2024). Generative artificial intelligence: A systematic review and applications. *Multimedia Tools and Applications*, 84(21), 23661–23700. https://doi.org/10.1007/s11042-024-20016-1

Sheng, H., Chen, X., Su, J., Rajagopal, R., & Ng, A. (2020). Effective data fusion with generalized vegetation index: Evidence from land cover segmentation in agriculture. *arXiv preprint arXiv:2005.03743*. https://arxiv.org/abs/2005.03743

Suganthi, S. U., Prinslin, L., Selvi, R., & Prabha, R. (2025). Generative AI in Agri: Sustainability in smart precision farming yield prediction mapping system based on GIS using deep learning and GPS. *Procedia Computer Science*, 252, 365–380. https://doi.org/10.1016/j.procs.2024.12.039

Vahdat, A., Kreis, K., & Kautz, J. (2021). Score-based generative modeling in latent space. *Advances in Neural Information Processing Systems*, 34, 11287–11299. https://github.com/NVlabs/LSGM

van den Oord, A., Kalchbrenner, N., & Kavukcuoglu, K. (2016). Pixel recurrent neural networks. In *Proceedings of the 33rd International Conference on Machine Learning* (pp. 1747–1756). https://proceedings.mlr.press/v48/oord16.html

Vaswani, A., Shazeer, N., Parmar, N., Uszkoreit, J., Jones, L., Gomez, A. N., ... & Polosukhin, I. (2017). Attention is all you need. *Advances in Neural Information Processing Systems*, 30, 5998–6008. https://10.48550/arXiv.1706.03762

Vogue Business. (2023, June 5). *How AI-Powered Satellite Data Could Boost the Organic Cotton Industry*. Vogue Business. https://www.voguebusiness.com/sustainability/how-ai-powered-satellite-data-could-boost-the-organic-cotton-industry

Voleti, V., Jolicoeur-Martineau, A., & Pal, C. (2022). Masked conditional video diffusion for prediction, generation, and interpolation. *arXiv Preprint, arXiv:2205.09853*. https://arxiv.org/abs/2205.09853

Wu, C., Raghavendra, R., Gupta, U., Acun, B., Ardalani, N., Maeng, K., Chang, G., Behram, F. A., Huang, J., Bai, C., Gschwind, M., Gupta, A., Ott, M., Melnikov, A., Candido, S., Brooks, D., Chauhan, G., Lee, B., Lee, H. S., ... Hazelwood, K. (2021). Sustainable AI: environmental implications, challenges, and opportunities. *arXiv* (Cornell University). https://doi.org/10.48550/arxiv.2111.00364

Zhang, H., Goodfellow, I., Metaxas, D., & Odena, A. (2019). Self-attention generative adversarial networks. In *Proceedings of the 36th International Conference on Machine Learning* (pp. 7354–7363).

Zhang, Y., Wang, S., & Liu, J. (2022). A review and meta-analysis of generative adversarial networks and their applications in remote sensing. *International Journal of Applied Earth Observation and Geoinformation*, 108, 102734. https://doi.org/10.1016/j.jag.2022.102734

Zhao, Y., Wang, K., Liu, J., Li, M., & Zhang, Y. (2023). A transformer-based approach for early prediction of soybean yield. *Frontiers in Plant Science*, 14, 1173036. https://doi.org/10.3389/fpls.2023.1173036

Zhou, Y., Wang, Y., & Zhang, X. (2023). DiffiT: Diffusion Vision Transformers for image generation. *arXiv Preprint, arXiv:2312.02139*. https://arxiv.org/abs/2312.02139

Part III

Resources, Challenges, and the Future of Gen AI in Remote Sensing

14 Harnessing AI to Unveil the Future

Modelling and Forecasting Climate Change Effects

Priyanka P. Shinde, Anurag Wazarkar, Pratik Gunjalkar, Tanmay Sawant, and Sanket Babar

14.1 INTRODUCTION

One of the most urgent challenges facing human society in the 21st century is the escalating impact of climate change. Rising global warming, growing instability in climatic patterns, and increasing extreme events like wildfires, floods, and droughts emphasize the need for more advanced methods to comprehend, forecast, and counteract these effects (Othman, 2024). Even though conventional climate science has made significant advancements in describing the physical foundation of such phenomena, highly interconnected and complex nature of the Earth's climate system requires advanced analytical tools (Belnekar, 2025). The most hopeful among these tools is artificial intelligence (AI), specifically its subfields of machine learning (ML) and deep learning (DL), which provide the ability to analyse enormous, multidimensional datasets and develop highly adaptive, data-driven prediction models. Climate modelling has made tremendous strides from the early conceptual schemes to advanced numerical models. Physical climate models, including General Circulation Models (GCMs), have been at the heart of modelling interactions between the Earth's atmospheric, oceanic, terrestrial, and cryosphere systems. Based on physical laws, these models have traditionally underpinned climate science (Hamdan et al., 2024). But with increasingly complex climate phenomena and exponentially greater availability of data through satellite, remote sensing, and in-situ monitoring, their inability to address regional dynamics and nonlinear interactions is now well recognized (Alnajjar et al., 2024).

Traditional models are plagued by issues such as low spatial resolution, inability to simulate fine-grained extreme events, and computation inefficiencies. Additionally, most traditional models utilize linear approximations, which may over-simplify inherently nonlinear climate processes. (Belnekar, 2025) The second large-scale limitation is the absence of integration with interdisciplinary tools, particularly those

DOI: 10.1201/9781003616207-17

from AI, that would improve both accuracy and interpretability (LunarTech, 2025). This also requires embedding AI techniques as fundamental components of future climate research, as to how climate forecasting and policy modelling are addressed. Furthermore, limited efforts towards real-time interpretability, ethical governance, and energy-efficient AI modelling signal key areas in need of exploration (NSF, 2025). This study seeks to investigate how AI, especially ML techniques, can improve climate modelling and forecasting systems. The major objectives of the study include:

a. To analyse the limitations of traditional climate models and identify the challenges they face in simulating nonlinear climate behaviour;
b. To investigate how AI methods like neural networks, support vector machines, and generative models may be used for improving the accuracy of climate forecasts and interpretability of climate models;
c. To emphasize the importance of interdisciplinary cooperation between climate scientists and AI researchers in creating scientifically sound and ethically well-founded AI-based climate solutions; and
d. To examine the future impacts of AI-enriched climate models on sustainable policy decisions, resource allocation, and global climate adaptation measures.

Table 14.1 presents a chronological overview of the evolution of climate modelling techniques, highlighting key developments, methodologies, and advancements across different eras.

TABLE 14.1
Evolution of Climate Modelling Techniques

Era	Key Development	Description	References
Pre-1950s	Conceptual Models	Early qualitative understandings of climate based on astronomy, geography, and basic physics.	Othman, 2024
19th Century	Mathematical Models (Energy Balance, Radiative Transfer)	Quantitative models focusing on the balance between incoming solar radiation and outgoing terrestrial radiation, and the role of greenhouse gases.	Othman, 2024
1950s	First General Circulation Models (GCMs)	Computer-based models simulating global atmospheric dynamics based on fundamental physical equations, marking the beginning of numerical climate modelling.	Othman, 2024
1960s-1970s	Incorporation of Oceans and Other Earth System Components	Development of coupled atmosphere-ocean models and the inclusion of other components like sea ice and land surface processes, leading to more comprehensive Earth system models.	Othman, 2024

(*Continued*)

TABLE 14.1 (Continued)
Evolution of Climate Modelling Techniques

Era	Key Development	Description	References
1980s-1990s	Increased Resolution and Complexity	Advancements in computing power allowed for higher resolution models and the inclusion of more detailed physical and chemical processes, such as volcanic activity and basic atmospheric chemistry.	Hamdan et al., 2024
2000s-Present	Earth System Models (ESMs) and Enhanced Parameterizations	Development of ESMs that integrate biogeochemical cycles (e.g., carbon cycle) and more sophisticated parameterizations to represent unresolved processes like cloud formation, alongside the emergence of regional climate models.	Belnekar, 2025
Future	Integration of Artificial Intelligence and Machine Learning (AI/ML) Techniques	Today's frontier is applying AI and machine learning to enhance model precision, efficiency, capture nonlinear dynamics, and extract insights from massive climate data sets.	Alnajjar et al., 2024

14.2 THE DAWN OF AI IN CLIMATE SCIENCE: A NEW ERA OF PREDICTIVE MODELLING

AI, and its subdomains ML and DL, provides effective methods for analysing complex datasets and identifying non-linear relationships in the vast amount of information that defines climate science (Alnajjar et al., 2024). ML algorithms have an ability to detect patterns and correlations in climate data, often yielding insights that may remain hidden from traditional modelling techniques (LunarTech, 2025). The development of DL models, which are adept at learning from large amounts of data through multi-layered neural networks, has been particularly useful in further understanding of climate dynamics (Frazier, 2024). While traditional models tend to falter with unstructured information, AI shows a remarkable ability to process and understand such information, including visual and textual data from satellites and research publications (Murari, 2024). AI systems can efficiently process and interpret the continuous streams of data from satellites, the localized measurements from ocean buoys and weather stations, providing a more detailed and holistic view of the Earth's changing climate (Team EMB, 2024).

Neural networks, drawing inspiration from the complex architecture of the human brain, are particularly good at identifying subtle correlations and non-linear relationships within large datasets, capabilities that can significantly improve the predictive capabilities of climate models beyond what traditional methods can accomplish (Team EMB, 2024). Furthermore, AI systems are not just limited to analysis; they are also efficient in important preprocessing tasks like data cleaning, where inconsistencies and errors are identified and corrected, and feature extraction, where relevant variables and patterns are identified and isolated for model training (Team EMB, 2024). AI-powered tools can also easily perform data fusion, intelligently combining information from multiple sources to create a unified and more informative dataset for climate modelling (Team EMB, 2024). This capability to integrate diverse data can result in improved resolution in climate projections, providing more granular and localized forecasts that are critical for effective regional planning and disaster preparedness (LunarTech, 2025). Overall, AI's capacity to learn, predict, and adapt is transforming the way we tackle the complex challenges of climate modelling and forecasting. (Anderson, 2024)

Climate sciences are not yours alone; in all areas of science involving complex modelling and forecasting problems, there have been successful approaches to implementation, including weather forecasting, astrophysics, and genomics (Othman, 2024). In weather forecasting, AI-based models were shown to outperform the traditional numerical methods, in terms of both accuracy and speed of computation, especially for short-term predictions and forecasting of extreme weather events (Donback, 2025). The AI techniques used in weather forecasting ranged over a wide spectrum of machine-learning approaches, including different types of neural networks, such as convolutional neural networks (CNNs), for spatial data such as satellite imagery and recurrent neural networks (RNNs) for sequential data such as time series of weather measurements, and even more advanced DL models (Climavision, 2024). Just as astronomy uses AI as a must-have tool to handle large data sets and complex phenomena that accommodate reality. Applications span from identification and classification of celestial bodies such as galaxies and exoplanets to the forecasting of cosmic phenomena like supernovae, from interpreting intricate gravitational wave signals in the mergers of black holes to forming rich images of distant black holes (Restack, 2025). AI's ability to pick up faint patterns or anomalies from the massive volume of astronomical data has allowed scientists to unravel mysteries and carve deep insight into the universe. Genomics has also undergone a revolution, courtesy of AI. AI algorithms are increasingly being used to analyse complex genetic data, predict 3D protein folding, identify genetic markers of various diseases, and fast-track drug discovery (Appinventiv, 2024). These powerful application examples across disciplines stand as testimony to the innate ability of AI to efficiently manage huge and complex data sets, pick up subtle patterns invisible to many conventional means of analysis, and predict accurately with high success (Othman, 2024).

14.3 DECIPHERING EARTH'S SIGNALS: AI-DRIVEN CLIMATE CHANGE MODELLING

The performance of AI models for climate change modelling and prediction relies on the availability and quality of climate data used to train them. AI-based methods make use of a variety of climate data sources in order to capture the multidimensional nature of the Earth's climate system, for example, satellite data, weather station data, and the output produced by traditional climate models (Othman, 2024). Satellite observations play a pivotal role in the sense that they provide comprehensive climate data offering insights into critical variables both for the purpose of observing the ongoing changes in the climate of our planet and for rigorously examining the performance of climate models (European Space Agency (ESA), 2024). Complementing this global perspective, weather station observations offer reliable, in situ measurements of meteorological variables such as temperature, precipitation, and wind speed, providing essential ground-truth data for regional climate studies and model validation (Colston et al., 2018). In addition, the output from conventional climate models themselves can be a rich source of training data for AI models, allowing efficient AI emulators that can replicate the behaviour of complicated physical models at only a fraction of the cost of computation to be built, and making advanced downscaling methods available to improve the spatial resolution of climate projections (U.S. Department of Agriculture (USDA) Climate Hubs, 2024).

Prior to its effective use to train AI models, this raw climate data undergoes several preprocessing steps, ranging from cleaning to normalization and feature engineering (Othman, 2024). Data cleaning represents a necessary starting point that comprises the detection of missing values and their proper management, the detection and correction of erroneous or aberrant observations likely to bias model training, and the overall assurance of consistency and integrity in the dataset (Speedwell Weather Ltd, 2016). Normalization represents an important mechanism employed to standardize the data to a standard range, preferably between 0 and 1 or -1 and 1. This action serves to increase the speed of convergence of AI models during model training and is also capable of improving their general performance by suppressing variables with large magnitudes from dominating the learning process (Pandit, 2023). Lastly, feature engineering entails converting raw inputs into structured, relevant features that can largely improve the accuracy and predictive power of the AI models. (Statsig, 2024) This may comprise the generation of new variables out of existing variables, the discovery of relevant temporal or spatial trends, or encoding domain knowledge within the data representation.

A wide range of AI models is being utilized to address the multifaceted issues of climate change modelling and prediction. Among the most common are neural networks, support vector machines (SVMs), and DL architectures, such as CNNs, RNNs, and long short-term memory networks (LSTMs) (Alnajjar et al., 2024). Artificial neural networks that are based on the structure and functioning of the human brain have captured the complex, intricate patterns that describe the climate information (Othman, 2024). DL methods—a subtype of multiple-layered neural network—have displayed huge potential in analysing the spatial patterns of climate data, such as satellite data, through CNNs, whereas temporal sequences involved in climate data, such as time series of temperature and rainfall, are analysed through RNNs (Climavision, 2024). LSTMs, a type of RNN, are perhaps most suitable for climate data modelling, since they can deal with long-term dependencies and hold information for very extended time intervals, which is highly important for representing climate variability and trends (Liu et al., 2025). Another research line in climate modelling involves generative AI: diffusion models, for instance, can be used to generate climate projection ensembles that reflect a range of possible future outcomes.

These models can also be used to generate synthetic climate data for a variety of research purposes (Climavision, 2024). Statistical learning theory-based SVMs are very powerful tools for regression and classification in climate science, ranging from the prediction of rainfall occurrence to the subsequent evolution of temperature (Kumar et al., 2024). Finally, an important advancement known as physics-informed neural networks (PINNs) utilizes the known physical laws and constraints directly in the training of neural networks. This allows for the creation of AI models that are data-driven, yet bound by fundamental scientific principles, with a view towards more physically realistic and thus more credible climate prediction (Yuval et al., 2021). Table 14.2 presents the major AI models employed in climate change

TABLE 14.2
Common AI Models for Climate Change Modelling

AI Model	Core Principles	Applications in Climate Change Modelling	References
Neural Networks	Interconnected nodes learn complex patterns from data through weighted connections.	Modelling non-linear relationships in climate variables, improving the accuracy of climate predictions, and emulating complex climate processes.	Othman, 2024
Convolutional NN (CNN)	Specialized for processing grid-like data, using convolutional filters for feature extraction.	Analysing satellite imagery for weather pattern recognition, detecting deforestation, and extracting spatial features from climate model outputs.	Murari, 2024
Recurrent NN (RNN)	Designed for processing sequential data, maintaining an internal state to capture temporal dependencies.	Modelling time series data such as temperature and precipitation records, forecasting weather patterns over time, and analysing the evolution of climate variables.	Othman, 2024
LSTM	A type of RNN particularly effective at learning long-term dependencies in sequential data.	Predicting long-term climate trends, forecasting extreme weather events with longer lead times, and capturing the temporal dynamics of climate variables over extended periods.	Liu et al., 2025*
Support Vector Machine (SVM)	Finds an optimal hyperplane to separate or map data for classification and regression.	Predicting rainfall, forecasting temperature, classifying climate patterns, and downscaling climate model outputs.	Alnajjar et al., 2024
Generative AI	Learns the underlying data distribution to generate new, synthetic data samples.	Creating ensembles of future climate projections, generating synthetic climate data to augment training datasets, and accelerating climate simulations.	Belnekar, 2025
Physics-Informed NN (PINN)	Integrates physical laws and constraints into the neural network training process.	Developing climate models that are not only data-driven but also adhere to known physical principles, potentially improving the realism and reliability of predictions.	Othman, 2024; Alnajjar et al., 2024

modelling, describing their fundamental principles and applications. It includes neural networks, CNNs, RNNs, SVMs, and emerging models like PINNs and generative AI, highlighting their roles in analysing climate data and enhancing prediction accuracy.

14.4 AI IN CLIMATE FORECASTING: ENHANCING PREDICTIONS AND MEASURING PERFORMANCE

Evaluating AI models in climate research involves rigorous performance testing using a suite of diverse metrics to evaluate how accurately, how reliably, and how effectively these models simulate and predict climate behaviour (Othman, 2024). The accuracy offers an overall measure of the frequency with which the model is correct for predicting all classes or outcomes (Organisation for Economic Co-operation and Development (OECD), 2024). For classification applications, precision measures the accuracy of positive predictions, reporting as a proportion those instances predicted positively that were found to be indeed positive (Andonians, 2023). Recall, synonymous with sensitivity, or the true positive rate, reports the ratio of the cases correctly identified by the model among all actual cases, emphasizing how complete the representation of the phenomenon of interest in the model (Andonians, 2023). For regression applications, root mean square error (RMSE) is another popular metric estimating the average scale of the discrepancy between the output of the model and the reality observed (Al-Mahdawi et al., 2023). In climate prediction, the anomaly correlation coefficient (ACC) examines the spatial association between the estimated anomalies (a departure from average climate state) and the estimated anomalies, highlighting the model's skill in reproducing the geographical climate variability patterns. For models that provide probabilistic predictions, for example, the probability a given temperature limit will be crossed, the Brier score of utility is its ability to appraise the consistency of these predicted probabilities with reference to the occurring frequency. Along with these quantitative estimates, explainable AI (XAI) scope encompasses a variety of methods and methods of interpreting outputs of AI models, particularly deeper learning models, to establish reasons behind their predictive outputs and explore implications for climatology. These XAI approaches are vital to establishing user trust in AI-based climate forecasts and constructing improved insights into the underlying physical mechanisms that the models have learnt (Othman, 2024).

AI is significantly improving the accuracy of short-term climate forecasts of severe weather and long-term forecasts of climate change. In short-term forecasting, AI is being used with great success to forecast phenomena such as hurricanes, floods, and droughts, leading to tremendous improvement both in the accuracy and lead time of these life-saving warnings (Othman, 2024). AI-driven models can process large volumes of real-time and historical data so that they can recognize fine-grained patterns and complex relationships that usually lead to extreme weather phenomena, hence more accurate and timely predictions (Othman, 2024). This high-end forecasting capability is crucial in enabling populations and authorities to better prepare and respond, finally mitigating the potentially devastating impact of these phenomena. Besides short-term forecasting, AI is now employed in the generation of long-term projections of climate change, such as future sea-level rise and temperatures (Othman, 2024). Furthermore, AI models are now being developed to predict the cascading impacts of climate change on various major sectors, such as agriculture, human health, and infrastructure, and their interesting implications for adaptation and long-term planning strategies (Othman, 2024). Generative AI models have emerged as a potential solution for accelerating the creation of long-term climate projections, potentially creating these complex simulations far faster than physics-based models today. Of particular significance is that AI approaches can be applied to enhance the spatial resolution of climate simulations, thereby enabling more accurate and localized forecasts and projections more useful at the regional and local level (LunarTech, 2025).

14.5 NAVIGATING THE LABYRINTH: CHALLENGES AND LIMITATIONS

While there is immense potential for AI to extend climate modelling and forecasting, its application in this complex domain is subject to significant technical and conceptual constraints (Othman, 2024).

- Availability and Quality of Data: Availability and quality of climate data are the primary concerns. AI models, particularly DL frameworks, often require large amounts of reliable and uniformly formatted datasets to learn efficiently, and gaps, inconsistencies, and potential biases in existing climate datasets can have a massive impact on the reliability and precision of AI-driven predictions (Belnekar, 2025).
- Energy and Computational Cost: The computational cost and associated energy expenditure required to train large-scale AI models that can be used in applications related to climate can be computationally intensive, demanding advanced processing capabilities and raising questions about the environmental viability of such AI endeavours (Alnajjar et al., 2024).
- Transparency and Interpretability: Transparency and interpretability are another major barrier, especially for DL models that can sometimes work as "black boxes," making it difficult for stakeholders to interpret decision logic behind their predictions and assess their reliability (Othman, 2024).
- Climate System Complexity: The inherent complexity and chaotic dynamics of the climate system itself pose a fundamental challenge to even the most sophisticated AI models to fully capture and predict with absolute accuracy (Othman, 2024).
- Generalization to Unseen Scenarios: The capacity of AI models trained on past data to extrapolate effectively to future climate conditions that might include novel conditions or severe events is still a central area of concern and ongoing research (Nowack & Watson-Parris, 2025).
- Ethical Considerations: Ethical concerns regarding the utilization of AI for climate modelling, including matters involving data secrecy, the prospect for algorithmic prejudice to adversely affect disadvantaged populations disproportionately, and ensuring fair access to the payoffs of these technologies, also need to be carefully managed (Belnekar, 2025).

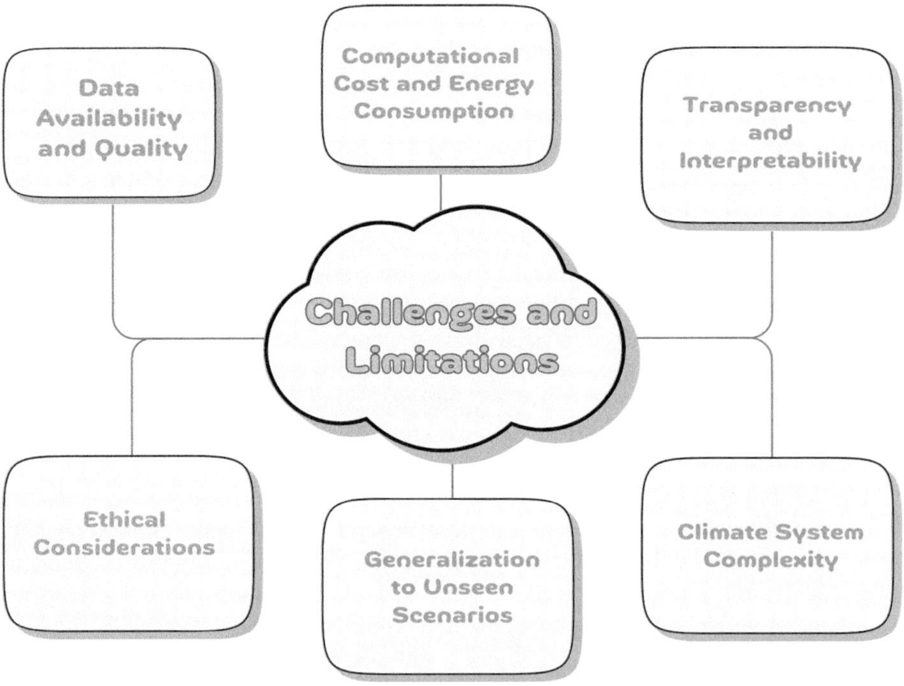

FIGURE 14.1 The primary challenges and limitations associated with applying AI in climate change modelling.

- Risk of Techno-Solutionism: Care must be taken not to fall into the trap of "techno-solutionism," whereby AI is viewed as a single solution to addressing climate change, and a balanced view of its promise within an overall framework of scientific knowledge and policy initiative must be preserved (Clutton-Brock et al., 2021).

Figure 14.1 illustrates the primary challenges and limitations associated with applying AI in climate change modelling. It highlights issues such as data quality, model interpretability, computational demands, generalization to novel scenarios, and ethical concerns, all of which must be addressed to ensure reliable and responsible AI-driven climate forecasting.

14.6 A FUTURE FORGED IN SILICON: THE PATH FORWARD

Unlocking AI's ability to transform the largely technical and sensitive domain of forecasting and modelling climate changes requires the consistent interdisciplinary collaboration of researchers and experts in both AI and climate (Othman, 2024). Building high-performing AI solutions for climate problems requires that the involved stakeholders possess knowledge of both the considerable environmental processes that govern our planet, as well as the sophisticated machine-learning techniques that permit AI to perform analyses and predictions (LunarTech, 2025). This collaborative integration guarantees that AI models are not just computationally efficient and technologically advanced but that they are also scientifically validated and able to deal with the true-to-life complexity of climate issues (LunarTech, 2025). Interdisciplinary teams composed of individuals from both the climate and community and AI research would be best suited to tackle the numerous problems present in the field, ranging from mediating dilemmas about data bias and quality, improving AI system transparency and interpretability, and balancing the ethical questions arising from their application. Also, such cross-fertilization of expertise could contribute to new AI methods and algorithms custom-designed to address the special problems and data features encountered in climate science (National Science Foundation (NSF), 2025). Owing to the overriding significance of this information, there are a number of workshops, research projects, and programmes connecting these two key fields in order to support joint research and mutual understanding leading to significant scientific advances (National Science Foundation (NSF), 2025).

The combination of AI and climate science offers many promising areas for future study to help improve our capacity to comprehend and act upon climate change. These encompass the development of ever-more advanced neural network architectures for learning more nuanced climate dynamics, advancements in physics-informed AI models bringing fundamental scientific physical laws into them to enhance predictive realism and confidence, and innovation of new AI methodologies for optimizing the quantification and communication of intrinsic uncertainties pertaining to climate predictions, as well as for improving complex AI model interpretability (Alnajjar et al., 2024). Additionally, it should be given higher priority for future research to explore the uses of AI towards attribution of climate change, working towards better appreciation of the proportionate contribution human activity is playing in causing selected aspects of change, and detailed local and regional scale impact evaluations to better inform selective adaptation (Othman, 2024). Because of the extreme computational efforts of training such large AI algorithms, research addressing developing more compute-efficient AI codes and methods by which overall power use by climate AI science could be minimized are equally critical issues towards ensuring those endeavours are more sustainable.

The growing involvement of AI in climate modelling and prediction also has profound policy implications. Governments and international institutions have a vital role to play in ensuring targeted support for research and development work at the interface of AI and climate science, as well as in putting in place clear regulations and ethical frameworks to govern the responsible application of AI technologies in this area (Kumar et al., 2024). Prioritizing the consideration of the ethical implications of AI, including the risk of bias in data and algorithms, and ensuring equitable access to the benefits of AI-facilitated climate solutions to all countries and communities, especially those most exposed to the impacts of climate change, should be high-policy priorities (Belnekar, 2025). Finally, supporting global collaboration and fostering free exchange of climate data, AI models, and research findings will be essential to speeding up progress in this important field and realizing the benefits of AI-facilitated climate science everywhere (Murari, 2024).

14.7 CONCLUSION

AI carries the transformative potential for climate modelling and forecasting. Being able to circumvent the fundamental limitations of classical climate models, mainly the inability to deal efficiently with non-linear dynamics and large-dimensional climate data, AI offers avenues to enhance accuracy, efficiency, and interpretability. Higher-level AI techniques, including deep neural networks, SVMs, and generative models, are now able to provide greater spatial and temporal resolution, richer predictions, and more control of uncertainty in climate projections (Alnajjar et al., 2024; Othman, 2024). To realize the promise of AI in this domain, however, requires more than just algorithmic complexity—it requires coordinated and sustained interdisciplinary collaborations with AI researchers, climate scientists, environmental policymakers, and ethicists. The partnership in place aims to have AI-based models that are not only technically sound but climate-science-based, ethically grounded, and socially responsive (LunarTech, 2025; National Science Foundation [NSF], 2025). In view of that, making sure that the AI sustainable development goes on requires eliminating issues relating to low-quality data, inherent bias in the AI algorithms, lack of explainability, and massive energy consumption due to these large-scale models.

Towards the future, a few research directions are particularly significant in this sector. These include physics-constrained AI models that embed scientific principles into learning methodologies, energy-efficient ML architectures to reduce computational overhead, and XAI methods that ensure trust through transparency. Additionally, AI could be helpful for regional-scale climate diagnostics, anthropogenic influence attribution, and climate adaptation strategy optimization (Belnekar, 2025; Kumar et al., 2024). In fact, the globalization of AI-based climate tools calls for urgent attention. Governments, universities, and world agencies will need to shadow in life through investment in open data-sharing infrastructures, building regulatory and ethical frameworks to govern the application of AI in climate research, and making the benefits of these technologies available for all, with particular emphasis on those vulnerable populations affected by climate change (Murari, 2024).

REFERENCES

Al-Mahdawi, H.K., Alkattan, H., Alhumaima, A.S., Al-Hadrawi, H.F., Abotaleb, M., Ali, G.K., Mijwil, M.M., Tofeek, A.K. and Helal, A.H., 2023. Analysis and prediction of rainfall using support vector machine (SVM) in the city of Najaf. *ResearchGate*, 9 September. Available at: https://www.researchgate.net/publication/383025386_Analysis_and_prediction_of_rainfall_using_support_vector_machine_SVM_in_the_city_of_Najaf

Alnajjar, M., Hamadaqa, M.H.M., Ayyad, M.N., Al-Nakhal, M.A., Abu-Nasser, B.S. and Abu-Naser, S.S., 2024. AI in climate change mitigation. *International Journal of Academic Engineering Research (IJAER)*, 8(10), pp. 31–37. Available at: https://philarchive.org/archive/ALNAIC

Anderson, K., 2024. How can artificial intelligence help tackle climate change? *Greenly Blog*, 13 February. Available at: https://greenly.earth/en-us/blog/industries/how-can-artificial-intelligence-help-tackle-climate-change

Andonians, V., 2023. Four key metrics for evaluating AI models. *Cognaize Blog*, 23 January. Available at: https://blog.cognaize.com/four-key-metrics-for-evaluating-ai-models

Appinventiv, 2024. How AI in genomics is transforming healthcare and research. *Appinventiv Blog*, 15 January. Available at: https://appinventiv.com/blog/ai-in-genomics/

Belnekar, M., 2025. AI for climate change modelling and environment sustainability. *International Research Journal of Modernization in Engineering Technology and Science*. [online] Available at: https://www.irjmets.com/uploadedfiles/paper//issue_1_january_2025/65848/final/fin_irjmets1735817316.pdf

Climavision, 2024. AI in weather forecasting: The future of weather prediction. *Climavision*, [online] Available at: https://climavision.com/resources/ai-weather-forecasting-guide/

Clutton-Brock, P., Rolnick, D., Donti, P.L., Kaack, L.H., Maharaj, T., Luccioni, A., Das, H.P., Miailhe, N., Chatila, R., Kwiatkowska, M., Dignum, V., Hodes, C., Rajotte, J., Teather, E. and Kirnberger, J.L., 2021. Climate change and AI: Recommendations for government action. *Global Partnership on AI (GPAI), in collaboration with Climate Change AI and the Centre for AI & Climate*. Available at: https://www.gpai.ai.projects/climate-change-and-ai.pdf

Colston, J.M., Ahmed, T., Mahopo, C., Kang, G., Kosek, M., de Sousa Junior, F., Shrestha, P.S., Svensen, E., Turab, A. and Zaitchik, B., 2018. Evaluating meteorological data from weather stations, and from satellites and global models for a multi-site epidemiological study. *Environmental Research*, 165, pp. 91–109. Available at: https://pmc.ncbi.nlm.nih.gov/articles/PMC6024078/

Donback, N., 2025. Traditional weather forecasting is slow and expensive. AI could help. *Grist*, 3 January. Available at: https://grist.org/extreme-weather/how-ai-could-help-predict-climate-fueled-extreme-weather/

European Space Agency (ESA), 2024. Modelling with climate data from space. [online] Available at: https://climate.esa.int/en/about-us-new/climate-change-initiative/modelling-with-climate-data-from-space/

Frazier, K., 2024. *Arctic Climate Data Science: The Role of Artificial Intelligence in Supporting Operational Decision Making*. Ted Stevens Center for Arctic Security Studies Special Report. Available at: http://tedstevensarcticcenter.org/wp-content/uploads/2024/01/18_TSC_SR_Arctic_Climate_Data_Science_FINAL.pdf

Hamdan, A., Ibekwe, K.I., Etukudoh, E.A., Umoh, A.A. and Ilojianya, V.I., 2024. AI and machine learning in climate change research: A review of predictive models and environmental impact. *World Journal of Advanced Research and Reviews*, 21(1), pp. 1999–2008. Available at: https://doi.org/10.30574/wjarr.2024.21.1.0257

Kumar, R., Goel, R., Sidana, N., Sharma, A.P., Ghai, S., Singh, T., Singh, R., Priyadarshi, N., Twala, B. and Ahmad, V., 2024. Enhancing climate forecasting with AI: Current state and future prospect [version 1; peer review: 1 approved]. *F1000Research*, 13, 1094. Available at: https://doi.org/10.12688/f1000research.154498.1

Liu, S., Liu, K., Wang, Z., Liu, Y., Bai, B., Zhao, R., 2025. Investigation of a transformer-based hybrid artificial neural network for climate data prediction and analysis. *Frontiers in Environmental Science*, 12, Article 1464241. Available at: https://www.frontiersin.org/journals/environmental-science/articles/10.3389/fenvs.2024.1464241/full

LunarTech, 2025. AI in climate modeling and environmental monitoring. *LunarTech Blog*, 18 January. Available at: https://www.lunartech.ai/blog/ai-in-climate-modeling-and-environmental-monitoring

Murari, H., 2024. AI for climate change: Innovative models for predicting environmental impact. *DATAVERSITY*, 15 October. Available at: https://www.dataversity.net/ai-for-climate-change-innovative-models-for-predicting-environmental-impact/

National Science Foundation (NSF), 2025. *Collaborations in Artificial Intelligence and Geosciences (CAIG)*. NSF Funding Opportunities. Available at: https://www.nsf.gov/funding/opportunities/caig-collaborations-artificial-intelligence-geosciences

Nowack, P. and Watson-Parris, D., 2025. Opinion: Why all emergent constraints are wrong but some are useful – a machine learning perspective. *Atmospheric Chemistry and Physics*, 25, pp. 2365–2384. Available at: https://doi.org/10.5194/acp-25-2365-2025

Organisation for Economic Co-operation and Development (OECD), 2024. Catalogue of Tools & Metrics for Trustworthy AI. [online] Available at: https://oecd.ai/en/catalogue/overview

Othman, A., 2024. AI in Climate Change Modeling and Prediction. [online] DOI:10.13140/RG.2.2.14900.97925

Pandit, B., 2023. Four most popular data normalization techniques every data scientist should know. *Dataaspirant*, 3 April. Available at: https://dataaspirant.com/data-normalization-techniques/

Restack, 2025. AI for space exploration: Astrophysics tools. *Restack.io*, 22 February. Available at: https://www.restack.io/p/ai-for-space-exploration-answer-ai-tools-astrophysics-cat-ai

Speedwell Weather Ltd., 2016. Introduction to Weather Data Cleaning. [pdf] Available at: https://www.speedwellweather.com/PDF/DataHome/Introduction%20to%20Weather%20Data%20Cleaning.pdf

Statsig, 2024. Feature engineering for time-series data. *Statsig Perspectives*, 25 November. Available at: https://www.statsig.com/perspectives/feature-engineering-timeseries

Team EMB, 2024. The role of AI in predicting climate change patterns. *EMB Global Blog*, 12 January. Available at: https://blog.emb.global/ai-in-predicting-climate-change-patterns/

U.S. Department of Agriculture (USDA) Climate Hubs, 2024. Basics of Global Climate Models. [online] Available at: https://www.climatehubs.usda.gov/hubs/northwest/topic/basics-global-climate-models

Yuval, J., O'Gorman, P.A. and Hill, C.N., 2021. Use of neural networks for stable, accurate and physically consistent parameterization of subgrid atmospheric processes with good performance at reduced precision. *Geophysical Research Letters*, 48(4), e2020GL091363. Available at: https://par.nsf.gov/servlets/purl/10300831

15 Challenges with Practical Solutions and Case Studies

Souvik Das and M Hemakumar Reddy

15.1 INTRODUCTION

Artificial intelligence (AI)-based technology will notably affect the agriculture sector by enabling the conventional way of farming to advance to digital agriculture (Shamshiri et al., 2024). The agricultural industry is challenged by its traditional methods of farming, due to labour-intensive methods, unstructured data sets, and uncertain weather conditions (Kalamkar et al., 2020; Morella et al., 2023). Also, AI has a broad impact on biodiversity conservation. Using generative adversarial networks (GANs) and machine learning-based models, AI can recreate virtual ecosystems and can predict the interactions between various species and the environment (Ayoola et al., 2024). Besides that, AI can help in smart waste management and green city planning using smart alternatives and innovative strategies (Xu et al., 2024). In the field of renewable energy, generative artificial intelligence (GenAI)-based technology enables smart forecasting and situation-based planning (Annapareddy, 2024). Climate change is a serious threat which can affect all entities of the environment, including mankind, so here GenAI-enabled smart technology forecasts the climate change in advance and helps to make adaptive moves (Deep & Verma, 2024).

Furthermore, productivity improvement requires better agricultural management, which helps to apply resources in a sustained manner and mitigate environmental impacts (Pretty, 2007). To fulfil the increasing demands of high-yielding variety development as well as to manage the natural resources, efficient crop management is required (Çakmakçı et al., 2024). So, the traditional methods of crop cultivation and AI-centred systems combinedly offer exciting possibilities for improved sustainability and efficiency in crop management (Gaitan et al., 2025). AI, ML, and deep learning (DL) are currently the three important innovations affecting new farming methods and environmental issues (Weraikat et al., 2024). Innovative and smart farming techniques are used with data-based insights to guide effective decision-making, which increases production and resource use efficiency (Elbasi et al., 2024). Farmers and all the ground-based workers could use AI to improve agricultural yield and manage fertiliser and water, and also they can predict weather patterns and disease-pest outbreaks easily in advance (Sidhu et al., 2021). The long-term benefits of soil health, like physical structure, composition and others, can be aided and

 DOI: 10.1201/9781003616207-18

monitored in real time by various AI-powered sensor mechanisms (Ogwu and Kosoe, 2025). Vertical farming with advanced Artificial General Intelligence (AGI)-based technologies could offer responsive greenhouse environmental control on air quality, temperature, humidity, and supply water, artificial lighting, and CO_2 consumption (Muhie, 2022). AI-driven technologies can have complex and exciting implications for farmers by providing more than dispensing information, as it provides the means of utilising AI technologies, such as the use of DL, together with Global Positioning System (GPS) technologies, and Geo Information Systems technologies, to provide practical precision farming (Senthil et al., 2025). In addition, GenAI-based platforms could ease knowledge transfer and increase stakeholder communication to induce real-time decision making (Rane et al., 2024). GenAI is a multifaceted area of AI that generates synthetic data, including text and images, among many types of data (Sengar et al., 2025). Rather than develop predictions about existing data like predictive AI systems do by assessing prior data to make predictions about a future outcome, GenAI generates entirely new information from learnt patterns and structures in large datasets (Kibria et al., 2018; Zhang et al., 2025).

Generative AI in agriculture extends its capabilities beyond data assessment and optimisation; it can also unlock a broad range of abilities, including simulating various weather scenarios, soil types, and plant growth stages (Sai et al., 2025). It can assist with yield predictions impacting outcomes, allowing farmers to take action and adapt their processes for real-world, profit-driven agricultural activity. Researchers and specialists have slowly recognised the capabilities that generative AI presents for agriculture. Nevertheless, we are now affected by the global distribution of generative AI that changes the way we work and interact (Feuerriegel et al., 2024). The term "Agriculture 4.0" (AG-4.0) has recently introduced the combination of GenAI with AI and precision agriculture and other wireless sensors and the Internet of Things (IoT) into the agricultural sector, producing new and valuable agricultural insights and solutions, through data analysis and interpretation (Sai et al., 2025). GenAI can accomplish a variety of task outputs depending on user prompts.

Generative AI is rapidly expanding in every sector, including addressing environmental issues. By using advanced ML techniques and training data, GenAI can analyse and gather a vast amount of data from satellites, ground-based sensors, biodiversity surveys and energy grids easily. GenAI uses this information as a trained data set to reveal the patterns and solve problems that were impossible to address earlier. GenAI is making a remarkable difference in the present day in various areas such as monitoring and mitigating pollution in air and water, modelling ecosystems for biodiversity conservation, optimising energy in a renewable way, designing greener and eco-friendly cities, improving waste management, and defending the climate change scenario by advancing modelling and adaptation planning.

GenAI is implemented through training deep neural networks (DNNs) to output data that produces specific outcomes. DNNs can also be applied to model another kind of data (Feuerriegel et al., 2024). Large language models (LLMs) are also another means of processing user inputs to train various GenAI-based applications (Yan et al., 2025). Some popular LLM-based models include Generative Pre-trained Transformer (GPT), created by OpenAI; Pathways Language Model (PaLM); and (Bidirectional Encoder Representations from Transformers) (BERT), developed by

Google (Feuerriegel et al., 2024). Additionally, Llama is an open-source tool that Meta built (Raiaan et al., 2024). Additionally, various DL models, such as Artificial Neural Networks (ANN) and Convolutional Neural Networks (CNN), are used to forecast weather parameters and soil components (Han & Wang, 2021). GenAI utilises algorithm-based models that work together to enhance the predictive accuracy and efficiency of systems (Goodfellow et al., 2020). In addition, the "You Only Look Once" (YOLO) performs well as a real-time object detection (e.g., fruit or panicle counting) model.

The farmers' traditional methods of production were insufficient to meet their overall needs. Crop management is a complex and time-consuming process; to accelerate this, incorporating GenAI with the traditional farming method is essential. Even in rural communities, modern technologies cannot replace traditional methods for the betterment of farming and crop management. Various environmental concerns are increasing day by day, such as pollution, waste management, and biodiversity conservation. So, the increased awareness and adoption of modern AI-powered technologies in our society are very much essential.

Therefore, it is vital to develop new advanced tools and techniques powered by GenAI, facilitating researchers to control this information for practical applications in solving real-world problems related to the environment and farming. Moreover, the adoption of AI-driven innovative farming solutions can be facilitated by a strong commitment to addressing ethical concerns related to data privacy, security, and labour displacement. Government policymakers, farmers, and developers must cooperate to design frameworks in a well-structured way that support innovation while maintaining the social and economic sustainability of rural communities to overcome these obstacles (Ayoub Shaikh et al., 2022; Rane et al., 2024).

This chapter provides a comprehensive overview of the mechanism, applications, challenges, and future directions of utilising generative AI in crop management and addressing other environmental problems related to biodiversity, waste management, and monitoring various parameters of the environment. In addition, we aim to identify and present the research gaps and recent case studies. The primary objectives of this chapter are to bridge the gaps between traditional and modern farming by utilising GenAI-based technology and delivering the latest insights on GenAI-based tools in agriculture and other sectors of the environment. Illustrate the mechanism, applications, future scopes, and limitations of GenAI in the agricultural sector.

15.2 MECHANISM BEHIND AI MODEL BUILDING

The use of technology within agriculture has significantly advanced due to the automation of processes using algorithms. To use AI effectively in agriculture, several key components are required. Collecting data from various sources, employing data preprocessing methods tailored to agricultural data, utilising modern architectures and model-building and validation frameworks, and ensuring a smooth transition to real-time interfacing for informed decision-making. This method helps farmers and scientists to utilise advanced technologies to tackle specific issues in crop management and environment-related issues (Mienye & Swart, 2024). A brief overview of

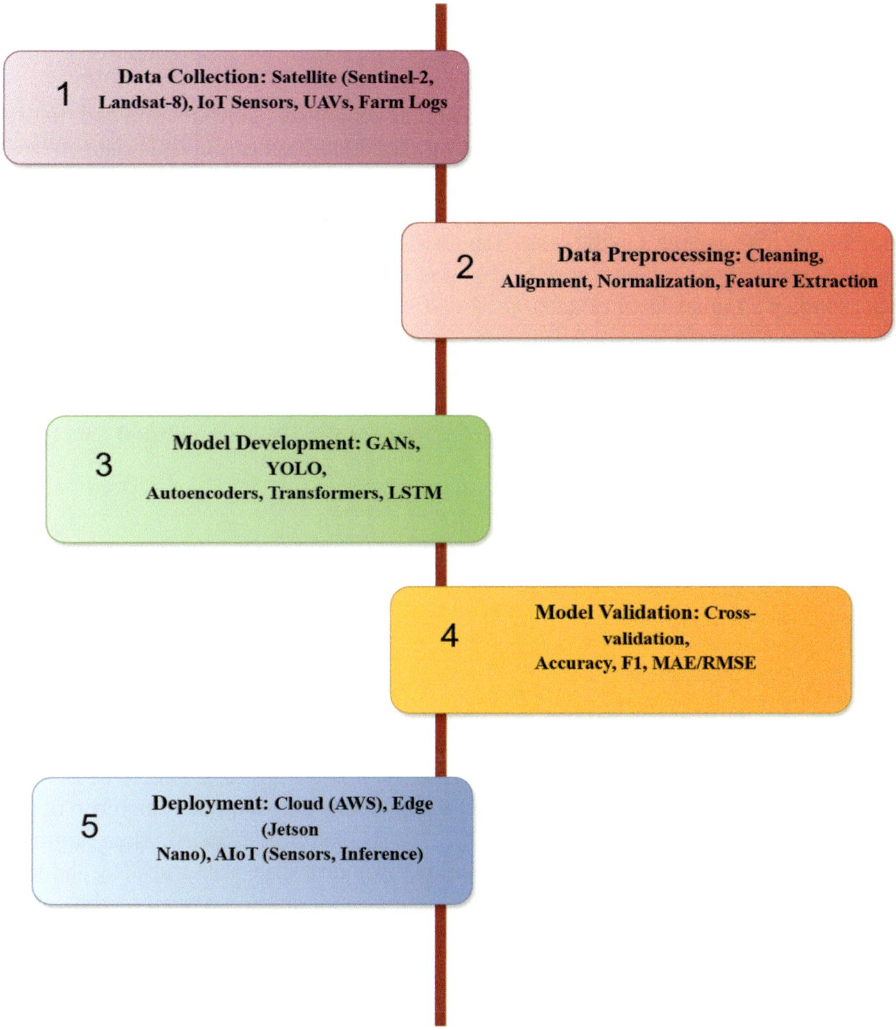

FIGURE 15.1 Diagram of mechanism behind AI model construction.

the mechanism is discussed next, and Figure 15.1 illustrates the mechanism behind AI model building in a simple diagram.

15.2.1 Data Collection

To build an AI model, first gather relevant data from the required sector for model training. In the case of the agriculture sector, relevant data sources are:

- Multispectral imagery from satellites such as Landsat-8 and Sentinel-2.
- IoT sensor readings for soil moisture, temperature, and pH.

- High-definition imagery from Unmanned Aerial Vehicles (UAVs) and drones.
- Collection of farm logs and mobile apps that record the crop conditions.

This multi-variate data collection supports complex pattern identification and predictive modelling. Liakos et al. (2018) told that the integration of multiple data sources will substantiate agroecosystem model accuracy. The Earth's observation satellites, such as Sentinel-2 and Landsat-8 data, can provide multispectral imagery crucial for crop health assessment. On the other hand, IoT networks offer real-time environmental measurements to enhance crop health, and UAV technology, when combined with ML, enables a higher level of accuracy in crop health analytics.

15.2.2 Data Preprocessing and Feature Engineering

This stage, called preprocessing, is vital for transforming raw agricultural data into a usable format. Key steps include:

- Handling missing values and outliers.
- Synchronising temporal data from varied sources.
- Normalisation and standardisation.
- Feature extraction and transformation.

Data from the agricultural sector is complex due to temporal and spatial variability (Fanelli, 2020). For example, weather data may be collected on a daily basis, whereas satellite images are typically updated weekly, and sensor readings are on a continuous basis, making time alignment critical and important (Thies & Bendix, 2011). Data extracted by computer vision using CNNs helps in detecting crop health and disease-pests (Kumar & Laxmi, 2022). GANs support synthetic data augmentation to address class imbalance, particularly for rare diseases, and also generate high-fidelity synthetic images, which help maintain model performance with sparse training datasets. (Lu et al., 2022).

15.2.3 Model Development

Model development involves training AI architectures to address agricultural challenges. Common architectures include GANs, CNNs, ANNs, YOLO, Autoencoders, Transformers, a Long Short-Term Memory (LSTM), and Diffusion Models. Table 15.1 describes the functional role of these models in agriculture.

15.2.4 Model Validation

Robust validation ensures AI models generalise well across regions and timeframes. Techniques include:

- Train-test splits and K-fold cross-validation.
- Metrics like accuracy, F1-score [classification], MAE (Mean Absolute Error), RMSE (Root Mean Square Error) [regression], and mAP (Mean Average Precision), IoU (Intersection over Union) [detection].

TABLE 15.1
Popular AI Models with Agricultural Functions and References

Model Name	Functional Role in Agriculture	References
GANs (Generative Adversarial Networks)	Synthetic data generation and crop disease image augmentation for improved training (e.g., soybean disease detection with 92% mAP@50).	(Akkem et al., 2024; Goodfellow et al., 2020; Lu et al., 2022; Rahman et al., 2024)
YOLO (You Only Look Once)	Real-time detection and classification of crops, weeds, and field objects from drone imagery.	(Badgujar et al., 2024; Deng et al., 2023)
Diffusion Models	High-quality denoising and image generation for diverse crop stress scenarios and synthetic field conditions.	(Muhammad et al., 2023; Yin et al., 2025)
Transformers & LSTMs (Long Short-Term Memory)	Sequence prediction for climate impact modelling, multi-modal integration for yield forecasting, and long-range yield prediction using temporal satellite, climate, or IoT data.	(Bi et al., 2023; Lu et al., 2022; Vaswani et al., 2023)
Autoencoders	Denoising satellite images and dimensionality reduction for anomaly detection in crop fields.	(Akkem et al., 2024; Picon et al., 2019)
Artificial Neural Networks (ANN)	Predicting crop yield, soil parameters, and disease risk using structured input data.	(Kamyshova et al., 2022; Kujawa & Niedbała, 2021; K. U. Singh et al., 2022)
Convolutional Neural Networks (CNN)	Image-based detection of diseases, pests, and plant phenotyping.	(G. Li et al., 2021; Sajitha et al., 2023)

CNN-LSTM (Convolutional Neural Network-Long Short-Term Memory) models have been validated for maize yield prediction across regions. Spatial validation uses diverse agroecological zones; temporal validation assesses performance over multiple growing seasons and climate extremes (Parashar et al., 2024).

15.2.5 Deployment Strategies and Real-Time Implementation

In the case of AI deployment, it faces challenges related to infrastructure and latency, especially in the agriculture sector (Senoo et al., 2024). The cloud services (AWS, Google Cloud) can handle heavy computation, while smart devices (e.g., Jetson Nano, Raspberry Pi) can perform real-time inference (Senoo et al., 2024). Besides, Jetson Nano achieves classification under 0.2 seconds per frame—suitable for field conditions, and IoT gateways enable real-time data fusion (Assunção et al., 2022; Mohan et al., 2021). A four-layer AI-IoT architecture (Alaba, 2024; Mrabet et al., 2020) includes the following:

1. **Perception layer**—Sensors gather environmental data
2. **Network layer**—Transmits data
3. **Processing layer**—AI models perform inference
4. **Application layer**—Interfaces for user decisions

These systems support and manage field monitoring, disease detection, irrigation control, and machinery.

15.3 KEY APPLICATIONS IN AGRICULTURE

Generative AI has vast uses in the field of agriculture, from crop monitoring to advanced crop breeding. In every aspect, AI will enrich this sector undeniably. The applications of "Generative Artificial Intelligence in Agriculture" are depicted in the form of a circular infographic in Figure 15.2.

15.3.1 Crop Monitoring

Generative AI plays an important role in crop monitoring by using DL and synthetic data generation methods. In more recent models, combining ANNs, GANs, and You

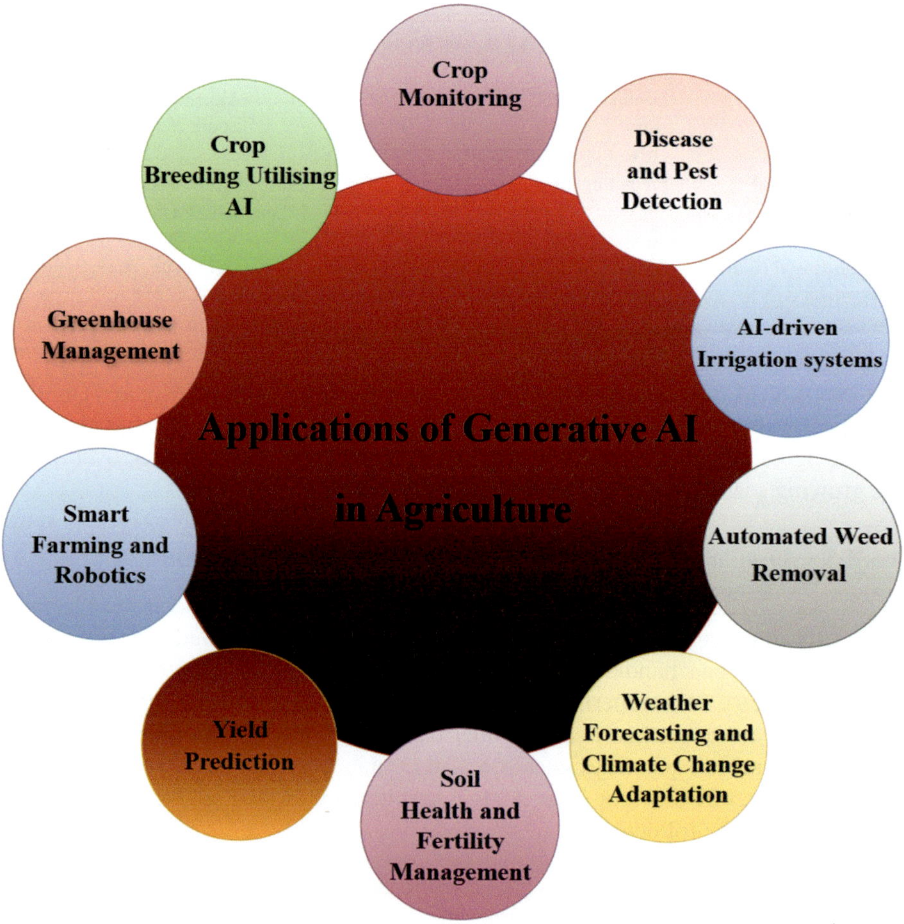

FIGURE 15.2 Applications of generative artificial intelligence in agriculture.

Only Look Once (YOLO) has been used for real-time crop health monitoring from satellite and drone images (Senthil et al., 2025). These platforms mimic the missing value or bad-quality data to fill in the field datasets, thereby enhancing the accuracy of yield prediction and field variability analysis. Due to the accurate GPS and GIS technology, site-specific recommendations can be made so that the application of fertiliser, irrigation, and pesticide is made in a manner that reflects the variability of the crop condition. Pihrt et al. (2024) addressed a significant problem of cloud interference in improved satellite-based crop monitoring using a DL model called UnCRtainPihrt (Uncertainty Quantification for Cloud Removal in Optical Satellite Time Series). This model integrates Sentinel-2's optical imagery with Sentinel-1's radar data to reconstruct the Normalised Difference Vegetation Index (NDVI), which is an important indicator of crop health.

15.3.2 Disease and Pest Detection

AI has made it possible for disease and pest diagnoses to be enhanced by image recognition models. Proposed tools, such as Plantix and PlantVillage Nuru, which are built using CNNs, enable farmers to take images of infected crops and receive immediate diagnostic feedback (Minhans et al., 2025). Drones with high-resolution cameras, powered by AI, fly over broad acres, quickly spotting signs of infestations or disease early to provide real-time treatment. Pan et al. (2023) developed a method called RiceNet, a two-stage approach for identifying different rice diseases using these predictive models: YOLOX and Siamese Network. The Siamese Network achieved an identification accuracy of 99.03%, while YOLOX achieved a 95.58% mAP (Mean Average Precision). Mahadevan et al. (2024) detected and analysed approximately 10,080 rice plant images at various stages using $DSGAN^2$ (Deep Spectral Generative Adversarial Neural Network). Khanramaki et al. (2021) developed a diverse ensemble of CNNs (AlexNet, VGG16, ResNet50, InceptionResNet-v2), and they achieved 99.04% accuracy to classify citrus pests (Leafminer, Sooty Mould, Pulvinaria) using transfer learning and data augmentation in real field conditions.

15.3.3 AI-driven Irrigation Systems

AI-based irrigation systems are introduced as a mechanism to tackle the water shortage in agriculture. Such systems process real-time data from soil sensors and weather stations to calculate when irrigation is required. Kashyap et al. (2021) developed the Deep Learning Neural Network-based Internet of Things (IoT)-enabled intelligent irrigation system for precision agriculture (DliSA). This IoT-enabled innovative irrigation system employs LSTM (extended short-term memory network) networks to forecast soil moisture and optimise water distribution, achieving approximately 30% water savings compared to existing models. A predictive irrigation mechanism developed by Adeyemi et al. (2018) using Dynamic Neural Network models was evaluated against AQUACROP simulations for irrigation applications in potato crops.

15.3.4 Automated Weed Removal

AI-based automatic weed detection systems have transformed field management. Automated robots like Blue River Technology's "See & Spray" rely on AI to spot and target weeds with precision-sprayed herbicide, decreasing chemical use by more than 90% (Li et al., 2025). These systems utilise DL and real-time vision processing to distinguish crops from weeds, leading to less labour and less environmental damage. Its accuracy contributes to increased crop yields and less soil pollution. A GenAI pipeline that has Segment Anything Model (SAM) combined with Stable Diffusion generates synthetic images under real-world conditions to detect weeds (Modak & Stein, 2025). A semi-supervised diffusion generative network (SSDGN) model was used to improve automated weed detection in digital agriculture, which achieved 94% precision, better than Detection Transformer (DETR) and YOLOv10. (Li et al., 2025).

15.3.5 Weather Forecasting and Climate Change Adaptation

Generative AI allows climate-resilient agriculture by simulating the effect of weather fluctuations on crop yields. These predictive systems, such as those used by Descartes Labs and The Climate Corporation, combine past weather data with satellite images and AI-powered forecasting models so owners can know in advance when their fields are likely to experience droughts, floods, or heatwaves (Jeyalakshmi et al., 2024). A farmer can employ the information to take adaptive measures, such as growing drought-tolerant crops or changing sowing dates for risk reduction. DL, particularly hybrid DL methods (e.g., variational autoencoders which incorporate domain knowledge), allows for robust evaluation of how agricultural and ecological drought is assessed. Convolutional neural random fields (CNNs) and transformers utilised multiple heterogeneous climate datasets to predict heatwaves. Variable Rate Technology (VRT) is a revolutionary tool in precision agriculture, used to inspect various field conditions, including different soil properties, topography, moisture variations, yield, biomass, and soil types (Sharma et al., 2025).

15.3.6 Soil Health Management

AI soil scanners such as SoilOptix use ML to determine soil structure and available nutrients (Mbah & Pentoś, 2025). Such systems produce customised fertilisation plans to help minimise excess fertilisation and protect the soil from degradation. GenAI can also infer missing data related to the soil's condition based on sensor data, including real-time nutrient deficiencies or imbalanced pH readings. Such a strategy would promote soil-based fertilisation, with no concerns for leaching, and at the same time increase yields. Models such as Random Forests (RF), Support Vector Machines (SVMs), ANNs, and DNNs provide powerful tools in interpreting complex, multi-dimensional agro-ecological data to support optimised soil nutrient management, thereby enabling farmers to uncover deficiencies and improve crop productivity (Barathkumar et al., 2025).

15.3.7 Yield Prediction

Combining GenAI with DL models has significantly improved yield prediction. GAN generated a synthetic dataset that simulates various weather and soil conditions, similar to natural ones, and can be used to train forecasting models more effectively. “Prospera” serves as an example of practical applications of these technologies in yield prediction and planning (Amin & Amin, 2025). A hybrid model based on DL for crop yield predictions using the soybean dataset of 395 features. This model was tested using the XGBoost algorithm for comparison, along with CNN-DNN (Convolutional Neural Networks-Deep Neural Networks), CNN-XGBoost, CNN-RNN (Convolutional Neural Networks-Recurrent Neural Networks), and CNN-LSTM (Convolutional Neural Networks-Long Short Term Memory) (Oikonomidis et al., 2022). The best performer overall was the CNN-LSTM model, followed by the XGBoost model. Another study was conducted on five unique DL architectures to identify an optimal hybrid architecture of an LSTM network and one-dimensional Convolutional Neural Networks (1D-CNN) for predicting rice yield accurately, approximately two months before harvest (Jeong et al., 2022).

15.3.8 Smart Farming and Robotics

AI drives the next generation of intelligent farming robotics, automating sowing, spraying, and harvesting devices. Companies like Agrobot and FarmWise have founded robots that handle delicate harvesting, using AI vision systems to identify when fruits are ripe (Kantha et al., 2024). To reduce input waste, AI-powered devices are used for weeding, fertilising, and spraying pesticides accurately to address labour shortages and improve efficiency by Agrawal and Arafat (2024). Gupta and Gupta (2024) evaluated robotics and AI in the field of precision agriculture (PA), focusing on food crops to know their role in farming systems. The technologies surveyed include uncrewed ground vehicles (UGVs) and autonomous tractors, and their applications span multiple growing tasks and decisions, including planting, crop cultivation tasks (e.g., weeding, pruning), crop scouting using GPS and sensors, harvesting, and utilising UGVs for post-harvest processing. Examples of controlled environment agriculture (CEA) technologies discussed include traditional greenhouse systems, in vitro culture (IVC), and gene banks.

15.3.9 Greenhouse Management

In greenhouses, AI can lead the way in climate control. These AI-driven climate control systems can autonomously control the temperature, light, humidity, and CO_2 levels exactly to the readings from the sensors, with the goals of optimising plant growth while minimising energy consumption and stabilising crop production. Vijayasuganthi et al. (2025) created the Fusion Enhanced Smart Farming System (FESFS), consisting of a mixture of IoT, ML, precision agriculture, and big data automation. The FESFS, which is built on the advances from combining Temporal Fusion Transformers (TFT), Graph Neural Networks (GNN), and GAN, showed yield improvements of 42%, decreased water use by 30%, improved soil health by

25% and increased operating efficiency by 31%. Lee et al. (2024) developed the AI-GECS (AI-powered greenhouse environmental control system), an ideal set of gridded weather forecasts, crop physiological indicators, micro-climatic forecasts, and automated greenhouse operations. The Multi-Model Super Ensemble (MMSE) forecasting framework is a hybrid DL model, CLSTM-CNN-BP, which generates accurate weather forecasts hourly. Different models were trained to generate synthetic data using the KASPRO (greenhouse climate simulation model), along with a modified version of the INTKAM cucumber crop model that incorporated high-wire cucumber cultivation (Hoseinzadeh & Astiaso Garcia, 2024).

15.3.10 Crop Breeding Utilising AI

AI accelerates crop breeding by analysing genetic properties and suggesting combinations that yield robust and high-yielding crops. Genotype X environment (G X E) integration is the principle of feeding traits to the ML algorithms, allowing for the identification of the best combination of attributes to get a stress-resistant variety. "CropGPT" is an integrative intelligence precision design-breeding (IPDB) proposal that relies on teamwork across many disciplines, including seed selection (for high-quality germplasm resources), batch cloning of functional genes using ML solutions, high-throughput genotyping, and phenotyping assessment, and is an open platform that can provide IPDB services. It can access shared research results from the worldwide genetics and breeding literature related to crops, integrate fundamental theoretical constructs with phenotypic information from crops of various species, and create a knowledge map that supports crop breeding work within breeding units (Zhu et al., 2024, 2025). Speed breeding expedites crop breeding by utilising controlled light and temperature to reduce growth cycles. Speed breeding brings together marker-assisted selection and gene editing to transform this process for early modification of key traits. Recently, AI has been used to understand plant biology as influenced by environmental settings and to comprehend complex OMICS data, which will accelerate the process of speed breeding as we develop high-yielding, adaptable crops (Rai, 2022).

15.4 GENAI USED FOR ENVIRONMENTAL SOLUTIONS

15.4.1 GenAI for Water and Air Pollution Monitoring and Mitigation

Generative AI is revolutionising monitoring systems for different types of pollution, which deeply influence the environment. Advanced AI systems now provide unprecedented accuracy in detecting and predicting pollution levels, water quality assessment, and air quality prediction in a more efficient way. Recently, many AI-driven systems, such as adaptive neuro-fuzzy inference systems (ANFIS) and DNNs, can detect and forecast any anomalies or contamination present in water (Balakumar et al., 2025; Gulati, 2024). AI-based systems can predict water pollution and contaminants with 94% accuracy while reducing field sampling costs by using satellites (Al-Khafaji et al., 2025). In the case of wastewater management, AI-driven systems and sensors can estimate different parameters present in it (Maqbool et al., 2025).

On the other hand, these ML-based algorithms have demonstrated exceptional performance in monitoring and managing air quality. 98.2% accuracy has been achieved in measuring air quality by ML method—RF (Chadalavada et al., 2025).

Whereas Xie et al. (2021) developed a hybrid model (DL-CTEM) for monitoring PM2.5, PM10, O3, CO, NO2, and SO2 with an accuracy of 95.6%. Li et al. (2023) also used RF on an hourly basis to evaluate the fuel combustion with an accuracy of 94% for predicting PM2.5. The integration of advanced satellites with ground-based sensors and various IoT-enabled devices has evolved a new domain in the field of technology. Combining all the approaches with Gen AI, we can now determine various environmental conditions, including temperatures, humidity, and many toxic compounds, across the globe, to monitor and mitigate the pollution efficiently.

15.4.2 Biodiversity Conservation through GenAI-Driven Ecosystem Modelling

GenAI can transform the way of biodiversity conservation by enabling sophisticated ecosystem modelling that can predict the relationship of species with environmental changes and find complex ecological systems. Large-scale ecological data are used to generate virtual ecosystem representations that capture key spatial and statistical patterns of real-world systems using generative models such as GANs and diffusion models.

ML and DL algorithms acquired data from different devices and sensors, such as remote sensing, motion sensors, and cameras. Mainly, GenAI helps to improve the efficacy of monitoring by integrating with advanced devices. Drones can identify endangered species because of the AI algorithms equipped with them (Jiang & Zhu, 2022). It improves the ability to track and protect vulnerable populations. In the case of biodiversity conservation, one of the major tools is predictive modelling and species distribution mapping (Ayoola et al., 2024; Rodríguez et al., 2007). To analyse the critical insights, generative AI is playing a pivotal role by improving the accuracy of these tools. Disease and pest outbreaks in forest animals can be predicted by the GenAI with the integration of advanced surveillance devices (Dong et al., 2025). Disease infestation in coral reefs has been successfully predicted by AI-enabled models with an accuracy rate of 85% (Ayoola et al., 2024). In this recent era, GenAI-based ecosystem modelling supports the development of strategies that can respond to the unpredictable environmental conditions. The integration of AI and big data analytics provides scalability around different ecological conditions from forest to underwater marine environments. These GenAI-based models continuously learn from training data and improve their predictive ability to address the ongoing biodiversity crisis in this changing environment.

15.4.3 Optimising Renewable Energy Systems with GenAI

GenAI-based technologies are also popular in the field of renewable energy optimisation by enabling smart forecasting, adaptive control, and situation-based planning. According to Gupta and Gupta (2024), GenAI-enabled models can generate simulated data and predict complex energy production and consumption patterns. It also allows dynamic optimisation of the grid performance and energy storage management by training data and generative simulation frameworks. GenAI also synthesises synthetic operational data from sparse historical datasets for solar and wind systems (Surathunmanun et al., 2024). A new GAN-based model (solarGAN) was designed to forecast household weather, rooftop solar generation time series, and household electricity demand (Annapareddy, 2024). Combining all these studies, it could be concluded that GenAI has a transformative role in creating resilient, adaptive, and data-efficient renewable infrastructures that have the capability of balancing fluctuating supply-demand cycles and accelerating the transition to net-zero energy ecosystems. Upcoming advancements in GenAI may result in decreased carbon emissions, for instance, through optimising the location of computing to use regions that have surplus renewable energy, or using fine-tuned models rather than those trained from first principles to reduce computation load (Richards et al., 2024).

15.4.4 GenAI for Urban Environmental Planning and Green Infrastructure

Generative AI (GenAI) is also reshaping urban environmental planning and design for green infrastructure today by bringing together and analysing multidimensional data to facilitate smarter, more sustainable cities (Xu et al., 2024). Through scenario modelling, GenAI assists urban planners by allowing for the better allocation of land use, modelling the potential impact of proposed policies, and designing green infrastructure solutions, such as parks, rain gardens, and urban forests, along a spectrum of ecological, social, and economic factors (Moreno-Ibarra et al., 2024). These AI-based techniques will assist with sustainable green city development and disaster response, as well as enhance city services. Chen et al. (2025) designed an integrated GAN, which is a stable diffusion multimodal large-scale image pre-training model that can generate green space plans. Furthermore, a GenAI-based traffic management system helps to reduce emissions and also improve smooth transportation service (Satpathy et al., 2025). These advanced algorithm-based systems can predict future disasters, the pollution level of the city, and emergency evacuation plans (Zhu et al., 2022). Above all, GenAI provides substantial benefits for urban environmental planning and green infrastructure by facilitating sustainability and creating new possibilities for innovation.

15.4.5 GenAI for Waste Management and Circular Economy

Recently, the waste management system and circular economy practices have transformed their approach from traditional to advanced systems by using intelligent GenAI systems. It promotes resource recovery, reduces contamination, and closes material loops for effective management (Sikander, 2024). Above 95% accuracy is achieved by GenAI-enabled waste sorting systems to improve recycling rates and the diversion of solid waste from landfill, using advanced computer vision capabilities and ML algorithms (Yevle & Mann, 2025). GenAI can predict effective waste collection routes, reducing fuel consumption and providing opportunities to adjust and reconfigure routes based on traffic and waste levels. By using a smart waste management digital twin, an optimised, improved, and quality strategy can be made for sustainable urban or city planning (Barth et al., 2023). GenAI promotes the circular economy by calculating recovered material lifecycles to identify high-waste commodity streams for reuse in marketplaces for manufacturers to swap out excess material in the form of new commodities (Caroli, 2025). However, the environmental costs of GenAI itself are a challenge because computational infrastructure could generate 1.2–5.0 million tonnes of e-waste in 2030 across 187 countries (Wang et al., 2024). Circular economy strategies across the GenAI value chain could mitigate existing e-waste and also make a smart, sustainable waste management plant for a better environment.

15.4.6 Climate Change Scenario Modelling and Adaptation Planning

Generative AI transforms climate change research by simulating modelling that can predict the complex interactions within Earth's climate system. Various GenAI-based models have the ability to analyse vast datasets, including oceanographic data, meteorological data, and many environmental components, to provide more reliable and accurate predictions at finer spatial and temporal scales (Deep & Verma, 2024). A generative AI-powered model can easily identify non-linear and complex relationships of climate that traditional model lacks (Ukoba et al., 2025). Recent developments of physically constrained GANs have significantly improved the pattern of precipitation in Earth systems models using simulation data sets (Hess et al., 2022). These models can open up advanced monitoring systems towards climate change, such as natural calamities—flood, tsunami, drought, earthquake, and also assess the impact on agriculture, water resources, animal habitats, and human settlements. This can help to make better adaptation planning and support targeted interventions like improved infrastructure design, early warning systems, and safe, sustainable land use planning that promote community resilience.

15.5 CHALLENGES AND ETHICAL CONSIDERATIONS FOR USING GENAI

The real-life implementation of AI-powered models and technologies faces different challenges and ethical considerations. Detailed challenges are discussed later in the chapter, and Table 15.2 gives a brief overview of the applications of these models and technologies in the format of case studies from recent research.

15.5.1 Data Privacy and Security Challenges

In this era of technological advances, data is the ultimate source of information wealth. Data acquisition and utilisation for training purposes and generating simulated data may be concerning if sensitive data is exposed on unsecured online platforms (Halder et al., 2025). A competitive party can misuse this data for its own use and would be a threat to national security. Different third-party organisations taking part as mediators between the farmers' community and buyers may misuse potential data related to land utilisation models, financial transactions, workforce utility

TABLE 15.2
Case Studies Highlighting Recent Agricultural Uses of AI Models and Technologies

Case Study Category	Models/Technologies Used	Applications	Citations
Smart Farming & Robotics	IoT, Big Data, Custom AI, AI Vision, UGVs (uncrewed ground vehicles), Autonomous Tractors, Robotics	To do real-time crop performance monitoring, sowing, spraying, harvesting, and post-harvest processing; robotics for scouting, weeding, pruning, and precision operations.	(Agrawal & Arafat, 2024; Gupta & Gupta, 2024; Kantha et al., 2024)
Crop Monitoring & Precision Farming	ANN, GAN, YOLO, VRT, GPS, GIS, Satellite/Drones, Piezo-MEMS, GenAI	Monitoring crop health via UAV/satellite, real-time field variability mapping, and site-specific input application; synthetic image enhancement; leaf water content tracking using piezo-MEMS sensors with AI; variable rate application based on topography and soil variability.	(Senthil et al., 2025; Roy et al., 2025; Sharma et al., 2025)
Crop Recommendation Systems	XGBoost, AdaBoost, Random Forest, IoT Sensors, VAE, GAN	Personalised SMS-based crop suggestions (94% accuracy); Synthetic data augmentation for model training under low data regimes.	(Akkem et al., 2024; G. Singh & Sharma, 2025)
Disease & Pest Detection	CNN (Plantix, Nuru), YOLOX, Siamese Network, DSGAN2, Ensemble CNNs, Transfer Learning	Image-based diagnosis of rice and citrus diseases, drone-assisted pest detection, and early warning alerts; high-accuracy multi-model detection (e.g., RiceNet), data augmentation using deep GANs.	(Khanramaki et al., 2021; Mahadevan et al., 2024; Minhans et al., 2025; Pan et al., 2023)
AI-Driven Irrigation Systems	LSTM (DliSA), CNN, GAN, Dynamic NN, IoT Sensors	Predictive irrigation scheduling	(Adeyemi et al., 2018; Kashyap et al., 2021; Laha et al., 2023)
Automated Weed Removal	Deep Learning, YOLOv10, SSDGN, DETR, SAM + Stable Diffusion, Synthetic Image Generation	Robotic weed detection and herbicide targeting with >90% chemical reduction; real-time vision classification of weeds vs crops; synthetic image generation to reduce data dependence.	(R. Li et al., 2025; Modak & Stein, 2025; San & Kakani, 2025)

Weather Forecasting & Climate	VAE, CNN, Earthformer, Explainable AI (ANN, Neuron Integrated Gradients), Spatio-Temporal Transformers	Forecasting droughts, floods, and heatwaves; climate bias correction; multimodal modelling using satellite, weather, and soil indices; rare event prediction using latent-space simulation and XAI models.	(Jacques-Dumas et al., 2022; Jeyalakshmi et al., 2024)
Soil Health & Fertility Management	RF, SVM, ANN, DNN, ML, DL, SoilOptix	Real-time nutrient mapping.	(Barathkumar et al., 2025; Mbah & Pentoś, 2025)
Yield Prediction	GAN, CNN-DNN, CNN-XGBoost, CNN-RNN, CNN-LSTM, LSTM+1D-CNN, DDM	Hybrid models for forecasting yields (soybean, rice); use of synthetic datasets simulating various weather and soil conditions; improved pre-harvest prediction of crop yield.	(Amin & Amin, 2025; Jeong et al., 2022; Oikonomidis et al., 2022)
Greenhouse Management	TFT, GNN, GANs (FESFS), DRL, CNN, RNN, DDPG, MMSE, CLSTM-CNN-BP, INTKAM, KASPRO	Automated environmental control and Greenhouse crop modelling using AI; yield increased by 42%, water saved by 30%; Climate simulation and AI weather forecasting using hybrid models and sensor integration.	(Hoseinzadeh & Astiaso Garcia, 2024; Lee et al., 2024; Vijayasuganthi et al., 2025)v
Crop Breeding Utilising AI	ML for Genotype × Environment, OMICS, CropGPT, IPDB, AI + Marker-Assisted Selection, Speed Breeding	Rapid breeding for drought/disease resistance; genome-to-phenotype prediction; high-throughput genotyping and phenotyping; open-access AI breeding platforms (e.g., CropGPT); OMICS data integration	(Rai, 2022; Zhu et al., 2024, 2025)
Conversational Farming AI	Large Language Models (LLMs), ChatGPT, NLP (Natural Language Processing)	Interactive advisory systems integrated with farming platforms; real-time guidance on irrigation, pest control, and planting; increases access to expert knowledge among rural farmers.	(Rane et al., 2024)

details, and production parameters. Especially in rural regions where technology and awareness are not so widespread, primary field-level data from cultivators is highly vulnerable and easily accessible to misusers. To prevent exploitation of these data, a definite legal framework should be implemented to ensure that the farmer community does not lose control over the necessary crop production and management-related information (Awais et al., 2025).

15.5.2 Job Replacement

Excessive use of AI-related technology in farming procedures, especially crop management practices, may displace traditional cultivation practices. It can affect the socio-economic backbone of society, specifically the labour community, which can lose its jobs, leading to widespread socio-economic disruption due to job displacement (Lazaroiu & Rogalska, 2023). It is worth noting that an AI generative model can provide you with all services based on the data. If the GenAI-based models capture any misinformation, it can destroy the whole farming practice from the sowing to the harvesting stage. So, we should limit our use to such an extent that it may assist you in smartly doing traditional farming.

15.5.3 Algorithmic Biases

GenAI-based learning models are trained by vast datasets that can contain biased information (Banh & Strobel, 2023). They can generate simulated data based on the trained data fed into the algorithms. Suppose biased information provided to farmers by GenAI-based systems for crop management can result in a considerable loss in production, as well as a lost season for this crop cultivation (Rane et al., 2024). If any biased information is present, it can affect the accurate and uniform decisions produced by AI-based models. Small-scale farmers who earn a marginal profit may be disadvantaged by limited access to quality data compared to the quantity of data. Data quality can be compromised by inadequate resources for quality data, which will affect the social and economic sectors by creating disparities in the farming industry. Also, a lack of efficient training may cause farmers to lag behind tech-based firms that capture the customer base with advanced technology. To ensure the farming community provides quality data, it is mandatory to track and tackle this through a scientific procedure, and also modify the algorithms so that they can process diverse data and identify misinformation from biased datasets.

15.5.4 Transparency and Accountability in AI Decision-Making

AI-, ML-, and DL-based models can provide information at a surface level, and they sometimes lose their transparency and accountability (Feuerriegel et al., 2024; Rane et al., 2024). They may not consistently deliver precise decisions based on sudden environmental factors. The lack of clarity in ensuring data reliability and accountability to the farmers may lead to a loss of trust in AI-based models and technologies. To mitigate the ambiguity in transparency and accountability of the

AI-based information, it is mandatory to form a regulatory body that can manage the accessibility of the interpretable side of GenAI-based systems, so that farmers can get a guaranteed support system for authentic and transparent service of AI-driven systems.

15.5.5 Environmental Impact

Excessive use of technology can also negatively impact our environment. Every year, lots of e-waste (electronic waste) is generated by using AI-powered devices, such as defective computers and their parts. Wrong predictions or misleading data can create major problems during smart green and sustainable city planning. Biodiversity conservation, such as monitoring endangered species by advanced GenAI-enabled drones, can also disturb the habitat of common animals and birds. Smart GenAI-based sensors are sometimes unable to detect the correct value of pollutants present in the environment due to a lack of training data. Besides, anomalies have been noted in the case of AI-based waste sorting devices. Above all, the technology itself requires a smart management system to make a more sustainable environment.

15.6 FUTURE DIRECTIONS AND RESEARCH OPPORTUNITIES OF GENERATIVE AI IN SMART FARMING AND SHAPING A BETTER ENVIRONMENT

GenAI can have a significant impact on farming practices, introducing innovative approaches to traditional farming through data-driven decisions, predictive analysis, and creative, sustainable approaches. To bridge the genotype-phenotype gap, GenAI-driven approaches can be a smart alternative to the conventional method, enabling next-generation plant breeding. From facilitating generative farm infrastructure design to providing personalised crop treatment through real-time environmental information and producing synthetic data to train inference models, GenAI-based systems can be a paradigm-shifting innovation. Subsequent studies could emphasise improving data augmentation techniques through VAEs, GANs, and various generative models to produce high-quality synthetic crop datasets. This might include improving the model structures, fine-tuning hyperparameters, and increasing the diversity of data to represent real-world scenarios more accurately. Advancements in data augmentation techniques are anticipated to enhance the resilience of recommendation systems. Connections with robotics, wireless sensors, and IoT-based devices give precise decisions, and ethical concerns promote openness and equitable access in AI-driven farming. Green city development, traffic management, waste management, biodiversity conservation, and measuring pollution from the atmosphere, soil surface, and underwater_all are bright avenues of this advanced GenAI technology. Holistic and judicious application of these technologies can change the environment and promote a better future for us. Future research should emphasise both the socio-economic impacts and technological advancements to promote "zero hunger" and maintain rural prosperity.

15.7 CONCLUSION

GenAI has transformed agricultural practices by shifting from traditional, labour-intensive methods to effective smart farming practices. This technology also enables efficient and smart solutions for the management of various environmental problems, such as optimising renewable energy, maintaining biodiversity, pollution and waste management, and planning a smart city. GenAI provides effective decisions to fix real-world problems by using synthetic data generation, prediction capabilities and multimodal data analysis. Also, various cooperative platforms can provide real-time consultancy to farmers and field workers by addressing current knowledge gaps and increasing accessibility using generative AI, mainly in remote locations. Greenhouse climate control and advanced farm robots using this technology maintain stable environmental conditions for crop production. In the field of crop breeding, AI using these algorithm-based tools can also accelerate the finding of beneficial markers, reduce breeding cycles, and develop more stress-tolerant varieties. Furthermore, effective waste management practices, making a circular economy, promoting surveillance towards endangered species, optimising renewable energy production, and providing adaptive management strategies for environmental disasters are some of the major aspects of this GenAI-enabled technology. GenAI has integrated with IoT devices, wireless sensors, and robotics, responding to environmental and socio-economic challenges in every sector. In light of this context, research and innovation should focus not only on the technical development but also on the social integration of AI-related tools to facilitate the evolution of inclusive, ethical, and effective agricultural and environmental systems. Generative AI is not a displacement but a reinforcement of traditional knowledge and skills, improved with a bit of modern AI tools. It solves major environmental-related issues, ultimately promoting growth in the agricultural sector, securing natural habitats, providing better human resilience, and managing food security.

REFERENCES

Adeyemi, O., Grove, I., Peets, S., Domun, Y., & Norton, T. (2018). Dynamic Neural Network Modelling of Soil Moisture Content for Predictive Irrigation Scheduling. *Sensors (Basel, Switzerland)*, *18*(10), 3408. https://doi.org/10.3390/s18103408

Agrawal, J., & Arafat, M. Y. (2024). Transforming Farming: A Review of AI-Powered UAV Technologies in Precision Agriculture. *Drones*, *8*(11), Article 11. https://doi.org/10.3390/drones8110664

Akkem, Y., View Profile, Biswas, S. K., View Profile, Varanasi, A., & View Profile. (2024). A Comprehensive Review of Synthetic Data Generation in Smart Farming by Using Variational Autoencoder and Generative Adversarial Network. *Engineering Applications of Artificial Intelligence*, *131*(C). https://doi.org/10.1016/j.engappai.2024.107881

Alaba, F. A. (2024). IoT Architecture Layers. In F. A. Alaba (Ed.), *Internet of Things: A Case Study in Africa* (pp. 65–85). Springer Nature Switzerland. https://doi.org/10.1007/978-3-031-67984-1_4

Al-Khafaji, M. S., Abdulameer, L., Al-Shammari, M. M. A., Maimuri, N. M. L. A., Dulaimi, A., & Al-Jumeily, D. (2025). Revolutionizing Water Quality Monitoring with Artificial Intelligence: A Systematic Review. *Journal of Studies in Science and Engineering*, *5*(1), 358–385. https://doi.org/10.53898/josse2025528

Amin, R., & Amin, S. (2025). Machine Learning Techniques for Increasing the Productivity of High-Quality Food Products. In *Artificial Intelligence in the Food Industry*. CRC Press.

Annapareddy, V. N. (2024). *Harnessing AI Neural Networks and Generative AI for Optimized Solar Energy Production and Residential Battery Storage Management* (SSRN Scholarly Paper No. 5203702). Social Science Research Network. https://doi.org/10.2139/ssrn.5203702

Assunção, E., Gaspar, P. D., Mesquita, R., Simões, M. P., Alibabaei, K., Veiros, A., & Proença, H. (2022). Real-Time Weed Control Application Using a Jetson Nano Edge Device and a Spray Mechanism. *Remote Sensing*, *14*(17), 4217. https://doi.org/10.3390/rs14174217

Awais, M., Wang, X., Hussain, S., Aziz, F., & Mahmood, M. Q. (2025). Advancing Precision Agriculture Through Digital Twins and Smart Farming Technologies: A Review. *AgriEngineering*, *7*(5), 137–137. https://doi.org/10.3390/agriengineering7050137

Ayoola, I. P. Idoko, S. O. Eromonsei, O. A., Apampa, A. R., & Oyebanji, O. S. (2024). The Role of Big Data and AI in Enhancing Biodiversity Conservation and Resource Management in the USA. *World Journal of Advanced Research and Reviews*, *23*(2), 1851–1873. https://doi.org/10.30574/wjarr.2024.23.2.2350

Ayoub Shaikh, T., View Profile, Rasool, T., View Profile, Rasheed Lone, F., & View Profile. (2022). Towards Leveraging the Role of Machine Learning and Artificial Intelligence in Precision Agriculture and Smart Farming. *Computers and Electronics in Agriculture*, *198*(C). https://doi.org/10.1016/j.compag.2022.107119

Badgujar, C. M., Poulose, A., & Gan, H. (2024). Agricultural Object Detection with You Only Look Once (YOLO) Algorithm: A Bibliometric and Systematic Literature Review. *Computers and Electronics in Agriculture*, *223*, 109090. https://doi.org/10.1016/j.compag.2024.109090

Balakumar, S., Mahesh, N., Kamaraj, M., & Aravind, J. (2025). Harnessing Artificial Intelligence for Sustainable Environmental Remediation a Review. *International Journal of Environmental Science and Technology*, *22*(13), 13189–13206. https://doi.org/10.1007/s13762-025-06528-9

Banh, L., & Strobel, G. (2023). Generative Artificial Intelligence. *Electronic Markets*, *33*(1), 63. https://doi.org/10.1007/s12525-023-00680-1

Barathkumar, S., Sellamuthu, K. M., Sathyabama, K., Malathi, P., Kumaraperumal, R., & Devagi, P. (2025). Advancements in Soil Quality Assessment: A Comprehensive Review of Machine Learning and AI-Driven Approaches for Nutrient Deficiency Analysis. *Communications in Soil Science and Plant Analysis*, *56*(2), 251–276. https://doi.org/10.1080/00103624.2024.2406484

Barth, L., Schweiger, L., Benedech, R., & Ehrat, M. (2023). From Data to Value in Smart Waste Management: Optimizing Solid Waste Collection with a Digital Twin-Based Decision Support System. *Decision Analytics Journal*, *9*, 100347. https://doi.org/10.1016/j.dajour.2023.100347

Bi, L., Wally, O., Hu, G., Tenuta, A. U., Kandel, Y. R., & Mueller, D. S. (2023). A Transformer-Based Approach for Early Prediction of Soybean Yield Using Time-Series Images. *Frontiers in Plant Science*, *14*. https://doi.org/10.3389/fpls.2023.1173036

Çakmakçı, S., Polatoğlu, B., & Çakmakçı, R. (2024). Foods of the Future: Challenges, Opportunities, Trends, and Expectations. *Foods*, *13*(17), Article 17. https://doi.org/10.3390/foods13172663

Caroli, N. (2025). *Evaluating the Sustainability and Economic Integration of Additive Manufacturing in Modern Production Lines: A Holistic Approach*. Laurea, Politecnico di Torino. https://webthesis.biblio.polito.it/36003/

Chadalavada, S., Faust, O., Salvi, M., Seoni, S., Raj, N., Raghavendra, U., Gudigar, A., Barua, P. D., Molinari, F., & Acharya, R. (2025). Application of Artificial Intelligence in Air Pollution Monitoring and Forecasting: A Systematic Review. *Environmental Modelling & Software*, *185*, 106312. https://doi.org/10.1016/j.envsoft.2024.106312

Chen, R., Yi, X., Zhao, J., He, Y., Chen, B., Liu, F., Yao, X., Jiang, X., Lian, Z., & Li, H. (2025). AI for Landscape Planning: Assessing Surrounding Contextual Impact on GAN-Generated Green Land Layouts. *Cities*, *166*, 106181. https://doi.org/10.1016/j.cities.2025.106181

Deep, G., & Verma, J. (2024). Deep Learning Models for Fine-Scale Climate Change Prediction: Enhancing Spatial and Temporal Resolution Using AI. In G. Tripathi, A. Shakya, S. Kanga, S. K. Singh, & P. K. Rai (Eds.), *Big Data, Artificial Intelligence, and Data Analytics in Climate Change Research: For Sustainable Development Goals* (pp. 81–100). Springer Nature. https://doi.org/10.1007/978-981-97-1685-2_5

Deng, J., Yang, C., Huang, K., Lei, L., Ye, J., Zeng, W., Zhang, J., Lan, Y., & Zhang, Y. (2023). Deep-Learning-Based Rice Disease and Insect Pest Detection on a Mobile Phone. *Agronomy*, *13*(8), Article 8. https://doi.org/10.3390/agronomy13082139

Dong, Y., Liu, L., Zhai, X., & Li, W. (2025). Artificial Intelligence in Agricultural Pest and Disease Management: Current Applications and Future Prospects. *Advances in Resources Research*, *5*(2), 971–986. https://doi.org/10.50908/arr.5.2_971

Elbasi, E., Mostafa, N., Zaki, C., AlArnaout, Z., Topcu, A. E., & Saker, L. (2024). Optimizing Agricultural Data Analysis Techniques through AI-Powered Decision-Making Processes. *Applied Sciences*, *14*(17), Article 17. https://doi.org/10.3390/app14178018

Fanelli, R. M. (2020). The Spatial and Temporal Variability of the Effects of Agricultural Practices on the Environment. *Environments*, *7*(4), Article 4. https://doi.org/10.3390/environments7040033

Feuerriegel, S., Hartmann, J., Janiesch, C., & Zschech, P. (2024). Generative AI. *Business & Information Systems Engineering*, *66*(1), 111–126. https://doi.org/10.1007/s12599-023-00834-7

Gaitan, N. C., Batinas, B. I., Ursu, C., & Crainiciuc, F. N. (2025). Integrating Artificial Intelligence into an Automated Irrigation System. *Sensors*, *25*(4), Article 4. https://doi.org/10.3390/s25041199

Goodfellow, I., Pouget-Abadie, J., Mirza, M., Xu, B., Warde-Farley, D., Ozair, S., Courville, A., & Bengio, Y. (2020). Generative Adversarial Networks. *Communications of the ACM*, *63*(11), 139–144. https://doi.org/10.1145/3422622

Gulati, S. (Ed.). (2024). *Application of Artificial Intelligence in Wastewater Treatment*. Springer Nature Switzerland. https://doi.org/10.1007/978-3-031-69433-2

Gupta, N., & Gupta, P. K. (2024). Robotics and Artificial Intelligence (AI) in Agriculture with Major Emphasis on Food Crops. In P. M. Priyadarshan, S. M. Jain, S. Penna, & J. M. Al-Khayri (Eds.), *Digital Agriculture: A Solution for Sustainable Food and Nutritional Security* (pp. 577–605). Springer International Publishing. https://doi.org/10.1007/978-3-031-43548-5_19

Halder, S., Rafiqul Islam, M., Mamun, Q., Mahboubi, A., Walsh, P., & Zahidul Islam, M. (2025). A Comprehensive Survey on AI-Enabled Secure Social Industrial Internet of Things in the Agri-Food Supply Chain. *Smart Agricultural Technology*, *11*, 100902. https://doi.org/10.1016/j.atech.2025.100902

Han, K., & Wang, Y. (2021). A Review of Artificial Neural Network Techniques for Environmental Issues Prediction. *Journal of Thermal Analysis and Calorimetry*, *145*(4), 2191–2207. https://doi.org/10.1007/s10973-021-10748-9

Hess, P., Drüke, M., Petri, S., Strnad, F. M., & Boers, N. (2022). Physically Constrained Generative Adversarial Networks for Improving Precipitation Fields from Earth System Models. *Nature Machine Intelligence*, *4*(10), 828–839. https://doi.org/10.1038/s42256-022-00540-1

Hoseinzadeh, S., & Astiaso Garcia, D. (2024). AI-Driven Innovations in Greenhouse Agriculture: Reanalysis of Sustainability and Energy Efficiency Impacts. *Energy Conversion and Management: X*, *24*, 100701. https://doi.org/10.1016/j.ecmx.2024.100701

Jacques-Dumas, V., Ragone, F., Borgnat, P., Abry, P., & Bouchet, F. (2022). Deep Learning-Based Extreme Heatwave Forecast. *Frontiers in Climate*, *4*. https://doi.org/10.3389/fclim.2022.789641

Jeong, S., Ko, J., & Yeom, J.-M. (2022). Predicting Rice Yield at Pixel Scale Through synthetic Use of Crop and Deep Learning Models with Satellite Data in South and North Korea. *Science of the Total Environment*, *802*, 149726. https://doi.org/10.1016/j.scitotenv.2021.149726

Jeyalakshmi, J., Vijay, K., Berna, I. E., & Sowmia, K. R. (2024). AI in Optimisation of Agricultural Output in Precision Agriculture. In *Artificial Intelligence for Precision Agriculture*. Auerbach Publications.

Jiang, M., & Zhu, Z. (2022). The Role of Artificial Intelligence Algorithms in Marine Scientific Research. *Frontiers in Marine Science*, *9*. https://doi.org/10.3389/fmars.2022.920994

Kalamkar, R. B., Ahire, M. C., Ghadge, P. A., Dhenge, S. A., & Anarase, M. S. (2020). Drone and its Applications in Agriculture. *International Journal of Current Microbiology and Applied Sciences*, *9*(6), 3022–3026. https://doi.org/10.20546/ijcmas.2020.906.363

Kamyshova, G., Osipov, A., Gataullin, S., Korchagin, S., Ignar, S., Gataullin, T., Terekhova, N., & Suvorov, S. (2022). Artificial Neural Networks and Computer Vision's-Based Phytoindication Systems for Variable Rate Irrigation Improving. *IEEE Access*, *10*, 8577–8589. https://doi.org/10.1109/ACCESS.2022.3143524

Kantha, P., Srivastava, D., Srivastava, S. K., Maakar, S. Kr., & Sah, B. (2024). Overview of Robotics in Agriculture. In *Smart Agritech* (pp. 29–56). John Wiley & Sons, Ltd. https://doi.org/10.1002/9781394302994.ch2

Kashyap, P. K., Kumar, S., Jaiswal, A., Prasad, M., & Gandomi, A. H. (2021). Towards Precision Agriculture: IoT-Enabled Intelligent Irrigation Systems Using Deep Learning Neural Network. *IEEE Sensors Journal*, *21*(16), 17479–17491. https://doi.org/10.1109/JSEN.2021.3069266

Khanramaki, M., Askari Asli-Ardeh, E., & Kozegar, E. (2021). Citrus pests Classification Using an Ensemble of Deep Learning Models. *Computers and Electronics in Agriculture*, *186*, 106192. https://doi.org/10.1016/j.compag.2021.106192

Kibria, M. G., Nguyen, K., Villardi, G. P., Zhao, O., Ishizu, K., & Kojima, F. (2018). Big Data Analytics, Machine Learning and Artificial Intelligence in Next-Generation Wireless Networks (No. arXiv:1711.10089). *arXiv*. https://doi.org/10.48550/arXiv.1711.10089

Kujawa, S., & Niedbała, G. (2021). Artificial Neural Networks in Agriculture. *Agriculture*, *11*(6), Article 6. https://doi.org/10.3390/agriculture11060497

Kumar, V., & Laxmi, V. (2022). Pests Detection Using Artificial Neural Network and Image Processing: A Review. *2022 International Conference on Sustainable Computing and Data Communication Systems (ICSCDS)*, 462–467. https://doi.org/10.1109/ICSCDS53736.2022.9760976

Laha, S. R., Pattanayak, B. K., Pattnaik, S., Mishra, D., Kumar Nayak, D. S., & Dash, B. B. (2023). An IOT-Based Soil Moisture Management System for Precision Agriculture: Real-Time Monitoring and Automated Irrigation Control. *2023 4th International Conference on Smart Electronics and Communication (ICOSEC)*, 451–455. https://doi.org/10.1109/ICOSEC58147.2023.10276266

Lazaroiu, G., & Rogalska, E. (2023). How generative artificial intelligence technologies shape partial job displacement and labor productivity growth. *Oeconomia Copernicana*, *14*(3), Article 3. https://doi.org/10.24136/oc.2023.020

Lee, M.-H., Yao, M.-H., Kow, P.-Y., Kuo, B.-J., & Chang, F.-J. (2024). An Artificial Intelligence-Powered Environmental Control System for Resilient and Efficient Greenhouse Farming. *Sustainability*, *16*(24), Article 24. https://doi.org/10.3390/su162410958

Li, G., Huang, Y., Chen, Z., Chesser, G. D., Purswell, J. L., Linhoss, J., & Zhao, Y. (2021). Practices and Applications of Convolutional Neural Network-Based Computer Vision Systems in Animal Farming: A Review. *Sensors*, *21*(4), Article 4. https://doi.org/10.3390/s21041492

Li, R., Wang, X., Cui, Y., Xu, Y., Zhou, Y., Tang, X., Jiang, C., Song, Y., Dong, H., & Yan, S. (2025). A Semi-Supervised Diffusion-Based Framework for Weed Detection in Precision Agricultural Scenarios Using a Generative Attention Mechanism. *Agriculture*, *15*(4), Article 4. https://doi.org/10.3390/agriculture15040434

Li, T., Zhang, Q., Peng, Y., Guan, X., Li, L., Mu, J., Wang, X., Yin, X., & Wang, Q. (2023). Contributions of Various Driving Factors to Air Pollution Events: Interpretability Analysis from Machine Learning Perspective. *Environment International*, *173*, 107861. https://doi.org/10.1016/j.envint.2023.107861

Liakos, K. G., Busato, P., Moshou, D., Pearson, S., & Bochtis, D. (2018). Machine Learning in Agriculture: A Review. *Sensors*, *18*(8), Article 8. https://doi.org/10.3390/s18082674

Lu, Y., Chen, D., Olaniyi, E., & Huang, Y. (2022). Generative Adversarial Networks (GANs) for Image Augmentation in Agriculture: A Systematic Review. *Computers and Electronics in Agriculture*, *200*, 107208. https://doi.org/10.1016/j.compag.2022.107208

Mahadevan, K., Punitha, A., & Suresh, J. (2024). A Novel Rice Plant Leaf Diseases Detection Using Deep Spectral Generative Adversarial Neural Network. *International Journal of Cognitive Computing in Engineering*, *5*, 237–249. https://doi.org/10.1016/j.ijcce.2024.05.004

Maqbool, A. I., Zaib-Un-Nisa, Kiayani S., & Safdar, A. (2025). Artificial Intelligence In Environmental Science And Engineering: A Review of Applications, Challenges, and Sustainable Futures. *Spectrum of Engineering Sciences*, *3*(6), 861–876. https://sesjournal.org/index.php/1/article/view/520

Mbah, J. T., & Pentoś, K. (2025). Predicting Soil Compaction from Soil Electrical Conductivity Based on Scanning Methods. A Review. *Agronomy Science*, *80*(1), Article 1. https://doi.org/10.24326/as.2025.5463

Mienye, I. D., & Swart, T. G. (2024). A Comprehensive Review of Deep Learning: Architectures, Recent Advances, and Applications. *Information*, *15*(12), Article 12. https://doi.org/10.3390/info15120755

Minhans, K., Sharma, S., Sheikh, I., Alhewairini, S. S., & Sayyed, R. (2025). Artificial Intelligence and Plant Disease Management: An Agro-Innovative Approach. *Journal of Phytopathology*, *173*(3), e70084. https://doi.org/10.1111/jph.70084

Modak, S., & Stein, A. (2025). Generative AI-Based Pipeline Architecture for Increasing Training Efficiency in Intelligent Weed Control Systems. *Journal of Systems Architecture*, *167*, 103464. https://doi.org/10.1016/j.sysarc.2025.103464

Mohan, H. M., Anitha, S., Chai, R., & Ling, S. H. (2021). Edge Artificial Intelligence: Real-Time Noninvasive Technique for Vital Signs of Myocardial Infarction Recognition Using Jetson Nano. *Advances in Human-Computer Interaction*, *2021*(1), 6483003. https://doi.org/10.1155/2021/6483003

Morella, P., Lambán, M. P., Royo, J., & Sánchez, J. C. (2023). Vertical Farming Monitoring: How Does It Work and How Much Does It Cost? *Sensors*, *23*(7), Article 7. https://doi.org/10.3390/s23073502

Moreno-Ibarra, M., Saldaña-Perez, M., Pérez Rodríguez, S., & Juárez Carbajal, E. (2024). Generative Artificial Intelligence in the Context of Urban Spaces. In M. F. Mata-Rivera, R. Zagal-Flores, & C. Barria-Huidobro (Eds.), *Telematics and Computing* (pp. 209–222). Springer Nature Switzerland. https://doi.org/10.1007/978-3-031-77290-0_13

Mrabet, H., Belguith, S., Alhomoud, A., & Jemai, A. (2020). A Survey of IoT Security Based on a Layered Architecture of Sensing and Data Analysis. *Sensors*, *20*(13), Article 13. https://doi.org/10.3390/s20133625

Muhammad, A., Salman, Z., Lee, K., & Han, D. (2023). Harnessing the Power of Diffusion Models for Plant Disease Image Augmentation. *Frontiers in Plant Science*, *14*. https://doi.org/10.3389/fpls.2023.1280496

Muhie, S. H. (2022). Novel Approaches and Practices to Sustainable Agriculture. *Journal of Agriculture and Food Research*, *10*, 100446. https://doi.org/10.1016/j.jafr.2022.100446

Ogwu, M. C., & Kosoe, E. A. (2025). Integrating Green Infrastructure into Sustainable Agriculture to Enhance Soil Health, Biodiversity, and Microclimate Resilience. *Sustainability*, *17*(9), Article 9. https://doi.org/10.3390/su17093838

Oikonomidis, A., Catal, C., & Kassahun, A. (2022). Hybrid Deep Learning-based Models for Crop Yield Prediction. *Applied Artificial Intelligence*, *36*(1), 2031822. https://doi.org/10.1080/08839514.2022.2031823

Pan, J., Wang, T., & Wu, Q. (2023). RiceNet: A Two Stage Machine Learning Method for Rice Disease Identification. *Biosystems Engineering*, *225*, 25–40. https://doi.org/10.1016/j.biosystemseng.2022.11.007

Parashar, N., Johri, P., Khan, A., Gaur, N., & Kadry, S. (2024). An Integrated Analysis of Yield Prediction Models: A Comprehensive Review of Advancements and Challenges. *Computers, Materials & Continua*, *80*(1), 389–425. https://doi.org/10.32604/cmc.2024.050240

Picon, A., Alvarez-Gila, A., Seitz, M., Ortiz-Barredo, A., Echazarra, J., & Johannes, A. (2019). Deep Convolutional Neural Networks for Mobile Capture Device-Based Crop Disease Classification in the Wild. *Computers and Electronics in Agriculture*, *161*, 280–290. https://doi.org/10.1016/j.compag.2018.04.002

Pihrt, J., Šimánek, P., Kovalenko, A., Kvapil, J., & Charvát, K. (2024). *AI-Based Spatiotemporal Crop Monitoring by Cloud Removal in Satellite Images*. 485–492. https://doi.org/10.15439/2024F5446

Pretty, J. (2007). Agricultural Sustainability: Concepts, Principles and Evidence. *Philosophical Transactions of the Royal Society B: Biological Sciences*, *363*(1491), 447–465. https://doi.org/10.1098/rstb.2007.2163

Rahman, Z. U., Asaari, M. S. M., Ibrahim, H., Abidin, I. S. Z., & Ishak, M. K. (2024). Generative Adversarial Networks (GANs) for Image Augmentation in Farming: A Review. *IEEE Access*, *12*, 179912–179943. https://doi.org/10.1109/ACCESS.2024.3505989

Rai, K. K. (2022). Integrating Speed Breeding with Artificial Intelligence for Developing Climate-Smart Crops. *Molecular Biology Reports*, *49*(12), 11385–11402. https://doi.org/10.1007/s11033-022-07769-4

Raiaan, M. A. K., Mukta, Md. S. H., Fatema, K., Fahad, N. M., Sakib, S., Mim, M. M. J., Ahmad, J., Ali, M. E., & Azam, S. (2024). A Review on Large Language Models: Architectures, Applications, Taxonomies, Open Issues and Challenges. *IEEE Access*, *12*, 26839–26874. https://doi.org/10.1109/ACCESS.2024.3365742

Rane, J., Kaya, Ö., Mallick, S. K., & Rane, N. L. (2024). *Smart farming using artificial intelligence, machine learning, deep learning, and ChatGPT: Applications, opportunities, challenges, and future directions*. Deep Science Publishing. https://doi.org/10.70593/978-81-981271-7-4

Richards, D., Worden, D., Song, X. P., & Lavorel, S. (2024). Harnessing Generative Artificial Intelligence to Support Nature-Based Solutions. *People and Nature*, *6*(2), 882–893. https://doi.org/10.1002/pan3.10622

Rodríguez, J. P., Brotons, L., Bustamante, J., & Seoane, J. (2007). The Application of Predictive Modelling of Species Distribution to Biodiversity Conservation. *Diversity and Distributions*, *13*(3), 243–251. https://www.jstor.org/stable/4539918

Roy, K., Sim, D., Wang, L., Zhang, Z., Guo, X., Zhu, Y., Swarup, S., & Lee, C. (2025). A Generative AI-Assisted Piezo-MEMS Ultrasound Device for Plant Dehydration Monitoring. *Advanced Science*, e04954. https://doi.org/10.1002/advs.202504954

Sai, S., Kumar, S., Gaur, A., Goyal, S., Chamola, V., & Hussain, A. (2025). Unleashing the Power of Generative AI in Agriculture 4.0 for Smart and Sustainable Farming. *Cognitive Computation*, *17*(1), 63. https://doi.org/10.1007/s12559-025-10420-6

Sajitha, P., Diana Andrushia, A., Mostafa, N., Younes Shdefat, A., Suni, S. S., & Anand, N. (2023). Smart Farming Application Using Knowledge Embedded-Graph Convolutional Neural Network (KEGCNN) for Banana Quality Detection. *Journal of Agriculture and Food Research*, *14*, 100767. https://doi.org/10.1016/j.jafr.2023.100767

San, C. T., & Kakani, V. (2025). Smart Precision Weeding in Agriculture Using 5IR Technologies. *Electronics*, *14*(13), Article 13. https://doi.org/10.3390/electronics14132517

Satpathy, I., Nayak, A., & Khang, A. (2025). Transformative Impact of Generative Artificial Intelligence (Gen AI) on Smart Transportation System. In A. Khang (Ed.), *Driving Green Transportation System Through Artificial Intelligence and Automation: Approaches, Technologies and Applications* (pp. 563–579). Springer Nature Switzerland. https://doi.org/10.1007/978-3-031-72617-0_29

Sengar, S. S., Hasan, A. B., Kumar, S., & Carroll, F. (2025). Generative Artificial Intelligence: A Systematic Review and Applications. *Multimedia Tools and Applications*, *84*(21), 23661–23700. https://doi.org/10.1007/s11042-024-20016-1

Senoo, E. E. K., Anggraini, L., Kumi, J. A., Karolina, L. B., Akansah, E., Sulyman, H. A., Mendonça, I., & Aritsugi, M. (2024). IoT Solutions with Artificial Intelligence Technologies for Precision Agriculture: Definitions, Applications, Challenges, and Opportunities. *Electronics*, *13*(10), Article 10. https://doi.org/10.3390/electronics13101894

Senthil, G. A., Suganthi, S. U., Prinslin, L., Selvi, R., & Prabha, R. (2025). Generative AI in Agri: Sustainability in Smart Precision Farming Yield Prediction Mapping System Based on GIS Using Deep Learning and GPS. *Procedia Computer Science*, *252*, 365–380. https://doi.org/10.1016/j.procs.2024.12.038

Shamshiri, R. R., Sturm, B., Weltzien, C., Fulton, J., Khosla, R., Schirrmann, M., Raut, S., Basavegowda, D. H., Yamin, M., & Hameed, I. A. (2024). Digitalization of Agriculture for Sustainable Crop Production: A Use-Case Review. *Frontiers in Environmental Science*, *12*. https://doi.org/10.3389/fenvs.2024.1375193

Sharma, V., Vaddevolu, U. B. P., Bhambota, S., Ampatzidis, Y., Bayabil, H., & Singh, A. (2025). Variable Rate Technology and Its Application in Precision Agriculture: AE607, 1/2025. *EDIS*, *2025*(1), Article 1. https://doi.org/10.32473/edis-ae607-2025

Sidhu, K. S., Gill, A. S., Arora, A., Singh, R., Singh, G., Verma, M. K., & Kaur, B. (2021). Advancements in Farming and Related Activities with the Help of Artificial Intelligence: A Review. *Environment Conservation Journal*, *22*(SE), 55–62. https://doi.org/10.36953/ECJ.2021.SE.2206

Sikander, A. (2024). Artificial Intelligence and the Circular Economy: How AI Advances Waste Reduction. *Green Environmental Technology*, *1*(2), 23–34. https://www.researchcorridor.org/index.php/get/article/view/214

Singh, G., & Sharma, S. (2025). Enhancing Precision Agriculture Through Cloud Based Transformative Crop Recommendation Model. *Scientific Reports*, *15*(1), 9138. https://doi.org/10.1038/s41598-025-93417-3

Singh, K. U., Kumar, A., Raja, L., Kumar, V., Singh Kushwaha, A. K., Vashney, N., & Chhetri, M. (2022). An Artificial Neural Network-Based Pest Identification and Control in Smart Agriculture Using Wireless Sensor Networks. *Journal of Food Quality*, *2022*(1), 5801206. https://doi.org/10.1155/2022/5801206

Surathunmanun, S., Ongsakul, W., & Singh, J. G. (2024). Exploring the Role of Generative Artificial Intelligence in the Energy Sector: A Comprehensive Literature Review. *2024 International Conference on Sustainable Energy: Energy Transition and Net-Zero Climate Future (ICUE)*, 1–11. https://doi.org/10.1109/ICUE63019.2024.10795598

Thies, B., & Bendix, J. (2011). Satellite Based Remote Sensing of Weather and Climate: Recent Achievements and Future Perspectives. *Meteorological Applications*, *18*(3), 262–295. https://doi.org/10.1002/met.288

Ukoba, K., Onisuru, O. R., Jen, T.-C., Madyira, D. M., & Olatunji, K. O. (2025). Predictive Modeling of Climate Change Impacts Using Artificial Intelligence: A Review for equitable Governance and Sustainable Outcome. *Environmental Science and Pollution Research International*, *32*(17), 10705–10724. https://doi.org/10.1007/s11356-025-36356-w

Vaswani, A., Shazeer, N., Parmar, N., Uszkoreit, J., Jones, L., Gomez, A. N., Kaiser, L., & Polosukhin, I. (2023). Attention Is All You Need (No. arXiv:1706.03762). *arXiv*. https://doi.org/10.48550/arXiv.1706.03762

Vijayasuganthi, K., Sudharson, K., Janaki, L., SureshKumar, A., Devi, K. K., & Mathiyalagan, P. (2025). Management Practices for Sustainable Agriculture in the Age of Smart Farming. *2025 IEEE International Students' Conference on Electrical, Electronics and Computer Science (SCEECS)*, 1–7. https://doi.org/10.1109/SCEECS64059.2025.10941589

Wang, P., Zhang, L.-Y., Tzachor, A., & Chen, W.-Q. (2024). E-waste Challenges of Generative Artificial Intelligence. *Nature Computational Science*, *4*(11), 818–823. https://doi.org/10.1038/s43588-024-00712-6

Weraikat, D., Šorič, K., Žagar, M., & Sokač, M. (2024). Data Analytics in Agriculture: Enhancing Decision-Making for Crop Yield Optimization and Sustainable Practices. *Sustainability*, *16*(17), Article 17. https://doi.org/10.3390/su16177331

Xie, X., Zuo, J., Xie, B., Dooling, T. A., & Mohanarajah, S. (2021). Bayesian Network Reasoning and Machine Learning with Multiple Data Features: Air Pollution Risk Monitoring and Early Warning. *Natural Hazards*, *107*(3), 2555–2572. https://doi.org/10.1007/s11069-021-04504-3

Xu, H., Omitaomu, F., Sabri, S., Zlatanova, S., Li, X., & Song, Y. (2024). Leveraging Generative AI for Urban Digital Twins: A Scoping Review on the Autonomous Generation of Urban Data, Scenarios, Designs, and 3D City Models for Smart City Advancement. *Urban Informatics*, *3*(1), 29. https://doi.org/10.1007/s44212-024-00060-w

Yan, B., Li, K., Xu, M., Dong, Y., Zhang, Y., Ren, Z., & Cheng, X. (2025). On Protecting the Data Privacy of Large Language Models (LLMs) and LLM Agents: A Literature Review. *High-Confidence Computing*, *5*(2), 100300. https://doi.org/10.1016/j.hcc.2025.100300

Yevle, D. V., & Mann, P. S. (2025). Artificial Intelligence-Based Waste Management: A Review of Classification, Techniques, Issues, and Challenges. *WIREs Data Mining and Knowledge Discovery*, *15*(2), e70025. https://doi.org/10.1002/widm.70025

Yin, J., Li, W., Shen, J., Zhou, C., Li, S., Suo, J., Yang, J., Jia, R., & Lv, C. (2025). A Diffusion-Based Detection Model for Accurate Soybean Disease Identification in Smart Agricultural Environments. *Plants*, *14*(5), Article 5. https://doi.org/10.3390/plants14050675

Zhang, J., Song, W., & Liu, Y. (2025). Cognitive Bias in Generative AI Influences Religious Education. *Scientific Reports*, *15*(1), 15720. https://doi.org/10.1038/s41598-025-99121-6

Zhu, W., Han, R., Shang, X., Zhou, T., Liang, C., Qin, X., Chen, H., Feng, Z., Zhang, H., Fan, X., Li, W., & Li, L. (2024). The CropGPT project: Call for a Global, Coordinated Effort in Precision Design Breeding Driven by AI Using Biological Big Data. *Molecular Plant*, *17*(2), 215–218. https://doi.org/10.1016/j.molp.2023.12.015

Zhu, W., Li, W., Zhang, H., & Li, L. (2025). Big Data and Artificial Intelligence-Aided Crop Breeding: Progress and Prospects. *Journal of Integrative Plant Biology*, *67*(3), 722–739. https://doi.org/10.1111/jipb.13791

Zhu, Y., Li, H., Wang, Z., Li, Q., Dou, Z., Xie, W., Zhang, Z., Wang, R., & Nie, W. (2022). Optimal Evacuation Route Planning of Urban Personnel at Different Risk Levels of Flood Disasters Based on the Improved 3D Dijkstra's Algorithm. *Sustainability*, *14*(16), 10250. https://doi.org/10.3390/su141610250

16 Future Trends and Innovations in AI and Remote Sensing

Rupinder Singh, Jarnail Singh, Amanpreet Singh, and Jaswinder Singh

16.1 INTRODUCTION

16.1.1 Global Environmental Imperatives and the Drive for Sustainability

The world is subject to unprecedented environmental pressures that are driven by population growth, industrialization, and climate change. It is estimated that the world will grow at a phenomenal rate, with the demand for food growing by up to 56 % by 2050. This population transformation requires a radical change in the way we use our planet's resources to make the most output with the least amount of waste and damage to the planet (Akkem et al., 2025). Traditional environmental management practices are often unable to cope with these increasingly rising demands because of their complex nature and the factors that are beyond their control. Among such pressures, there are the massive impact of climate change, decreasing supply of such valuable resources as water and fertile soils, and the distribution of attacks of pests and plant diseases. When manifested beyond the agricultural sector, these pressures are experienced in the accelerated loss of biodiversity, strained water systems, damaged ecosystems, and increased pollution in urban centers (Benos et al., 2021).

This situation raises a grave issue: the interrelation between human prosperity and environmental degradation. The quest to extract additional food production in the conventional and typically resource-rich nature of agriculture may often be the cause of environmental issues. The resource-intensive activities that have led to the environmental challenges have been the basis of such economic growth. Conversely, environmental degradation, that is, soil depletion or severe weather conditions as a result of climate change, has direct effects on human well-being and economic stability. It introduces a paradoxical situation where the traditional solutions are likely to introduce a tradeoff between environmental health and productivity. The utilization of AI in the environmental industry is not merely a way to rationalize the current procedures, but it may turn out to be one of the potentially promising ways toward the achievement of these rather contradicting goals in a world in which the development should not contradict the ecological stability but rather coexist with it. This

 DOI: 10.1201/9781003616207-19

creates the pretext concerning AI as a significant technology that can lead to a new era of sustainability in which efficiency and environmental friendliness could become a reality with the assistance of technological innovation (Bawa et al., 2026).

16.1.2 The Foundational Role of Crop Rotation in Sustainable Agriculture

Crop rotation is a systematic activity of cultivating various crops on the same piece of agricultural land every year or on a similar piece of agricultural land in various seasons. The method is one of the fundamental permaculture principles, which makes a garden, as well as an entire field, much healthier and productive. Unlike monoculture (the continuous use of agricultural land as producer of a single crop, which often leads to serious degradation of agricultural lands and an increased risk of pests and diseases), crop rotation provides new habitat diversity. It is used to disrupt the crop cycles of the pests, diseases, and the weeds to promote soil fertility and sustainability in agriculture as a whole (Turchetta et al., 2024), and historical crop rotations have been seen as one of the oldest and most practical means of improving soil health and minimizing the use of synthetic chemical fertilizers and pesticides (Turchetta et al., 2024).

As we can show from its principles, crop rotation is considered a biological form of sustainability. The soil health, pest management, and decreased chemical inputs play a strong part in why this practice is being recognized as a biological solution, working with nature's cycles instead of against them. In the pursuit of long-term sustainability, the approach uses intrinsic ecology extensively. This is only reinforced by the comparison with the generally linked monoculture and its overexploitation of chemicals, which causes soils to be degraded. It comprises the creation of a more enhanced soil system with a diverse root system and biological partnerships, which is federally important for the intrinsic soil resistance (Chellaswamy et al., 2020).

16.1.3 The Dawn of Artificial Intelligence in Agricultural Transformation

Artificial intelligence (AI) is changing the face of environmental management through site-specific and data-driven decision-making using information made available by devices, satellites, drones, sensors, and weather stations. In contrast to the traditional methods that relied on generic and historic recommendations, AI takes volatile factors such as soil heterogeneity, weather, crippling, illnesses, industry needs, and farmer inclinations into account and makes very specific recommendations. Due to this change, the AI in the environmental technology market has been exploding; for example, the agricultural AI market alone had an estimated value of 2.14 billion dollars as of May 2021, where the projected annual growth rate is 23.06%, and the projected value in 2035 is 20.96 billion dollars. The flexibility in planning generated by the AI, offering hyper-personalized plans, will allow for the translation of scientific knowledge into real and sustainable practices that can be applied on a much larger, smoother, and more comprehensive scale than current advanced methods (Chen et al., 2018).

16.1.4 Objectives, Scope, and Structure of the Chapter

The purpose of this chapter is to cover in detail the various aspects of how AI and machine learning (ML) technologies can be used in a broad range of environmental solutions to make human activities more productive and less environmentally demanding. The topics that are covered are the principles of sustainable practices, the background concepts of ML, and their diverse use cases in agriculture, climate monitoring, water management, biodiversity conservation, and urban planning. The chapter also reviews calculable advantages, practical case stories, existing limitations, and further research. The chapter is organized as follows: Section 16.2 outlines traditional rotational crops; Section 16.3 defines the basics of ML; Section 16.4 addresses the AI tools and benefits; Section 16.5 provides case studies; Section 16.6 addresses difficulties; Section 16.7 provides the graph of the future scope; and Section 16.8 concludes the paper.

16.2 TRADITIONAL CROP ROTATION: PRINCIPLES, BENEFITS, AND INHERENT LIMITATIONS

16.2.1 Core Principles and Practices of Sustainable Crop Rotation

Fundamentally, crop rotation aims to transform systematically the nature of crops grown in a particular field to curb soil erosion and check the spread of pests and diseases. It differs from monoculture in that this agricultural practice involves planting crops in a particular piece of land in one season or year, rather than in one planting.

- **Nutrient Cycling and Soil Health Enhancement:** Sustainable crop rotation is essentially based on alternate use of crops like heavy feeders, such as nightshades (tomatoes, peppers, eggplant), and light feeders and legumes (beans, peas, clover). The strategic sequencing of legumes is used to take advantage of the special relationship legume plants have with rhizobia bacteria, which colonize the roots of a legume plant and convert atmospheric nitrogen to ammonia available to the legume plants. As the root systems of legumes break down, they return this fixed nitrogen to the soil and provide a natural, renewable source of nutrients to future crops (Kalichkin et al., 2025).
- **Integrated Pest and Disease Management:** Susceptibility to common pests and diseases tends to be shared between members of the same plant family, for example, legumes, brassicas, or nightshades. The pest and pathogen mechanisms with systematic crop rotation are interrupted because they are trapped in soils, and pathogens will not get their preferred hosts, hence decreasing transmission through soil and reducing the use of chemical pesticides. This strategy is not only advantageous toward Integrated Pest Management (IPM), but it also offers environmental and economic benefits. Crop rotation is likewise an adaptive crop health strategy by preventing the increase of pests, disease, and weeds from taking over because they cannot adapt and their changes will not be able to arise to fit the environment. This is done through constant variation of the cropping environment. This kind of biodiversity-based prevention includes herbicide-resistant weeds

and reduces chemical inputs, and is long-term sustainable (Botero-Valencia et al., 2025).

- **Weed Control and Biodiversity Promotion:** Crop rotation provides added flexibility in plant defense because less of the chemicals is required to control the weeds. It benefits the entire ecosystem services as it cultivates a diverse range of crops, which include legumes, cereals, and plants that grow deeply. This results in increased nutrient cycling, natural control of pests and pollination, and storage of carbon. Thus, the system becomes more stable in the long term. Crop rotation not only offers short-term solutions but also makes farms more resilient to such an extent that they are more able to cope with climate shocks and disease outbreaks. This is in line with the ideas of agroecology, which aims to create stability in natural systems.

16.2.2 Established Benefits of Implementing Crop Rotation

- The implication of wise crop rotation incorporates a variety of indirect advantages in the context of promoting agricultural sustainability and farm productiveness: it is also the fact that the soil is good and fertile due to the crop rotation in that organic matter, microorganisms, water retention capability, and equilibrium of the nutrients of the soil are improved, and these are significant in the health of the soil.
- Natural Pest and Disease Management: Altering crops disrupts the natural cycle of pests and diseases, and thereby it takes fewer chemical pesticides.
- Weed Management: Diversity royally checks weeds, reducing the chances of developing any resistance to herbicides and the usage of chemical herbicides as a result.
- Greater Bio-diversity: Diversity in crops improves nature helpers in the ecosystem, such as pollination and natural pest control, so farms are more resilient.
- Water Conservation: Rotating, especially deep-rooted crops, improves the efficiency of water utilization and reduces erosion.

16.3 ENVIRONMENTAL OPTIMIZATION USING MACHINE LEARNING

16.3.1 Machine Learning and Deep Learning Environment Science Paradigms

AI has been a game-changer in the environmental sector, introducing many features, with ML being the most important. This enables rural farmers to achieve precise predictions, location detection, and automated farming processes. ML helps process large and complex data, such as soil conditions, weather changes, indicators of crop health, and market trends, or similarly, ocean salinity, atmospheric carbon levels, wildlife migration patterns, and urban energy consumption bring out patterns people may overlook, making all predictions more accurate and the management of resources more intelligent (Dubrovin et al., 2024).

A more sophisticated ML method, deep learning (DL), founded on artificial neural networks, performs excellently in identifying complex patterns in vast amounts of data. As used in environmental science, DL has been particularly useful in diagnosing crop diseases, forecasting crop yields, and optimizing irrigation timetables, as well as classifying land cover from satellite imagery, monitoring deforestation, and predicting air quality in urban areas. Combining ML with agricultural Big Data and high-performance computing helps to create even more knowledge, transforming raw data into strategic, actionable results to pursue efficient and sustainable agriculture.

16.3.2 Machine Learning Models and Techniques for B. Crop Rotation Optimization

Some ML techniques can be effectively applied to optimize crop rotation, providing different sets of abilities to predict and make relevant decisions.

16.3.2.1 Supervised Learning Methods (e.g., Support Vector Machines, Random Forests, Neural Networks, Gradient Boosting, XGBoost)

Agricultural optimization uses supervised learning models that learn using labeled past data to make forecasts. Such models can include Support Vector Machines (SVMs), Random Forests (RF), or Artificial Neural Networks (ANNs), and can be trained with high amounts of data, possibly of past crop behavior, complicated soil properties, dynamic weather patterns, etc., with respect to related variables. With this kind of training, they can also foresee the harvest of crops and schedule more complicated crop rotation patterns (Hao et al., 2019). The models are used to seek complex relationships between other variables and project the performances of varying combinations of crops in a specific piece of fabric under various conditions. As an illustration of this, an ML would be capable of making predictions of corn production after soybeans based on historical data, including some conditions of soil and the predicted weather.

These algorithms have found application empirically in agricultural work. A comprehensive comparison of the output of six algorithms that are available to predict the crop rotation productivity, that is, Decision Tree (CART), RF, Bootstrap Aggregating (Bagging), Gradient Boosting, Extreme Gradient Boosting (XGBoost), and ANN, showed that ANN, Gradient Boosting, and XGBoost are the most powerful ones with the coefficient of determination (R2) ranging between 0.90 and 0.93. Among them, XGBoost performed better than them and obtained an R2 of 0.93, the Root Mean Square Error (RMSE) of 2.34, and the Mean Absolute Error (MAE) of 1.81.7. Moreover, an XGBoost has been known to be markedly more accurate, with an accuracy of 99.7 in predicting the most suitable crop to plant given the specific parameters of the soil, namely, the pH, temperature, level of humidity, and the quantity of nutrients (Meng et al., 2021). These empirical experiments attest to the effectiveness of ML tools and XGBoost in particular, in offering practical knowledge in agricultural planning rather than theorizing the models themselves, by offering field-scale decision-support services (Zhang et al., 2020).

16.3.2.2 Deep Learning Architectures (e.g., Convolutional Neural Networks, Long Short-Term Memory Networks)

DL involves artificial neural networks to find intricate patterns in massive datasets. The two architectures in agriculture are as follows.

Convolutional Neural Networks (CNNs) are particularly effective in analyzing images. They identify crop diseases based on photos of leaves, soil classification, and pictures of weeds and crops, as well as guidance on selecting varieties. Such models as VGG-16, ResNet, and AlexNet demonstrate high accuracy in the detection of leaf-disease (Konduri et al., 2020) Long short-term memory (LSTM) networks effectively handle time-series data, including weather patterns and crop growth cycles, enabling accurate yield prediction and resource planning." They fuel the prediction of crop yields and the well-being of the land to enhance the planning of resources. Collaboratively, CNNs and LSTMs extract previously obscured visual and temporal signals—sensing the early onset of disease based on slight leaf manifestation, and making projections on yields, by correlating soil, weather, and crop health data—all guesses that can now be made faster and more confidently (Tang et al., 2020).

16.3.3 Essential Data Requirements and Diverse Data Sources for AI Models

The efficacy and reliability of AI models designed for crop rotation planning are fundamentally contingent upon the quality, volume, and diversity of the input data (Berger et al., 2019). The basis of the yield prediction models is data; the more accurate and detailed the data, the higher is the strength and reliability of the resultant predictions.

16.3.3.1 Detailed Soil (Nutrient Content, pH Levels, Texture, Moisture)

Although soil analysis is the most important in the field of precision farming, the concept is applicable in other fields, such as geospatial data. This may involve seabed composition to examine marine ecosystems, urban soil sealing data to plan a hydrological approach, or landslide risk assessment by geological survey. The datasets that are available are the labeled images that are classified as specific soil types, which are Alluvial, Clayey, Laterite, Loamy, Sandy Loam, and Sandy Soil, and which are automatically connected to the soil texture. Moreover, satellite observations, including those of NASA SMAP (Soil Moisture Active Passive), provide moisture throughout the world.

There are also labeled images that can be categorized into soil types, like Alluvial, Clayey, Laterite, Loamy, Sandy Loam, and Sandy Soil, which are inherently correlated with soil texture. Additionally, satellite-based data, like the one offered by NASA SMAP provide data regarding moisture on an Earth-scale. Although this remote sensing information is not the usual visual images, it is of great use in DL applications, especially when it comes to moisture mapping and prediction at a large scale. Also, it is essential to state that it has been difficult to predict soil chemical composition based on typical visual photographs taken in the RGB color system; to conduct accurate prediction, the application of hyperspecialized imageries or spectroscopes is needed, which offers more comprehensive chemical data that cannot be identified by the human eye (Balaska et al., 2023).

The consideration of the multi-dimensionality of the soil data is important. Physical (soil texture, soil moisture), chemical (soil pH, soil nutrients), and biological (microbial activity, soil organic matter) aspects make up the soil. To obtain this multidimensionality properly, the combination of the traditional methods of soil sampling, ground sensors, and the sophisticated methods of remote sensing is required. The fact that forecasting nutrients based on RGB pictures is challenging also highlights the need to use specific sensors, for example, hyperspectral cameras, to collect the needed chemical data.

16.3.3.2 Moving Data on Weather and Climate

The artificial intelligence systems are coded to synthesize both actual-time weather forecasts and historical weather data with specific system requirements, either crop water requirement or city storm-drain capacity, to generate highly precise and dynamic irrigation plans. The analysis of weather data is also significant in order to forecast the conditions that are favorable to the growth of plant pathogens, along with the forecasting of the emergence of extreme weather conditions that can influence the health and crop harvest. Climate risk analysis plays an important role, as it is AI-driven and allows identifying the most suitable crops and the most suitable rotation pattern to reduce the susceptibility of agricultural systems to climatic risks (Ben Ayed & Hanana, 2021).

The variability of environmental inputs, especially weather and climate data, brings about a lot of variability in the agricultural systems. The ability of AI to analyze real-time information, as well as to forecast in advance, is a requirement of adaptive management, allowing one to move past historical averages. This is in direct response to the problem of "climatic variability," which may lead to changes in yield, and hence traditional forecasting would be challenging. This functionality highlights the importance of AI in creating agriculture that is resistant to climate change to enable farmers to decide the date to plant their crops, the types of crops to use, and the allocation of resources based on altered climate patterns.

16.3.3.3 Past Crop Performance and Yield Data

The data of past crops is an essential base to train ML models to predict future crop yields accurately and optimize the complicated crop rotation patterns. Although yield history data give an approximation level, its univariate use is restricted, given the fact that changes in weather, soil status, and pest pressure may make such a procedure unreliable. In turn, AI systems are created to assess information about the previous yields, as well as the most recent environmental and crop health information, and produce firmer and more precise predictions regarding the harvest outcomes (Bhatnagar and Chandra, 2020).

An essential principle in this regard is learning the lessons of the past and adjusting to the new future. The past yield data provides important trends and a background to ML model. Nevertheless, it is only useful when combined with real-time dynamic environmental data because the past performance does not ensure the future in the ever-changing agricultural systems. The power of AI is exactly in the fact that it will unite these inert historical patterns with dynamic real-time inputs and will, therefore, be able to make predictions stronger and more reliable. This explains the cyclical learning of AI models, in which past data is used as a basis. Nevertheless, with a constant inflow of real-time data, one can refine and adapt and make predictions, which will become more accurate.

16.3.3.4 Remote Sensing and Satellite Imagery (e.g., Sentinel-2, MODIS, SMAP)

Remote sensing (e.g., Sentinel-2, MODIS), satellite-based and drone (e.g., Sentinel-2), allows scalable and multi-scale monitoring of crop health, soil moisture, deforestation, glacial retreat, and urban sprawl with the assistance of AI. By integrating satellite (macro), drone (meso), and ground sensor (micro) data, the AI will be able to distinguish between macro trends and micro issues and make timely interventions, choose more favorable management tools, and climate-sensitive decisions on a larger geographical scale.

16.3.3.5 Internet of Things (IoT) Sensor Data

The IoT devices are also important in this case as they constantly monitor the soil moisture level, which is directly related to reducing overwatering and saving valuable water resources. These sensors will also capture real-time information about the state of affairs of the general soil, weather, and health specifics of crops, which is essential data to input into AI models. On the one hand, there are systems of interconnected sensors constantly measuring dozens of parameters not only because they determine the overall health of the farm, but also because the indicators can allow determining whether it is vulnerable to insect pests and vermin, whether it is agile enough, whether the equipment it uses is working properly even whether their farm animals are healthy or not devices are not only on the farm and the sensor ring has become a standard practice, but also everywhere on its territory (Du et al., 2018). The ML algorithms can then process this flow of information to identify any anomalies to be solved as soon as possible, identify new trends that indicate something has gone wrong, automatically respond to varying circumstances, draft working recommendations that can be transmitted to human operators, and make continual improvements to the feedback loop system performance.

The IoT sensors facilitate dynamic intervention as the sensors are real-time and granular. Such sensors are capable of delivering an unlimited supply of extremely localized information and allow AI systems to detect slight changes and ambiguity before they manifest as issues. This will facilitate proactive, systematic, and precise responses as exemplified by targeted irrigation, which is a great innovation compared to the regular manual appraisals that are done after intervals. The synergistic capability of big remote sensing data and the granular IoT data provides a powerful, multi-layered monitoring framework that will streamline the utilization of resources and reduce wastage (Du et al., 2022).

16.3.3.6 Market Dynamics and Economic Indicators

AI algorithms allow considering dynamic market needs and particular farmers' preferences in a sophisticated process of crop rotation planning. It is possible to compute the most desirable month to collect diverse crops with the help of ML models that analyze past prices, season performance, and other aspects that relate to the market (Ennouri et al., 2021). Proper prediction of the yield of crops with the assistance of these AI tools also leads to more accurate market prediction, which in turn helps to stabilize food prices and avoid their sudden changes. The agribusiness, farmers, and consumers are some of the stakeholders who benefit most from the stability.

The agribusiness and agronomy merge is one of the merits. Planning crop rotation has been customarily a choice and not an agronomically oriented approach. However, the most effective use of real-time market data enabled by AI transforms it into a holistic agribusiness decision that is able not only to maximize the biological yield but also the profitability as a whole and adaptability to the market changes (Tran et al., 2022). This change transcends the biological maximization to the economic one. This integration also gives farmers a stronger hand to make better strategic decisions, not only in planning their production but also in selecting sources and determining the harvesting time, which reduces the risk of financial losses and hence supply-demand management in the food supply chain as a whole.

16.3.4 Data Preprocessing and Feature Engineering for Robust Agricultural Datasets

High-quality, quantity diversity is the key to the quality and reliability of the predictions generated by AI models. Agricultural data present significant challenges; they are heterogeneous, missing, error-prone, often incomplete and do not conform to any uniform standard; they are sometimes irregular and unpredictable, which can introduce regular yield fluctuations and complicates the prediction; and, even when measuring the same thing, they are often heterogeneous, even variance in yield and they cannot easily be combined to manage the production of a variety of sources, such as soil sensors (Michishita et al., 2014).

As such, specific preprocessing procedures are necessary to ensure raw farmland-based data is set up to work with ML algorithms. With image-based data, this typically means resizing and normalization, along with image augmentation and noise reduction, to offer consistency and quality specifications. The process of feature engineering is no less important either, and the challenge is extracting and converting the most useful characteristics of the raw data set into the features that can be presented to maximize the model's abilities. A case of feature engineering features that have been used on datasets predicting crops is adding engineered input features, such as the pH of soil, temperature, humidity, and the concentration of particular nutrients to the model: efficient feature scaling and multivariate imputation by chained equations.

Even though sophisticated algorithms are sophisticated, they do have basic limitations in the exercise of their performance on high-quality input data. The lack of data issues that are increased by heterogeneity, information error, and the overall absence of standardization suggest that the agricultural sector can be underdeveloped on the solid data basis that highly sophisticated AI applications demand. This is the major bottleneck that must be broken before maxizing the enjoyment of the sophistication of the algorithms. That is why it is important to invest heavily in data infrastructure,standardize the data standards, and develop common data-sharing models across the agricultural ecosystem to make the most of AI (Son et al., 2014).

16.4 AI TOOLS FOR OPTIMIZING ENVIRONMENTAL SOLUTIONS: APPLICATIONS AND QUANTIFIABLE IMPACT

AI applications are radically changing the way environmental management is conducted beyond conventional ways to facilitate accuracy, efficiency, and sustainability in different fields. These tools are not only hypothetical but also demonstrated by quantifiable gains in essential performance indicators.

16.4.1 Artificial Intelligence (AI) Solutions to the Environment: Uses and Measurable Impact

AI and remote sensing are essential in the perception and reaction to climate change. ML algorithms can be used to process large amounts of satellite data, including Sentinel-2 and MODIS, to track the quality of important climate metrics such as polar ice cap depth, deforestation, and ocean surface temperatures with great accuracy. AI (deep learning) can identify trends in sophisticated weather information to enhance the precision of long-range weather predictions and extreme weather phenomena such as hurricanes and droughts. In addition, high-resolution climate simulations are being generated with the help of generative AI, enabling researchers to simulate the possible effects of various emissions scenarios and policy measures, and offering invaluable resources in the field of global climate negotiations and adaptation planning.

As seen in Figure 16.1, predictable and accurate yield, enabled by AI, will make farming management more resourceful (e.g., water, fertilizer) and pest control-oriented, which will increase environmental sustainability, cost reduction, and overall efficiency. It assists in informed decision-making when it comes to the selection of right crop varieties, sourcing and harvesting of these crops, and the reduction of risks and maximization of profits. Moreover, accurate yield forecasting can enable financial institutions and insurers to evaluate agricultural loans and crop insurance in order to customize their services to reduce risks to both the farmers and the banks. At a greater level, the factual plausibility of yields on regional and national levels can improve the food market forecasts and stabilize food value and prevent drastic fluctuations, which is beneficial to farmers, businesses, and consumers (Chen and Guestrin, 2016). Other data on historical yields include high plot/field detail, weather data history, soil quality measurements and composition, management practices and inputs, remote sensing vegetation indices, market, and economics.

Such transformation is associated with the substitution of the reactive mode of agricultural management with the proactive mode. The traditional methods of estimating yields using the manual sampling process and the subjective eye estimate are vulnerable to human error and usually ignore the natural heterogeneity of a field. Unlike AI-powered predictive analytics, disparate datasets can be incorporated and deployed with a complex model to come up with an accurate data-driven forecasting approach. The new paradigm will empower farmers toward a more proactive attitude

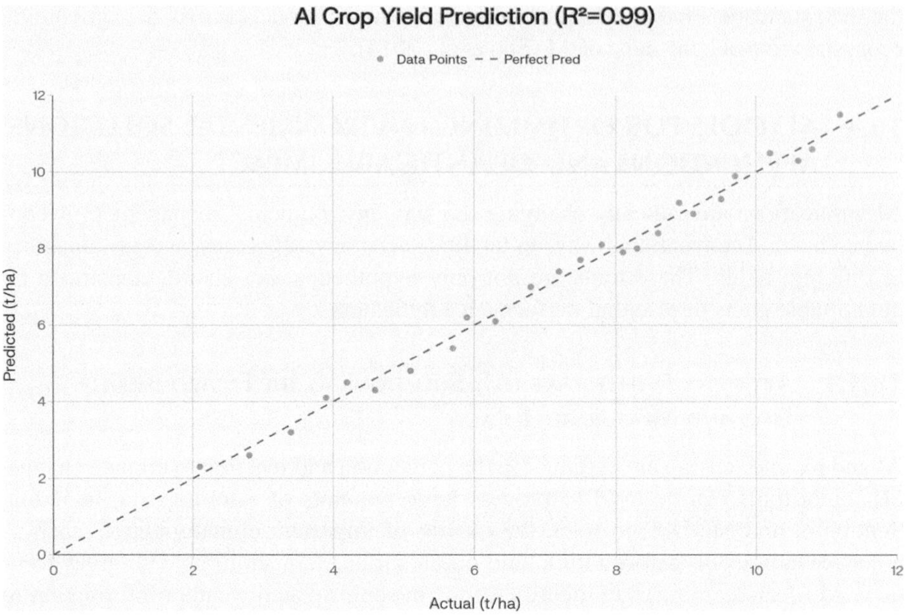

FIGURE 16.1 Predicted versus actual crop yields with AI.

in planning their agricultural value chain. This capability is necessary for the entirety of food security in the world as the more effective resource distribution will eliminate large volumes of unnecessary food and present more stable food markets (Stepanov et al., 2022).

16.4.2 Water Resource Assessment and Management

Water management is vital to the health and survival of the human population. Based on the satellite imagery and the data provided by the IoT sensors, AI systems examine the state of a particular reservoir, the depletion of groundwater, and the presence of the snowpack, which can allow forecasting water availability more precisely. In cities, AI can be used to construct water distribution networks to identify leakage and minimize waste. To protect the environment, ML algorithms can be used to operate a hyperspectral image of drone or satellite to determine the presence of chemical spills or algal growth within rivers and lakes to respond promptly during pollution incidents. This is a data-driven method that can be used to guarantee water security and conserve aquatic ecosystems.

Water Management: AI systems are deployed and combined to use the correct balance of weather forecasts, the content of soil moisture (as measured by the ground sensor and IoT), and the specific crop water needs. Indeed, with this detailed examination, they can generate precise and dynamic irrigation plans and successively eliminate overwatering, diminish soil erosion, and achieve 30% savings of water. Such platforms as Farmonaut apply further optimization of water consumption based on data provided by the satellites (Gardezi et al., 2024). Figure 16.2 shows this.

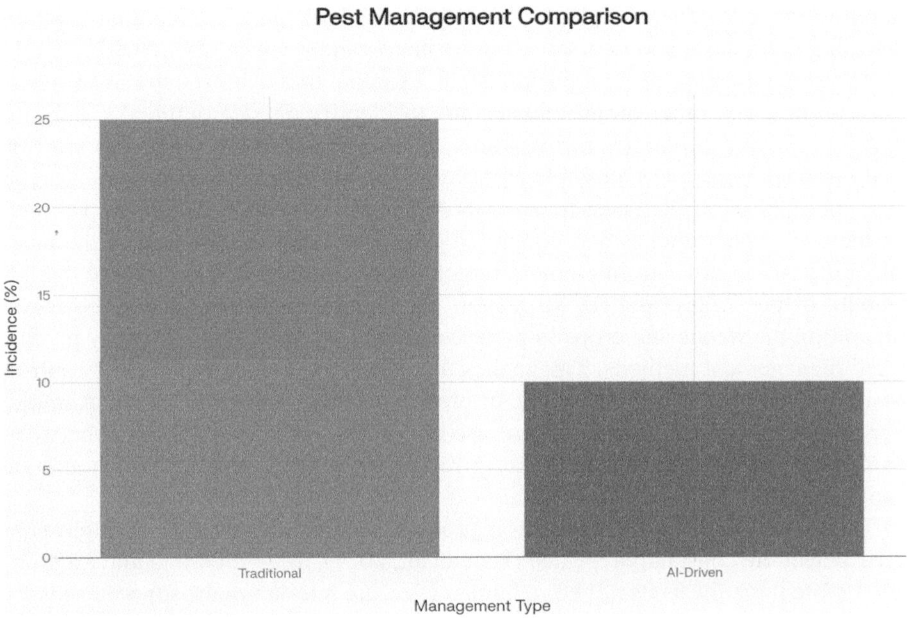

FIGURE 16.2 Reduction in water usage with AI-driven irrigation.

Fertilizer Application: Utilizing AI models to analyze the complicated nutrient profile of the soil and monitor crop development very closely. In this analysis, they propose the precise amounts of water and fertilizer needed, and their suggestions significantly reduce the nutrient runoff into waterways, decrease the amount of nutrients removed from the soil, and decrease agricultural emissions of greenhouse gases. It is an approach that will directly lower the input costs to farmers as well as minimize the environmental impact of the farming practice.

Application of Pesticides: AI-enabled precision agriculture can minimize the total environmental footprint of chemical pesticides, given the particular application patterns that such technologies make possible.

Appreciable co-benefits to the environment will be evident due to the AI-based optimization of the inputs. The constant emphasis in different studies on the optimization of water, fertilizers, and pesticides usage by AI shows that the optimization is not necessarily a process of cost-mongering by farmers (Gill et al., 2024).

16.4.3 Advanced Pest and Disease Surveillance and Early Detection

Pest and disease management is being revolutionized by the use of AI-powered image recognition systems, which are commonly operated through drones to identify diseases in plants and pest infestation in theirearly stages. This enables early measures that would greatly minimize the losses of crops. The current AI systems have the ability to detect minor differences in crop coloration, the shape of the leaf, and the growth and development patterns weeks before humans see them, which is a crucial time to intervene. There are high accuracy rates of AI models, including 95% on detecting apple scab, and in practice, the scale of yellow rust in wheat fields.

The sophistication of AI algorithms can examine massive data stored by satellites, drones, and other on-ground sensors to identify disease outbreaks, nutrient deficiencies, and anomalies in irrigation. In many cases, these diagnoses are carried out even before any possible symptoms can be seen by the human eye. The concept of AI-controlled pest traps, that is, the ones created by Trapview, is based on the use of pheromones and integrated cameras to detect and forecast the presence of pests in order to implement targeted interventions and reduce the overall amount of pesticides applied. Moreover, DL algorithms are capable of correctly separating the weeds and crops in pictures and help farmers deploy highly precise weed management methods. This makes the weeds more effective in their control, besides significantly lowering usage of broad-spectrum pesticides. As an example, See and Spray robots of John Deere use computer vision to distinguish between crops and weeds, which subsequently results in a 90 % decrease in herbicide application.

This is a high-tech protection that is proactive against crop health. Such terms as early detection, constantly repeated (Hara et al., 2021), the ability to identify a problem before it manifests itself before human eyes, are relevant to the superior sensory and analytical abilities of the AI over those of a human. This allows the active preventative control to the passive treatment, which is not as effective and, in some cases, consumes more chemical inputs. Another aspect through which the efficiency of this ability can be assessed is the precision of disease detection, which should be of high quality (e.g., in case of Apple Scab, 95% precision). The comparison of the pest management is checked in Figure 16.3. This is in direct relation to the philosophy of IPM since localized controls can be made with precision and reduce the instances where the broad-spectrum chemicals are deployed and the following occurrence of resistance, which is a major problem of the conventional crop rotation practices.

16.4.4 Sophisticated Off-the-Soil Smart Soil Health Observing and Recommendations Nutrient Management

AI is important to research the condition of the soil and crops, and to make recommendations based on the data, which assist in maintaining the management practice and eventually enhance yield. The AI systems assist farmers to make use of the instruments of conservation farming (minimal tilling, the use of cover crops, and organic inputs), however, according to real-time soil details and satellite photos. AI will be capable of providing the most optimal crop rotation depending on an in-depth investigation of soil structure, presence of pests and diseases, and past crop yields. This will minimize chances of disease and a greatly favoring the natural processes of managing the pests.

Besides, the DL algorithms would be able to scan crop pictures and help farmers identify particular nutrient deficiencies in the soil. By means of such visual analysis, more precise fertilizing plans can be created and eventually result in healthier soil and high-quality crop production (Nautiyal et al., 2025).

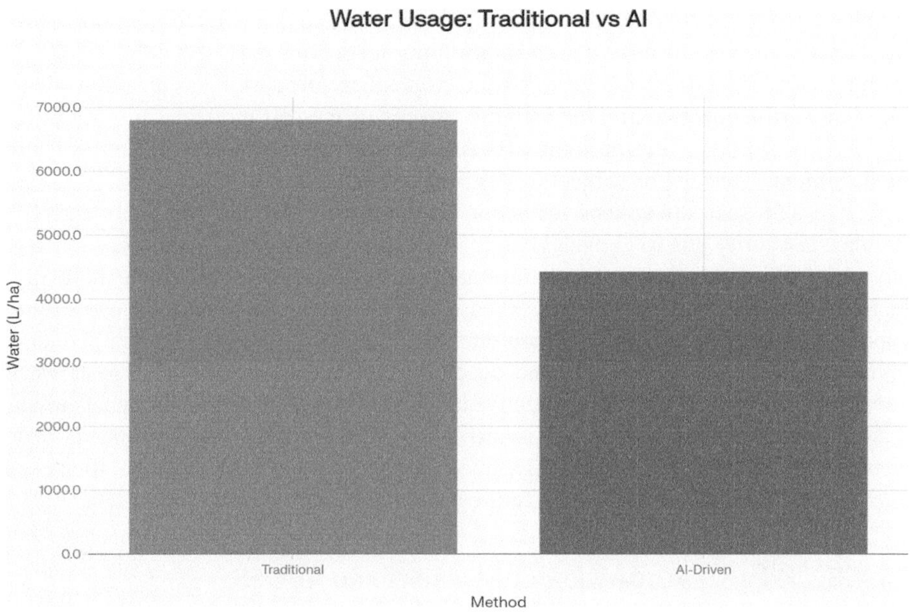

FIGURE 16.3 Impact of AI on pest and disease reduction.

The creation of a subterranean ecosystem was an important move. Besides monitoring, AI would provide implementable soil management and nutrient application suggestions. This marks the beginning of the end of passive, inactive guardianship of the soil itself, that is not a sterile medium upon which to culture plants, but an active and living ecosystem. This soil care method is further enhanced by the direct linkage with conservation agriculture, marketing of organic inputs, etc. This plays an important role in the sustainability of farming in the long term, where healthy soil is the foundation of viable and robust farming.

16.4.5 Climate Change Resilience, Adaptation, and Evolution of Cropping System

AI is a necessary technology in farming due to climate change. The AI that works on forming the most suitable crop and laying the most effective rotation patterns is founded on the complicated analysis of the climate risk factors. The AI systems are capable of analyzing complex weather forecasts, analyzing the prevailing conditions of the earth, and determining the reaction of the crop to weather stimuli. With such inclusive information, they will be able to prescribe such flexibility measures that will be oriented at containing the risks associated with the change of climate change. With the help of the statistical analysis of huge amounts of climate data, DL models are able to predict extreme weather events so that farmers can implement advanced, proactive, and adaptive methods. This is done through the adaptation of sowing seasons, planting crops more resistant to the conditions of agricultural growth, and better resource distribution based on the anticipated climatic changes (Kumar et al., 2023).

The data that is obtained through remote sensing and drone technology provides the farmer with useful information on the long-term climate aspects and the short-term impacts of weather conditions on the crops, therefore serving as an aid in adapting their farms. Specifically, the satellite images of the soil moisture are urgent for the early warning of the drought and simplifying the position of effective water management over regions subject to water stress.

It is a skill that strengthens resilience attributes in a turbulent environment. The weather insecurity due to climate change is an issue that does not give precedence to any of the agricultural systems and thus traditional-based planning would be more ineffective with time. This could be resolved by AI using the potential of its capacity to capitalize on complex climate systems and integrate it with live environmental data to allow proactive adaptation plans and develop intrinsic resilience in crop systems. It becomes a desperate shift from a reactive to a proactive reaction on the impacts of climate change. AI can thus be used to improve food security of the world, despite climate change that is likely to change rapidly, since farmers would understand what they need to make decisions depending on the future state of the environment (Khan et al., 2021).

16.4.6 Farming Automation of Machines and AI-DSS

The concept of the farming industry using AI-directed robots and autonomous vehicles will transform the manner in which farms are run, with the machines being able to carry out the duties of planting, weeding, and harvesting. This automation is not only rendering complete dependence on the manual labor force unnecessary but also boosting efficiency tremendously. These advanced machines are based on AI algorithms that combine with GPS and sensor files to produce excellent performance in a list of activities regarding the agricultural process. This is done in regards to the depth and adequate spacing of seeds, fertilizer in the appropriate locations, and appropriate, and harvesting at the appropriate maturity of crops. We can have the AI systems that will examine the real-time crop health and environmental data along with historic yield data, which will determine outcomes related to harvests, hence providing valuable planning information in terms of storage, logistics, and markets. Moreover, the conquest of AI by the decision support systems provides farmers with more data to make the most suitable decision of planting, fertilizing, and harvesting time because all available information is based on accurate weather and soil parameters. It is one of the inventions in human potential development and is a more effective operation. Machines imply that there are fewer manual laborers and more precision in repetitive tasks. This allows a man to channel his effort into more complicated scenarios of decisions and management. The other point is that AI is advantaged by offering specific and actionable information and directions in accordance with the location. On that account, AI is a complex decision support system, which supplements the intuitive knowledge of farmers with information intelligence. This boosts the overall working efficiency and minimizes the possibility of human error, which leads to improved utilization of the resources and a high yield (Khodjaev et al., 2025). Nevertheless, one should also speak about the fact that this development also produces concerns about the risk of unemployment (which will be examined in more detail in Section 16.6).

16.4.7 Economic Profitability and Return on Investment (ROI) AI Integration

AI approaches have since then recorded revolutionary developments in fundamental farming. Indicators—crop yield, pest control, and resource control. These inventions are all leading to high resiliency, high sustainability, and productivity of the farming systems (Kukar et al., 2019).

Beyond these general metrics, specific applications of AI demonstrate direct impacts on marketable yield and profitability, as shown in Table 16.1:

- **Crop Yield:** Over 50% of farmers who adopt AI tools report increased crop yields within the first two seasons of implementation. AI can boost overall crop productivity by approximately 20% to 150%.
- **Water Usage:** AI-driven irrigation systems can achieve water savings of up to 30%.
- **Pest Management Efficiency:** AI models can detect crop diseases with 95% accuracy.
- **Herbicide Use:** AI-driven robots, such as John Deere's See & Spray, can reduce herbicide use by up to 90%.
- **Harvest Timing Optimization:** Precision maturity monitoring enabled by AI can increase marketable yield by 15–25%.
- **Overall Marketable Yield:** Comparative field trials of the 2023 to 2024 growing seasons indicated that crops monitored by using professionally annotated AI systems had an average of 16.4% higher marketable harvest than crops monitored using generic monitoring systems. This variation translated to more than $175 per acre of extra revenue from high-value crop. The payback of professional annotation services that drive these AI systems, as illustrated in Figure 16.4, is calculable using measures like

TABLE 16.1
Comparison of Traditional and AI-Driven Farming Metrics

Farming Aspect	Traditional Approach (Estimated Values)	AI-Driven Approach (Estimated Values)	Improvement (AI vs. Traditional)
Crop Yield (tons/hectare)	2.8–3.5	3.8–5.2	Until 46% growth [24]
Water Usage (liters/hectare)	12,000–18,000	8,000–12,000	Less than 30% cut-off [24].
Pest Management Efficiency (% Reduction)	20%–30%	50%–70%	Up to 50% increase [24]
Resource Optimization (Estimated Savings)	Low—Moderate	High—Up to 30% reduction	Significant savings [24]
Time for Monitoring (hours/week)	10–15	2–4	Maximum 80% reduction

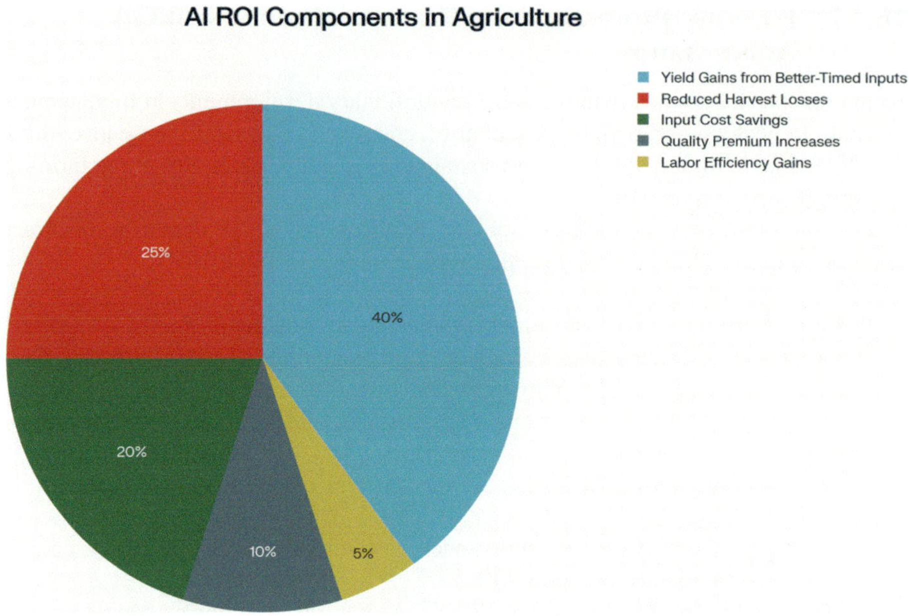

FIGURE 16.4 Economic return on investment (ROI) from AI integration.

bushels/tonnage saved by previous intervention at disease cycles, quality gains as a result of more precise harvest timing, gains in yields as a result of more precise timing of inputs at critical growth stages, and harvest losses as a result of more precise equipment scheduling.

These are quantitative measures that give a good, analytically supported case of the economic feasibility of AI in agriculture. The emphasis on the marketable yield and the extra profit directly connects the implementation of AI to the bottom line of the farmer. This is one of the main economic incentives behind the high-paced development of AI in the agriculture market, which implies that sustainability can be ensured not only by regulatory decisions or philanthropic intentions but also by economically reasonable judgments.

16.5 CASE STUDIES AND REAL-WORLD IMPLEMENTATIONS

The hypothetical advantages of AI in sustainable crop rotations are proven by the fact that the various real-life applications and platforms show calculable effects in various agricultural environments.

16.5.1 Crop Rotation AI Platforms and Solutions

A significant tendency is the development of agri-tech ecosystems, such as Sairone and Farmonaut. Sairone is a developed AI and ML service that is a specific service in the agricultural sector designed by Saiwa. It uses the technology of drones to offer extremely precise data about crop recognition and estimation of nitrogen levels. This is directly what helps farmers to maximize crop sequences and soil management based on accurate, data-driven decisions that respond to the realities of current-day agriculture.

Another innovative platform is Farmonaut, which combines multispectral satellite imagery, AI, and blockchain technology. The main aim of it is to provide practical recommendations, live tracking, and resource management to both small-scale and high-scale agribusinesses. Farmonaut democratizes information and makes it affordable and readily available on Android, iOS, and web applications through satellite-driven insights, and with an enormous Application Programming Interface (API) that allows them to fit into any third-party system in a seamless manner. They have devices built to perfect and track resources and land farming activities based on satellite information, even to measure carbon imprints.

These examples lead to an observation about increasingly integrated platforms that combine several technologies, such as AI, ML, drones, satellites, blockchain, and IoT, in order to present holistic solutions, rather than separated tools. Such a mixture seems to be essential to the uninterrupted flow and processing of the information, the demonstration of an overall interface, and the accessibility of such solutions, which are the main conditions of a successful scale and maximum implementation of the synergistic beneficial effects of AI in agriculture (Zhen et al., 2023).

16.5.2 Special Implications and Effects That Can Be Measured

Besides universal applications, they also have a variety of special applications to demonstrate the usefulness of AI in the sustainability of agriculture: Boni Rob (Autonomous Weeding Robot): The modern AI robot is a representation of an innovation since it identifies and removes weeds manually. Boni Rob applies ML scripts in order to distinguish between plants and weeds, making it possible to remove the crop successfully. The adoption of Boni Rob in farms has presented outstanding outcomes in terms of a potent decrease in the use of herbicides, and therefore serves as a direct reflection of the growing interest in sustainable farming across continents.

- Agrovech (AI-powered Drones to Monitor Crops): Aggrotech is a company leverages advanced drone-based technologies for monitoring the crops that grow in the field using drones, which are powered by AI technology. The technology of these drones is modern in terms of imaging and environmental measurements that yield data points that are vital in monitoring moisture levels, the health of the plant, and nutrient deficiencies. The innovation can measure crop health with unmatched accuracy and granularity, a level of detail that has never been achieved with conventional methods. The information gathered by such drones enables farmers to easily identify problems in the field, allowing them to make targeted interventions that prevent extensive crop disasters.
- Smart Farm Agri tech (AI to Optimize Resources in rice and palm oil): The firm has been at the top of agricultural technology regarding the integration

of AI in improving the use of resources when cultivating staples like rice and palm oil. They have come up with AI systems, which rely on collected data by sensors to determine how much water and fertilizer a particular area of a crop requires. In precision farming using AI in Smart farm agrotech, the company has realized a lot of optimization in the usage of resources and the production of crops. The technology has undergone over trial runs demonstrated itself to be able to cut down on water usage by up to 25%, at the same time hiking (or at least not reducing) crop yields. Such resource use precision also minimizes runoff and contamination, thus creating healthier ecosystems in farms. Organic Orchard (AI in Disease Detection and Decision Support).

- Organic Orchard is an established working system that practices sustainable agriculture and enjoys the leadership in this field but has implemented the use of AI technologies to optimize its agricultural processes, while also paying close attention to preserving the existing knowledge on the topic of farming. The major uses of AI in Organic Orchard are in disease detection and decision support. The farm uses sensors and imaging technologies that run on AI to detect the early signs of a sick plant that cannot be detected by human eyes. A system of this kind is able to come up with timely interventions and significantly reduce the need to utilize chemical treatment. The application of AI as a strategic tool in Organic Orchard has already had positive effects, including the decline of the dependency of the farm on the utilization of chemical fertilizers and pesticides, which eventually positively influenced the soil condition and reduced the impact of the business on the environment.
- Bihar Agricultural University (BAU) and Centre of Development of Advanced Computing (CDAC), India:In this context, the two institutions have joined forces to investigate the prospects of Agricultural Informatics and AI in precision farming in India. Their interests are in utilizing innovation to optimize farm produce by offering a stable agricultural ecosystem and therefore maximizing the incomes of farmers. This would redefine the farming practice through timely implementation, field testing, and capacity building exercises among the farmers to turn Bihar into a technology giant in Agri-tech in the country. All these case studies demonstrate the diversity of AI applications and their flexibility according to a situation. Their range of applications—automated weeding and years-long monitoring of crops to the optimization of resources and their ability to detect diseases early enough—is quite wide, and they can apply to a wide variety of crops (e.g., rice, palm oil), as well as different levels (individual farms to state-wide programs).

The example of Organic Orchard is particularly illuminating, as it showcases AI supporting and enhancing traditional farming wisdom rather than replacing it. This empirical evidence reinforces the practical viability and transformative potential of AI across the agricultural spectrum, providing valuable blueprints for future adoption and development (Suguna et al., 2024).

16.6 CHALLENGES AND LIMITATIONS OF AI IN SUSTAINABLE CROP ROTATIONS

Although AI in sustainable environmental management promises to become a powerful tool, it also has several major obstacles. Among these is the high cost of acquiring sensors, drones, and AI-acquired machinery, which can be prohibitive for smaller organizations, developing nations, and small and medium-sized farms, which form the majority. Such a financial divide threatens to lead to increased disparities between well-funded entities and smaller stakeholders unless the new business models, subsidies, or low-cost AI solutions are made available.

The other potential weakness is data quality, availability, and interoperability. Environmental data is often non-standardized, incomplete, or inconsistent, making it challenging to train trustworthy AI models. Combining the data of soil sensors, drones, and satellites is especially difficult, and the problem of ownership, privacy, and access adds further complications to sharing.

Technical skills and the black box problem are also barriers to AI adoption. A significant number of practitioners and land managers lack the expertise to apply AI, and the black-boxed AI predictions are often perceived as distrustful, especially when they contradict established field knowledge and instinct. It is even further complicated by integration with pre-existing systems and equipment, as legacy infrastructure largely incompatible with modern AI tools necessitates increasingly expensive upgrades or replacements, breaking previously established operations (Fedorova & Aseeva, 2022).

Lastly, socio-ethical problems are of great concern. Automation prompted by AI could jeopardize jobs in sectors like the farming sector since it might render low-skilled farmers disposable in the agricultural sector. There is a threat to privacy caused by the accumulation of large volumes of environmental and farm data, and even unrepresentative or biased datasets may result in black box outcomes that discriminate against some regions or communities. The mitigation of these problems must be based not only on technical solutions but also on policy actions, ethical guidelines, and multi-layered partnerships to advance AI in the environmental sector in a way that is both participatory and democratic, as well as fair and reliable.

16.7 FUTURE SCOPE AND EMERGING RESEARCH DIRECTIONS

The future of AI-based sustainable environmental science needs to focus on the development of new multi-modal systems and federated learning, mixing dissimilar data (e.g., images, weather, soil, genomics), with little cost to data privacy and little infrastructure overheads; expanding Explainable AI (XAI) methods (e.g., LIME, SHAP, Grad-CAM) to generate user trust and trust through transparent, human-understandable advice; and using generative AI in more powerful predictive modeling. In data-scarce settings, the generative models may produce widely realistic artificial datasets to be used to train more resilient AI systems. They can also produce future environmental conditions, for example,forecasting the effect of urbanization on the local biodiversity or the proliferation of wildfires in various climatic conditions, and offer powerful decision-making tools. One more important area is developing digital twins of ecosystems, for example,a rainforest or watershed, where AI will simulate complex interactions and experiment with strategies to manage an ecosystem in a virtual environment and test them before real-world implementation.

16.8 CONCLUSION

Using ML techniques such as XGBoost and DL, is a game changer for environmental management. Although this has been promising, there are still major barriers to adoption, such as high start-up costs, data quality challenges, technical expertise requirements, and the black box design of certain models. To overcome these issues, it is necessary to pay attention to future trends such as XAI to establish trust, research on new technologies such as generative AI, and develop intuitive instruments available to any community. Finally, the future of AI needs to continue being researched, invested in, and collaborate among many stakeholders to make the most of AI in a sustainable global future.

The chapter also introduces several fundamental ML algorithms, supervised learning algorithms, such as XGBoost or Random Forest, and DL-based algorithms, such as CNNs or LSTMs. It also presents the latest reinforcement learning models by digital twins that illustrate their profound applicability in the optimization of agriculture. It has been emphasized that different quality data presented through soil sensors, remote sensors, IoT machines, and historical data is a key factor, and efficient pre-processing of these data is necessary; the practical results of the AI application are more significant crops, less water, more efficient control of pests, and more optimized resources, which translate to the improved financial performance of the farms, and the investments payback can be evaluated. Such technologies are applicable in the farming environment of any kind, as examples such as autonomous weeding robots, AI-managed crop monitoring, and similar case studies are pointed out as real-life examples.

Although such an opportunity is so large, there are no problems with the large-scale adoption of AI in farms. Expensive capital entry, data quality, data access and data interoperability, technical skills that are specialized to use the technology, and the fact that it is impossible to explain how specific AI-based models operate (black box) are major challenges. Such issues as job displacement and ethical concerns concerning data privacy must also be treated very seriously. In the future, enhancing multi-modal AI systems or adopting the concept of federated learning should be considered to improve data integration and privacy. XAI has become a crucial requirement for gaining trust and adoption among farmers, as AI decisions will be transparent and understandable. Moreover, with the development of quantum computing and synthetic biology, even more complex optimization will occur in the future, and crops can be created that will help them in the future conditions of new environments. Most importantly, a conducive environment must be created by having policy frameworks, strengthening the development of digital infrastructure, and partnerships. Designing AI tools that simplify technology will be a priority in enabling more people, especially smallholders, to access life-changing tools.

To sum up, AI tools not only enhance the efficiency of an object but also drive a paradigm shift toward a more intelligent, sustainable, and efficient agricultural future. Further research, careful investment, and cooperation from all stakeholders should play a crucial role in overcoming current impediments and unlocking AI's full potential to ensure global food security without compromising our ability to leave the planet in a healthy state for future generations.

REFERENCES

Akkem, Y., Biswas, S. K., & Varanasi, A. (2025). Role of explainable AI in crop recommendation technique of smart farming. *International Journal of Intelligent Systems and Applications*, 17(1), 31–52.

Balaska, V., Adamidou, Z., Vryzas, Z., & Gasteratos, A. (2023). Sustainable crop protection via robotics and artificial intelligence solutions. *Machines*, 11(8), 774.

Bawa, S. S., Singh, H., & Singh, R. (2026). A study on how artificial intelligence is shaping. In G. Madaan et al. (Eds.), *Generative AI in food systems: Predictive demand, smart supply chains, and sustainable service futures* (pp. 67–76). Hershey, PA: IGI Global Scientific Publishing. https://doi.org/10.4018/979-8-3373-5182-7.ch005

Ben Ayed, R., & Hanana, M. (2021). Artificial intelligence to improve the food and agriculture sector. *Journal of Food Quality*, 2021, 1–7.

Benos, L., Tagarakis, A. C., Dolias, G., Berruto, R., Kateris, D., & Bochtis, D. (2021). Machine learning in agriculture: A comprehensive updated review. *Sensors*, 21, 3758. https://doi.org/10.3390/s21113758

Berger, A., Ettlin, G., Quincke, C., & Rodríguez-Bocca, P. (2019). Predicting the normalized difference vegetation index (NDVI) by training a crop growth model with historical data. *Computers and Electronics in Agriculture*, 161, 305–311.

Bhatnagar, V., & Chandra, R. (2020). IoT-based soil health monitoring and recommendation system. *Internet of Things and Analytics for Agriculture*, 2, 1–21.

Botero-Valencia, J., García-Pineda, V., Valencia-Arias, A., Valencia, J., Reyes-Vera, E., Mejia-Herrera, M., & Hernández-García, R. (2025). Machine learning in sustainable agriculture: Systematic review and research perspectives. *Agriculture*, 15, 377. https://doi.org/10.3390/agriculture15040377

Chellaswamy, C., Geetha, T. S., Venkatachalam, K., & Mulla, M. A. (2020, July). Soil test based smart agriculture management system. In *2020 7th International Conference on Smart Structures and Systems (ICSSS)* (pp. 1–6). IEEE.

Chen, T., & Guestrin, C. (2016). XGBoost: A scalable tree boosting system. In *Proceedings of the 22nd ACM SIGKDD International Conference on Knowledge Discovery and Data Mining* (pp. 785–794). ACM.

Chen, Y., Lu, D., Moran, E., Batistella, M., Dutra, L. V., da Silva, I., Huang, J., Luiz, A., & de Oliveira, M. (2018). Mapping croplands, cropping patterns, and crop types using MODIS time-series data. *International Journal of Applied Earth Observation and Geoinformation*, 69, 133–147.

Du, Y. D., Niu, W. Q., Gu, X. B., Zhang, Q., Cui, B. J., & Zhao, Y. (2018). Crop yield and water use efficiency under aerated irrigation: A meta-analysis. *Agricultural Water Management*, 210, 158–164.

Du, Z., Yang, L., Zhang, D., Cui, T., He, X., Xiao, T., … Li, H. (2022). Corn variable-rate seeding decision based on gradient boosting decision tree model. *Computers and Electronics in Agriculture*, 198, 107–118.

Dubrovin, K., Verkhoturov, A., Stepanov, A., & Aseeva, T. (2024). Multi-year cropland mapping based on remote sensing data: A case study for the Khabarovsk Territory, Russia. *Remote Sensing*, 16, 1633. https://doi.org/10.3390/rs16091633

Ennouri, K., Smaoui, S., Gharbi, Y., Cheffi, M., Ben Braiek, O., Ennouri, M., & Triki, M. A. (2021). Usage of artificial intelligence and remote sensing as efficient devices to increase agricultural system yields. *Journal of Food Quality*, 2021, 1–17.

Fedorova, T. N., & Aseeva, T. A. (2022). Changes in regional climatic characteristics of the Middle Amur Region and their impact on soybean yield. *Vestnik FEB RAS*, 3, 138–148.

Gardezi, M., Joshi, B., Rizzo, D. M., Ryan, M., Prutzer, E., Brugler, S., & Dadkhah, A. (2024). Artificial intelligence in farming: Challenges and opportunities for building trust. *Agronomy Journal*, 116(3), 1217–1228.

Gill, K. K., Bhatt, K., Akansha Setiya, P., Sandhu, S. S., & Kaur, B. (2024). Comparative evaluation of statistical and machine learning models for weather-driven wheat yield forecasting across different districts of Punjab. *Arabian Journal of Geosciences*, 17(10), 274.

Hao, P., Tang, H., Chen, Z., Yu, L., & Wu, M. (2019). High resolution crop intensity mapping using harmonized Landsat-8 and Sentinel-2 data. *Journal of Integrative Agriculture*, 18, 2883–2897.

Hara, P., Piekutowska, M., & Niedbała, G. (2021). Selection of independent variables for crop yield prediction using artificial neural network models with remote sensing data. *Land*, 10(6), 609.

Kalichkin, V. K., Fedorov, D. S., Maksimovich, K. Yu., & Riksen, V. S. (2025, January). Machine learning in predicting crop rotation productivity. *Agricultural Engineering (Russian:* Агроинженерия*)*, 27(1), 41–52.

Khan, R., Kumar, S., Dhingra, N., & Bhati, N. (2021). The use of different image recognition techniques in food safety: A study. *Journal of Food Quality*, 2021(1), 7223164.

Khodjaev, S., Bobojonov, I., Kuhn, L., & Glauben, T. (2025). Optimizing machine learning models for wheat yield estimation using a comprehensive UAV dataset. *Modeling Earth Systems and Environment*, 11(1), 15.

Konduri, V. S., Kumar, J., Hargrove, W. W., Hoffman, F. M., & Ganguly, A. R. (2020). Mapping crops within the growing season across the United States. *Remote Sensing of Environment*, 251, 112048.

Kukar, M., Vračar, P., Košir, D., Pevec, D., & Bosnić, Z. (2019). AgroDSS: A decision support system for agriculture and farming. *Computers and Electronics in Agriculture*, 161, 260–271.

Kumar, Y., Singh, R., & Moudgil, M. R., et al. (2023). A systematic review of different categories of plant disease detection using deep learning-based approaches. *Archives of Computational Methods in Engineering*, 30, 4757–4779.

Meng, S., Wang, X., Hu, X., Luo, C., & Zhong, Y. (2021). Deep learning-based crop mapping in the cloudy season using one-shot hyperspectral satellite imagery. *Computers and Electronics in Agriculture*, 186, 106188.

Michishita, R., Chen, J., & Xu, B. (2014). Empirical comparison of noise reduction techniques for NDVI time-series based on a new measure. *ISPRS Journal of Photogrammetry and Remote Sensing*, 91, 17–28.

Nautiyal, M., Joshi, S., Hussain, I., Rawat, H., Joshi, A., Saini, A., Kapoor, R., Verma, H., Nautiyal, A., Chikara, A., Ahmad, W., & Kumar, S. (2025). Revolutionizing agriculture: A comprehensive review on artificial intelligence applications in enhancing properties of agricultural produce. *Food Chemistry: X*, 29, 102748.

Son, N. T., Chen, C. F., Chen, C. R., Duc, H. N., & Chang, L. Y. (2014). A phenology-based classification of time-series MODIS data for rice crop monitoring in Mekong Delta, Vietnam. *Remote Sensing*, 6, 135–156.

Stepanov, A., Dubrovin, K., & Sorokin, A. (2022). Function fitting for modeling seasonal normalized difference vegetation index time series and early forecasting of soybean yield. *Crop Journal*, 10, 1452–1459.

Suguna, R., Raghuveer, K., Singh, R., Doddabasappa, N., & Kavin, R. (2024). Enhancing productivity through convolutional neural network-based classification and detection of strawberry plant diseases. In *Proceedings of the 3rd International Conference on Optimization Techniques in the Field of Engineering (ICOFE-2024)*.

Tang, L., Zhao, Z., Tang, P., & Yang, H. (2020). SURE-based optimum-length S-G filter to reconstruct NDVI time series iteratively with outliers' removal. *International Journal of Wavelets, Multiresolution and Information Processing*, 18, 2050001.

Tran, K. H., Zhang, H. K., McMaine, J. T., Zhang, X., & Luo, D. (2022). 10 m crop type mapping using Sentinel-2 reflectance and 30 m cropland data layer product. *International Journal of Applied Earth Observation and Geoinformation*, 107, 102692.

Turchetta, M., Corinzia, L., & Sussex, S. (2024, September 2). Fertile ground: How reinforcement learning can improve food sustainability. NCCR Automation. Retrieved from https://nccr-automation.ch/news/2024/fertile-ground-how-reinforcement-learning-can-improve-food-sustainability

Zhang, H., Kang, J., Xu, X., & Zhang, L. (2020). Accessing the temporal and spectral features in crop type mapping using multi-temporal Sentinel-2 imagery: A case study of Yi'an County, Heilongjiang province, China. *Computers and Electronics in Agriculture*, 176, 105618.

Zhen, Z., Chen, S., Yin, T., & Gastellu-Etchegorry, J.-P. (2023). Improving crop mapping by using bidirectional reflectance distribution function (BRDF) signatures with Google Earth Engine. *Remote Sensing*, 15, 2761.

Index

Note: Pages in *italics* refer to figures and pages in **bold** refer to tables.

N

O

P

R

S

V

W

Y

For Product Safety Concerns and Information please contact our EU representative GPSR@taylorandfrancis.com Taylor & Francis Verlag GmbH, Kaufingerstraße 24, 80331 München, Germany